Mathematical Research

Algebraic and Logic Programming

edited by
J. Grabowski · P. Lescanne
W. Wechler

Band 49

AKADEMIE-VERLAG BERLIN

In this series original contributions of mathematical research in all fields are contained, such as

— research monographs
— collections of papers to a single topic
— reports on congresses of exceptional interest for mathematical research.

This series is aimed at promoting quick information and communication between mathematicians of the various special branches.

In diese Reihe werden Originalbeiträge zu allen Gebieten der mathematischen Forschung aufgenommen wie

— Forschungsmonographien
— Sammlungen von Arbeiten zu einem speziellen Thema
— Berichte von Tagungen, die für die mathematische Forschung besonders aktuell sind.

Die Reihe soll die schnelle Information und gute Kommunikation zwischen den Mathematikern der verschiedenen Fachgebiete fördern.

Manuscripts

in English and German comprising at least 100 pages and not more than 500 pages can be admitted to this series. With respect to a quick publication the manuscripts are reproduced photomechanically. Authors who are interested in this series please turn directly to the 'Akademie-Verlag'. Here you will get more detailed information about the form of the manuscripts and the modalities of publication.

Manuskripte

in englischer und deutscher Sprache, die mindestens 100 Seiten und nicht mehr als 500 Seiten umfassen, können in diese Reihe aufgenommen werden. Im Interesse einer schnellen Publikation werden die Manuskripte auf fotomechanischem Weg reproduziert. Autoren, die an der Veröffentlichung entsprechender Arbeiten in dieser Reihe interessiert sind, wenden sich bitte direkt an den Akademie-Verlag. Sie erhalten dort genauere Informationen über die Gestaltung der Manuskripte und die Modalitäten der Veröffentlichung.

Algebraic and Logic Programming

Mathematical Research · Mathematische Forschung

Wissenschaftliche Beiträge
herausgegeben von der
Akademie der Wissenschaften der DDR
Karl-Weierstraß-Institut für Mathematik

Band 49

Algebraic and Logic Programming

Algebraic and Logic Programming

Proceedings of an International Workshop held in Gaussig (GDR), November 14–18, 1988

edited by

Jan Grabowski
Pierre Lescanne
Wolfgang Wechler

Akademie-Verlag Berlin 1988

Herausgeber:

Prof. Dr. Jan Grabowski
Humboldt-Universität zu Berlin

Prof. Dr. Pierre Lescanne
Centre de Recherche en Informatique de Nancy

Doz. Dr. Wolfgang Wechler (Chairman)
Technische Universität Dresden

Die Titel dieser Schriftenreihe werden vom Originalmanuskript der Autoren reproduziert.

ISBN 3-05-500665-8
ISSN 0138-3019

Erschienen im Akademie-Verlag Berlin,DDR-1086 Berlin,Leipziger Str. 3-4

Lizenznummer: 202·100/545/88
Printed in the German Democratic Republic
Gesamtherstellung: VEB Kongreß- und Werbedruck, 9273 Oberlungwitz
Lektor: Dr. Reinhard Höppner
LSV 1095
Bestellnummer: 763 976 8 (2182/49)

03400

P r e f a c e

Logic programming has made its way into professional software tools and serious applications. Inevitably there will emerge more and more systems that compete in enhancing the expressive power of logic programming,in a purely pragmatic way or - more desirably - in a way that is justified by theoretical considerations.

In particular, there is a continuing need for integrating the notion of a *function* into logic programming. Functions are the semantics of algebraic specifications. These are developing towards executability, bringing forth a new programming paradigm which may be called equational (or, in terms of our title, *algebraic*) programming.

Theoretical results achieved during the last five years helped us understand the common foundations of both issues. They provide the unified model-theoretic viewpoint from which to look at the whole scene, and to look for adequate inference techniques and system concepts.

The Gaussig workshop ought to be a place where people from both areas could study and discuss the achievements of both algebraic and logic programming, as well as the attempts to integrating them into unified systems.

Now that the volume has been prepared, we can already say that the idea worked. We have a nice collection of papers that reflect the current state of the field.

We are particularly happy that six authors accepted our invitation to give a main lecture at the workshop. Most of the invited papers arrived in time and could be included in the volume.

In spite of the workshop character of our meeting, we had decided that all submitted papers be subject to refereeing and selection by the programme committee. A number of them had to be rejected, while most of them were included after improvement.

The papers concentrate mostly on the following topics:

- integration resp. translation between functional and relational specification and programming paradigms
- term rewriting theory,
- equation solving techniques,
- modularity in algebraic specifications and logic programs.

We are indebted to the members of the programme committee and the referees for their careful work. This helped to achieve a satisfactory quality and timely publication of the proceedings. Thanks are also due to the publisher for his effective support.

If Gaussig will be a success, we have to owe it primarily to our colleagues from Dresden University who solved the organizational problems, especially to Jürgen Brunner. Thank you for this extraordinary engagement!

In promoting specialized scientific co-operation, our meeting should contribute to a more general aim as well establishing consensus on our common task in this world.

July 1988

Jan Grabowski
Pierre Lescanne
Wolfgang Wechler

PROGRAMME COMMITTEE

Leo Bachmair (Stony Brook, USA)
Peter Bachmann (Dresden, GDR)
Hans-Dieter Ehrich (Braunschweig, FRG)
Jan Grabowski (Berlin, GDR)
Stéphane Kaplan (Orsay, France)
Pierre Lescanne (Nancy, France)
Alberto Martelli (Torino, Italy)
Peter Padawitz (Passau, FRG)
Horst Reichel (Dresden , GDR)
Donald Sannella (Edinburgh, UK)
Magnus Steinby (Turku, Finland)
Wolfgang Wechler (Dresden, GDR) - *Chairman* -

Contents

INVITED PAPERS

SUBMITTED PAPERS

REFEREES

P.Bachmann	R.Hennicker	H.Reichel
F.Bellegarde	H.Hussmann	P.Réty
F.J.Brandenburg	M.Jokinen	G.F.Rossi
J.Brunner	C.Kirchner	M.Rusinowitch
J.Dick	M.K.F.Lai	G.Saake
K.Drosten	A.Lazrek	D.Sannella
H.D.Ehrich	P.Lescanne	M.Steinby
D.Galmiche	U.W.Lipeck	W.Struckmann
A.Geser	A.Martelli	M.Thomas
E.Giovannetti	Ch.Mohan	W.Wechler
M.Gogolla	P.Padawitz	
J.Grabowski	M.Penttonen	

ACP WITH SIGNALS

J.A. Bergstra
Programming Research Group, University of Amsterdam
Department of Philosophy, State University of Utrecht

ABSTRACT. New operators are introduced on top of ACP [BK 84] in order to incorporate stable signals in process algebra. Semantically this involves assigning labels to nodes of process graphs. The labels of nodes are called signals. In combination with the operators of BPA, a signal insertion operator allows to describe each finite tree labeled with actions and signals, provided the signals do not occur at leaves of the tree. In a merge processes can observe the signals of concurrent processes. This research was sponsored in part by ESPRIT under contract 432, METEOR.

1 INTRODUCTION

This paper is concerned exclusively with concrete process algebra in the sense of [BB 87]. Concrete process algebra is that part of process algebra that does not involve Milner's silent action τ [Mi 80] or the empty action ϵ due to Vrancken [Vr 86]. The advantage of concrete process algebra is that it admits a fairly clear operational intuition that can serve as a basis for finding algebraic specifications of new operators.

The new feature that this paper contributes to process algebra is the presence of explicit signals of a persistent nature. The original setup of process algebra inherited from Milner's CCS views process semantically as trees of actions. These actions are best thought of as atomic actions because otherwise the intuition behind the axioms becomes rather obscure. Now it was not claimed that the philosophical concept of a process has been analysed in full depth with the introduction of the concepts of process algebra. Some salient features remain unanalysed. For instance real time behavior and true concurrency. A mechanism of particular importance is the presence of visible aspects of the state of a process. Usually in process algebra the state of a process can only be understood (or observed) via the actions that can be performed from that state. In the setup that will be presented here some aspects of the system state are visible not so much through the actions that will follow but much more directly as signals that persist in time for some extended duration. A signal insertion operator will allow to put signals at the root of a process. Using the other operators of process algebra it is then possible to describe processes that have signals on intermediate nodes as well. The main technical complication lies in the signal observation mechanism. Probably I have not defined that mechanism in the full generality that it would admit. In full generality the mechanism should be so powerful as to admit the description of observing continously changing signals in real time.

The reader is supposed to have some familiarity with the axioms systems BPA, PA and ACP for process algebra. In comparison with the extended version [B 88] this discussion is significantly simplified by the absence of terminal signal insertion. Some remarks on the meaning of the operators of ACP can be found in the discussion at the and of this paper. In order to make the paper self contained all axioms of process algebra (ACP) from [BK 84] are repeated.

2 ADDING SIGNALS TO BASIC PROCESS ALGEBRA

The reader is supposed to have some familiarity wih the axioms of BPA. See the discusion for some comments on the meaning of the operators. The core system BPA_δ (basic process algebra with δ) has the following axioms.

A1 $x + y = y + x$

A2 $(x + y) + z = x + (y + z)$

A3 $x + x = x$

A4 $(x + y) \cdot z = x \cdot z + y \cdot z$

A5 $(x \cdot y) \cdot z = x \cdot (y \cdot z)$

A6 $x + \delta = x$

A7 $\delta \cdot x = \delta$

Four new operators and many new axioms will be presented. The first step is to add a finite new sort S of signals equipped with the binary operator & that describes signal combination and with a constant $\emptyset$ denoting the empty signal. Besides the combination operator there is a need for a filtering or intersection operator $\sqcap$. ATS is the small collection of axioms for signal algebra. It explains that & has the algebraic properties of set union and that $\sqcap$ corresponds to set intersection. Indeed the standard interpretation for signal algebra involves a finite set ATS of atomic signals and takes S to be the power set of ATS, identifying atomic signals with the corresponding singleton sets. The axioms of ATS are as follows:

ATS1 $u \& v = v \& u$ ATS2 $(u \& v) \& w = u \& (v \& w)$

ATS3 $u \& u = u$ ATS4 $u \& \emptyset = u$

ATS5 $u \sqcap v = v \sqcap u$ ATS6 $u \sqcap (v \sqcap w) = (u \sqcap v) \sqcap w$

ATS7 $u \sqcap u = u$ ATS8 $u \sqcap \emptyset = \emptyset$

ATS9 $u \& (v \sqcap w) = (u \& v) \sqcap (u \& w)$

ATS10 $u \sqcap (v \& w) = (u \sqcap v) \& (u \sqcap w)$

The principal new operator to be introduced here is the root signal insertion operator denoted with [.,.]. The intuition behind this operator is that it assigns labels (signals) to the states of processes. Root signal insertion places a signal at the root node of a process.

Whereas in process algebra one usually confines oneself to labeling the transitions and perhaps to some labeling of the nodes that is directly related to the mechanism of transition labeling, here it is intended to have labelings of states of processes withe the same status as the labelings of the state transitions by means of atomic actions. With some effort it turns out that an algebraic specification of the resulting notion of processes can be given that indeed constitutes a conservative enrichment of ACP (at least regarding identities between finite closed process expressions). The following equations are added to BPA thus obtaining BPAS (BPA with signals).

RS1 $[u, x] \cdot y = [u, x \cdot y]$

RS2 $[u, x] + y = [u, x + y]$

RS3 $[u, [v, x]] = [u \& v, x]$

RS4 $[\emptyset, x] = x$

The first axioms expresses the fact that the root of a sequential product is the root of its first component. Axiom RS2 can be given in a more symmetric form as follows:
[u, x] + [v, y] = [u & v, x + y]. This equation depends on the fact that the roots of two processes in an alternative composition are identified. Therefore signals must be combined. The third axiom expresses the fact that there is no sequential order in the presentation of signals. Of course one might image that a sequential ordering on signals is introduced, but preliminary investigations of this matter have shown that the introduction of sequential ordering is far from obvious. The combination of the signals is taking 'both' of them whereas x + y has to choose between x and y.

An interesting identity that follows with the presence of δ is : [u,x] = [u,δ] + x. This equation is indeed very useful for writing efficient process specifications mainly because it allows to a large extent to work with process algebra expressions that are not cluttered with signal insertions.

EXAMPLES. (i) A traffic light that changes color from green via yellow to red and back to green. As names for the traffic lights one may simply use the natural numbers. For the traffic light with number n, the signals are then green(n), yellow(n) and red(n), the only action is change(n).

TL(n) = [green(n), change(n)] · [yellow(n), change(n)] ·
        [red(n), change(n)] · TL(n)

(ii) An electric lamp with power switch. In the equations the names for signals and actions speak for themselves. I will assume that the lamp is indexed by a name n from NAMES = {hall1,hall2, hall3, kitchen1, kitchen2, living1, living2, living3, staircase, garage, cellar}. The lamp is parametrised by two additional parameters: L(n, x, y), x ∈ SWITCH = {on, off}, y ∈ STATUS = {defect, functioning}. The function sw permutes the elements of SWITCH and the actions switch(n) correspond to switching lamp n. For each n ∈ NAMES there are two signals: light(n) and dark(n). Together these signals constitute the atomic signals and the signals in the sense of signal algebra are just the finite subsets of this collection. It must be noticed however that the combined signal light(n) & dark(n) will not occur in any process in the intended interpretation. The function F: NAMES x SWITCH x STATUS → SIGNALS is defined as follows:

F(n,x,y) = <u>if</u> x = on <u>and</u> y = correct <u>then</u> light(n) <u>else</u> dark(n) <u>fi</u>

Recursion equations for the behavior of the lamp with name n are then as follows.

```
L(n, x, functioning) = [F(n, x, functioning), δ] +
                       switch(n) · L(n,sw(x), functioning) +
                       defect(n) · L(n, x, defect) +
                       get_new_lamp(n) · L(n, x, functioning)
```

$L(n, x, defect) = [F(n, x, defect), \delta] + switch(n) \cdot L(n, sw(x), defect) + get\_new\_lamp(n) \cdot L(n, x, correct)$

3 SIGNALS AND MERGE

We can extend the axioms to PAS⁻, including the free merge (without communication), the left merge with the usual axioms and adding axioms that handle the interaction between the signal insertion operators and the left merge. The superscript ⁻ indicates that there is no feature present that allows the observation of signals, addition of that feature is the subject of sction 4. (Of course the introduction to ACP will require a modification of the merge expansion axiom by adding an additional term for the communication merge, this will be described in section 5.)

M1	$x \parallel y = x ⌊⌊ y + y ⌊⌊ x$
M2	$a ⌊⌊ x = a \cdot x$
M3	$(a \cdot x) ⌊⌊ y = a \cdot (x \parallel y)$
M4	$(x + y) ⌊⌊ z = x ⌊⌊ z + y ⌊⌊ z$
MSI5	$[u, x] ⌊⌊ y = [u, x ⌊⌊ y]$

EXAMPLE. This example refers to the previous example that introduced a collection of lamps in a private house. The simultaneous behavior of the collection of lamps in the living is appropriately described by $L(living) = L(living1) \parallel L(living2) \parallel L(living(3)$

For the kitchen one obtains: $L(kitchen) = L(kitchen1) \parallel L(kitchen2)$.

For the hall we have $L(hall) = L(hall1) \parallel L(hall2)$

Composing these one obtains: $L(living/kitchen/hall) = L(kitchen) \parallel L(living) \parallel L(hall)$.

SIGNALS AND SYNCHRONOUS COMMUNICATION

The axioms system ACPS⁻ describes the addition of signals to processes with synchronous communication as modeled by ACP. The superscript ⁻ indicates that there is no communication by means of observation of signals. Addition of that feature will involve the addition of two more summands to the merge expansion axiom. The remaining axioms of ACPS⁻ can now be added without leading to any inconsistency, but it is necessary to add some equations that describe the interaction of left merge and communication merge and the node labels. The axioms CM1-4 replace the axioms M1-4.

C1	$a \mid b = b \mid a$
C2	$(a \mid b) \mid c = a \mid (b \mid c)$
C3	$a \mid \delta = \delta$
CM1	$x \parallel y = x ⌊⌊ y + y ⌊⌊ x + x \mid y$
CM2	$a ⌊⌊ x = a \cdot x$
CM3	$(a \cdot x) ⌊⌊ y = a \cdot (x \parallel y)$
CM4	$(x + y) ⌊⌊ z = (x ⌊⌊ z) + (y ⌊⌊ z)$

MSI1 $[u,x] ⌊⌊ y = [u, x ⌊⌊ y]$

CM5 $a \mid (b \cdot x) = (a \mid b) \cdot x$
CM6 $(a \cdot x) \mid b = (a \mid b) \cdot x$
CM7 $(a \cdot x) \mid (b \cdot y) = (a \mid b) \cdot (x \parallel y)$
CM8 $(x + y) \mid z = (x \mid z) + (y \mid z)$
CM9 $x \mid (y + z) = (x \mid y) + (x \mid z)$

MSI3 $[u, x] \mid y = [u, x \mid y]$
MSI4 $x \mid [u, y] = [u, x \mid y]$

D1 $\partial_H(a) = a$ if $a \notin H$
D2 $\partial_H(a) = \delta$ if $a \in H$
D3 $\partial_H(x + y) = \partial_H(x) + \partial_H(y)$
D4 $\partial_H(x \cdot y) = \partial_H(x) \cdot \partial_H(y)$
DSI1 $\partial_H([u,x]) = [u, \partial_H(x)]$

4 PROCESSES WITH MERGE AND SIGNAL OBSERVATION

In this section an observation mechanism is added to the features available in process algebra. In order to simplify the discussion, the communication mechanism is first left out in order to be added in section 5. It is assumed that some actions a are able to read the signals of processes that are put in parallel with the process executing a. To this end a subset OBS(A) of the atomic actions is introduced. This set contains the actions that allow non-trivial observations. The reduction of PAS as a special case of PAS$^-$ is found by taking OBS(A) empty. The deadlock constant is never in OBS(A). Moreover a signal observation function is introduced which takes two arguments, an action and a signal and it will return a finite sum of actions. Thus with A the collection of atomic actions including OBS(A), FS(A) is the collection of finite sums of actions in A. Now signal observation is a mapping of type A x SIG → FS(A).

I will assume that the boolean algebra of signals is prime. So every signal is the composition of finitely many indecomposable signals. The inspection of a signal will check whether or not some atomic signal is contained in it. Plausible axioms for the signal observation function (./.) are the following:

O1 $a / \emptyset = a$ if $a \in OBS(A)$
O2 $a / u = \delta$ if $a \notin OBS(A)$
O3 $a / (u \,\&\, v) = (a / u) + (a / v)$
O4 $(a / u) / v = a / (u \,\&\, v)$

Notice that axiom O1 is necessary in view of axiom O4, which in turn plays an essential role to ensure associativity of the merge. Notice that axiom O3 is incompatible with the option to have one single action a that observes several signals simultaneously. I have incorporated O3 for two reasons. It simplifies matters, and I did not find any natural examples that need the synchronous observation of multiple signals. Multiple observations are possible if one leaves out this axiom. Equipped with the signal observation function on atomic actions and signals it is possible to define the result of the observation function on processes with non-trivial (root) signals. The intuitive meaning of x / u is a process that behaves like x be it that the first action of x is viewed as an observation on the signal u.

O6 $(x + y) / u = (x / u) + (y / u)$

O7 $(x \cdot y) / u = (x / u) \cdot y$

O8 $[u, x] / v = [u, x / v]$

O9 $\langle x, u\rangle / v = \langle x / v, u\rangle$

THE ROOT SIGNAL OPERATOR

It is useful to extend the system with an operator S which determines the root signal of a process. If S(x) is Ø we say that x has a trivial root signal.

S1 $S(a) = \emptyset$

S2 $S(x + y) = S(x) \mathbin{\&} S(y)$

S3 $S(x \cdot y) = S(x)$

S4 $S([u, x]) = u \mathbin{\&} S(x)$

PAS: SIGNAL OBSERVATION AND FREE MERGE

Two new summands must be added to the merge expansion equation in order to obtain the recursion equation for a merge operator that takes signal observation into account as well. This completes the description of the axiom system PAS.

OM $x \parallel y = x ⌊\!⌊ y + y ⌊\!⌊ x + (x / S(y)) ⌊\!⌊ y + (y / S(x)) ⌊\!⌊ x$

In order to support the intuition for the observation mechanism I will now describe a model for the algebra of signals and the effect of the observation function connected to it.

The model takes a finite set Z of atomic signals as point of departure, z and z' range over Z. SIG(Z), the set of signals over Z is the power-set of Z. The observation actions all have the form test(z) with z an element of Z. These actions represent the intention to observe a signal z. There is a special observation action yes which is the result (confirmation) of a succesful observation. The signal observation function then works as follows.

test(z) / V = <u>if</u> z ∈ V <u>then</u> yes <u>else</u> test(z) <u>fi</u>.

yes / V = yes

One easily verifies the validity of all axioms in this case, for instance

(test(z) / {z, z'}) / {z} = yes / {z} = yes and

test(z) / ({z, z'} & {z}) = test(z) / {z, z'} = yes.

In connection with this observation function one will use a form of encapsulation which shields off all observations except yes. In this way seen from outside the encapsulation only succesful observations are present.

5 ACPS: SYNCHRONOUS COMMUNICATION AND OBSERVATION

Now I will combine the above specifications of ACPS⁻ and PAS to obtain ACPS (ACP with signals and signal observation) by adding the two expansion axioms for the merge. The purpose of the following equations is to ensure that all operators except $+$, $\cdot$, $[.,.]$ and $\langle.,.\rangle$ can be eliminated from finite process expressions by means of left to right term rewriting. Notice that the original case of ACP can be viewed similarly with the understanding that all operators except $+$ and $\cdot$ can be eliminated in that case.

Besides taking the axioms for ACPS⁻ and PAS together while taking the sum of their merge expansion axioms, only one axiom for the communication function is needed; this axiom says that observations will not be involved in any non-trivial communications.

O5 $\quad a \mid b = \delta \quad$ if $a \in OBS(A)$

The new expansion axiom for merge is then as follows:

OCM $\quad x \parallel y = x ⫦ y + y ⫦ x + x \mid y + (x / S(y)) ⫦ y + (y / S(x)) ⫦ x$

If the signal observation function yields δ for every atomic action, the sub-expressions $x / S(y)$ and $y / S(x)$ vanish (become equal to δ) and the interaction between processes works exactly as in ACP. ACP is therefore the special case of ACPS if the signal observation function vanishes everywhere. Similarly PA is the special case of ACP if the communication function vanishes everywhere. Obviously there is a situation where the communication function vanishes everywhere but the signal observation function may assume non-trivial values. In such cases one may omit the communication merge.

EXAMPLE. In this example I will elaborate once more on the example of the person in a house with various. The point here is that the person may want to inspect whether or not a light that was switched on indeed functions correctly. If not s/he will replace the bulb by a new one. The actions that must be introduced here are inspections of the signals of the various lamps. Recall that the signals are in this case just finite sets of atomic signals of the form light(n) and dark(n).

test(light(n)) / u = yes if light(n) ∈ u
test(dark(n)) / u = yes if dark(n) ∈ u.

The action pswitch(n) can now be replaced, whenever it is assumed to switch the lamp on, by a more involved non-atomic process as follows:

pswitch(n) · (test(light(n)) +
test(dark(n)) · put_new_lamp(n))

Of course the communication function has to be extended. In particular one needs a communication for the new lamp actions, for instance:

put_new_lamp(n) | get_new-lamp(n) = new_lamp(n)

Here is process that describes a person entering the house and focussed on repairing broken lights

person = enter_front_door · pswitch(hall1) ·
(test(light(hall1)) +
test(dark(hall1)) · put_new_lamp(hall1)) ·
enter_living · (pswitch(living1) ∥ pswitch(living2)) ·

```
pswitch(living3) ·
  ( (test(light(living1)) +
       test(dark(living1)) · put_new_lamp(living1)) ||
    (test(light(living2)) +
       test(dark(living2)) · put_new_lamp(living2)) ||
    (test(light(living3)) +
       test(dark(living3)) · put_new_lamp(living3)
  ) ·
enter_kitchen · pswitch(kitchen1) ·
(test(light(n)) + test(dark(n)) · put_new_lamp(n)) ·
leave_kitchen · pswitch(hall1) · enter_living
```

Composing this process `person` with `L(living/kitchen/hall)` in the system ACPS will indeed produce a process that shows appropriate interaction of the person with its environment. Notice however that there are many different plausible behaviours of the person. There seems to be no universal generic behavior for `person` even in this very simple case. Obviously other factors that bear no relation to the switching of light influence the behavior of the person. Nevertheless it is a fair hypothesis to assume that the person behaves for some short time as a process. Thus locally (in time) `person` is a process in the sense of process algebra, but globally a more complex stucturing mechanism is needed. One can imagine, however, that it is possible to formally incorporate in process algebra a small knowledge base which structures the decision taking process of the person.

6 DISCUSSION

Given a finite alphabet of atomic actions A provided with a commutative and associative communication function and a finite model of SA the initial algebras of BPAS, PAS and ACPS are process domains for finite processes with signals. Process equivalence in these models corresponds to bisimulation equivalence where it is understood that a bisimulation may only relate a pair of nodes in two process graphs if these carry the same signals (or are both terminal nodes in their respective graphs). It follows that the equations of ACPS can be read without hesitation as an algebraic specification for which the initial algebra semantics is the primary meaning.

Of course there is nothing new in the explicit description of process states. JSD [Ja 83] is based on processes but state oriented, Wieringa [Wi 88] uses terminology of knowledge representation as well as the state operator of [BB 87] to have a balanced approach to static and dynamic aspects of process descriptions. [Va 86] provides a process algebra semantics of the parallel programming language POOL which shows that ACP is sufficiently powerful to describe the semantics of a substantial imperative programming language. The relevance of signals is not so much that process algebra without signals lacks expressive power but that some systems can be described in a more natural way using signals.

ACPS attempts not primarily to combine processes and data, as it is done for instance in LOTOS ([Br 87]) for the case of CCS or in PSF [MV 88] for the case of ACP. In both cases an integration between a process oriented formalism and a formalism for algebraic data type specification is found but no new process concepts are developed. The use of the data types is mainly in facilitating the recursive definition of processes.

Without difficulties an operational semantics with action rules can be given for processes with signals, just as this was done in [vG 86] for ACP. I will collect the transition rules below. Before that I will give a short informal description of the meaning of the operators of ACP. Atomic actions are progams that have the effect of instantaneous events. The alternative composition $X + Y$ of processes X and Y is some process that can evolve like X or like Y. Nothing is presupposed about the mechanism involved in making the choice between X and Y. The product $X \cdot Y$ of X and Y is a process that starts with executing X and then proceeds with executing Y. The merge $X \| Y$ executes X and Y in parallel with arbitrary interleaving of atomic actions. The left merge $X \lfloor\!\lfloor Y$ is like the merge but now with the constraint that the first action of $X \lfloor\!\lfloor Y$ is chosen from X. The communication merge $X \mid Y$ is like the merge again but this time with the additional constraint that the first action must be a communication between an action of X and an action of Y. Left merge and communication merge are auxiliary operators used to find an initial algebra specification of merge. Communication at the elvel of individual atomic actions is given by the communication function. $a \mid b$ is the atomic action that results if atomic actions a and b are performed simultaneously (and therefore in communication). If both actions have nothing to communicate the result of $\mid$ is δ. δ is deadlock, a process which has no alternative to proceed. Notice that in a context $\delta + X$ alternatives are added and δ cannot represent deadlock. To finish the description of ACPS here are the action rules for an operational model for processes with signals.

$$a \rightarrow^{a} \surd \qquad\qquad X \rightarrow^{a} X' \Rightarrow [u, X] \rightarrow^{a} X'$$

$$X \rightarrow^{a} X' \Rightarrow X + Y \rightarrow^{a} X' \qquad\qquad X \rightarrow^{a} \surd \Rightarrow X + Y \rightarrow^{a} \surd$$

$$Y \rightarrow^{a} Y' \Rightarrow X + Y \rightarrow^{a} Y' \qquad\qquad Y \rightarrow^{a} \surd \Rightarrow X + Y \rightarrow^{a} \surd$$

$$X \rightarrow^{a} X' \Rightarrow X \cdot Y \rightarrow^{a} X' \cdot Y \qquad\qquad X \rightarrow^{a} \surd \Rightarrow X \cdot Y \rightarrow^{a} X$$

$$X \rightarrow^{a} X' \Rightarrow X \| Y \rightarrow^{a} X' \| Y \qquad\qquad X \rightarrow^{a} \surd \Rightarrow X \| Y \rightarrow^{a} Y$$

$$Y \rightarrow^{a} Y' \Rightarrow X \cdot\| Y \rightarrow^{a} X \| Y' \qquad\qquad Y \rightarrow^{a} \surd \Rightarrow X \| Y \rightarrow^{a} X$$

$$X \rightarrow^{a} X', \quad Y \rightarrow^{b} Y', \quad a \mid b = c \Rightarrow X \| Y \rightarrow^{c} X' \| Y'$$

$$X \rightarrow^{a} \surd, \quad Y \rightarrow^{b} Y', \quad a \mid b = c \Rightarrow X \| Y \rightarrow^{c} Y'$$

$$X \rightarrow^{a} X', \quad Y \rightarrow^{b} \surd, \quad a \mid b = c \Rightarrow X \| Y \rightarrow^{c} X'$$

$$X \rightarrow^{a} \surd, \quad Y \rightarrow^{b} \surd, \quad a \mid b = c \Rightarrow X \| Y \rightarrow^{c} \surd$$

$$X \rightarrow^{a} X', a \notin H \Rightarrow \partial_H(X) \rightarrow^{a} \partial_H(X') \qquad\qquad X \rightarrow^{a} \surd, a \notin H \Rightarrow \partial_H(X) \rightarrow^{a} \surd$$

$$\langle s \mid E \rangle \rightarrow^{a} Y \Rightarrow \langle X \mid E \rangle \rightarrow^{a} Y \qquad\qquad \langle s \mid E \rangle \rightarrow^{a} \surd \Rightarrow \langle X \mid E \rangle \rightarrow^{a} \surd$$

In the two above rules $\langle X \mid E \rangle$ is a guarded system of recursion equations that contains the equation $X = s$. $\langle s, E \rangle$ denotes a process as follows: s is a process expression with free variables in X, for these process variables the unique solution of the equations E is substituted. Last but not least there are the operational rules for signal observation.

$X \to^a X', \quad a/S(Y) = b \neq \delta \quad \Rightarrow \quad X \parallel Y \to^b X' \parallel Y$

$X \to^a \surd, \quad a/S(Y) = b \neq \delta \quad \Rightarrow \quad X \parallel Y \to^b Y$

$Y \to^a Y', \quad a/S(X) = b \neq \delta \quad \Rightarrow \quad X \parallel Y \to^b X' \parallel Y$

$Y \to^a \surd, \quad a/S(X) = b \neq \delta \quad \Rightarrow \quad X \parallel Y \to^b X$

Adress of the author: University of Amsterdam, Faculty of Mathematics and Computer Sciences, Programming Research Group, Kruislaan 409, 1098 SJ Amsterdam

REFERENCES

[BB 87] J.C.M.Baeten & J.A.Bergstra, Global renaming operators in concrete process algebra, Report P8709, University of Amsterdam, Programming Research Group (1987), to appear in Information and Computation

[B 88] J.A.Bergstra, Process algebra for synchronous communication and observation, University of Amsterdam, Programming Research Group, Report P88XX, (1988)

[BK 84] J.A.Bergstra & J.W.Klop, Process algebra for synchronous communication, Information and Control 60 (1/3), (1984) 109-137

[BCC 86] G.Berry, P.Couronne, & G.Gonthier, Synchronous programming of reactive systems: an introduction to ESTEREL, in Proc.first France-Japan Symposium on Artificial Intelligence and Computer Science, Tokyo, North-Holland, (1986)

[Br 87] (Ed. E. Brinksma) Information Processing Systems- Open Systems Interconnection- LOTOS- A formal description technique based on the temporal ordering of observational behavior, ISO/TC 97/SC 21/20-7-1987

[vGl 87] R.van Glabbeek, Bounded nondeterminism and the approximation induction principle in process algebra, in: proc. STACKS 87 (Eds. F.J.Brandenburg, G.Vidal-Naquet & M.Wirsing) LNCS 247 Springer (1987) 336-347

[MV 88] S.Mauw & G.J.Veltink, A process specification formalism, University of Amsterdam, Programming Research Group, Report XXX (1988)

[Mi 80] R.Milner, A Calculus of Communicating Systems, Springer LNCS, (1980)

[Ja 83] M.Jackson, System development, Prentice Hall, (1983)

[Va 86] F.W.Vaandrager, Process algebra semantics for POOL, Report CS-R8629, Centre for Mathematics and Computer Science, Amsterdam (1986)

[Vr 86] J.L.M.Vrancken, The algebra of communicating processes with empty process, Report FVI 86-01, Programming Research Group, University of Amsterdam, (1986)

[W 88] R.J.Wieringa, Jackson system development analysed in process algebra, Report IR-148, Free University, Amsterdam, Department of Mathematics and Computer Science, (1988)

FUNCTIONAL EXTENSIONS TO PROLOG: ARE THEY NEEDED?[1]

Laurent Fribourg[2]
L.I.E.N.S., 45 rue d'Ulm - 75230 Paris cedex 05 - France

Introduction

Prolog [6] is a logic language which, besides logical variables, manipulates relation and data constructor symbols. In the last few years, the enrichment of logic languages with function symbols and functional paradigms of computation has been extensively investigated (for a survey, see e.g. [2]). The main paradigm of functional computation is **reduction**, which applies when an expression matches the left-hand side of a program statement, and consists in replacing the expression by the corresponding right-hand side. The advantages of Functional Programming (FP) over Logic Programming (LP), in particular Prolog, are discussed in [2], and can be summarized as:

(1) *Notation*. The functional formalism is more readable than the relational one.

(2) *Control.*

(2.1) *Backtracking*. Since reduction replaces an expression by an equivalent one, it is backtracking-free (assuming the search strategy to be depth-first). It is also able to handle negative knowledge, and precipitate backtracking when reducing an expression to a form such as "true=false".

(2.2) *Evaluation strategy*. Owing to the nesting of functional subterms, a functional program *a priori* contains more control information than the corresponding logic program. In particular, an optimal strategy (lazy evaluation) can be used in order to avoid unnecessary computation and to handle infinite structures.

(3) *Higher-order features and Types.*

In this note we focus on point (2), and argue that control mechanisms similar to those of functional programs can be reproduced in the framework of Prolog programs without integrating functions.

1. Logic programming with functions: a bit of history

In order to enrich Logic Programming with functions, a natural approach is to consider the union of a set of Horn clauses H with a set E of equations as a program. In such a

[1] This title is inspired from D.H.D. Warren's paper: "Higher-order extensions to PROLOG: are they needed?" in *Logic Programming*, K.L. Clark and S.-A. Tarnlund, eds., Academic Press, 1982, pp. 441-454.

[2] This work has been partially supported by Laboratoires de Marcoussis and ESPRIT project 432.

framework, the essential property to be satisfied by the system of inference rules of the interpreter is completeness for first-order logic with equality.

The naive solution consists in limiting the inference rules of the interpreter to resolution and adding the equality axioms to the program. Unfortunately, this leads to a combinatorial explosion of the size of the search space. To remedy this defect, Plotkin has proposed instead to replace the standard unification involved in resolution by semantic unification modulo the theory defined by the equational system E [24]. Semantic unification of two terms modulo E corresponds to finding instantiations (solutions) such that the instantiated terms be equal modulo E. In [19] this approach is provided with a semantic foundation. The major drawback of this approach is that the process of semantic unification may itself loop forever and is essentially undecidable [16]. Another approach consists of amalgamating the LP paradigm of resolution with the FP paradigm of reduction. An early experiment in this trend was FUNLOG [28]. In FUNLOG the equational system E is made up of a **confluent** and **terminating** rewrite system, i.e. a functional program made of left-to-right oriented equations (rewrite system) such that reduction applied repeatedly always yields one irreducible form (termination property) and only one (confluence property). The FUNLOG interpreter applies resolution on predicates after having reduced their arguments as much as possible via the rewrite rules. Although FUNLOG combines several nice features of both programming styles (such as the logical variable of LP and the lazy evaluation of FP), the inference rules are not complete for first-order logic with equality.

In the field of automated theorem proving, a method, named **narrowing**, was especially conceived to ensure completeness for first-order logic with equality (in the case of a confluent and terminating system E) [27]. Narrowing is the natural extension of reduction to incorporate unification. It consists in applying the minimal substitution to an expression in order to make it reducible, and then to reduce it. The minimal substitution is found by unifying the expression with the left-hand side of a rewrite rule. A step of narrowing can also be followed by a sequence of reductions leading to an irreducible form (normalized narrowing). The idea of integrating LP with FP by adding (normalized) narrowing to resolution was first proposed in [14] (EQLOG). Narrowing can be used by itself without resolution (except for resolution with the reflexive axiom X=X) in order to semantically unify two terms modulo the confluent and terminating rewrite system E. Fay [9] and Hullot [17] have proved the completeness of narrowing to that purpose (in the sense that all the normalized solutions are enumerated). Hullot's proof is based on the parallel that can be established between the sequence of reductions applied to an instance of a term and a sequence of narrowings applicable to that term. This result is the counterpart of the "lifting lemma" in the theory of resolution [26]. Such results have been extended to conditional rewrite systems [21][18][13], and to non-terminating systems [32][13]. It thus becomes possible to consider the set H of Horn clauses as a special case of conditional rewrite rules and to incorporate it in *E*. The inference rule of resolution (except for resolution with X=X) is then subsumed by the rule of narrowing, which is used as a computation paradigm for first-order logic with equality [8].

2. Narrowing for Theories with Constructors

Unfortunately, as it is pointed out, e.g., in [11][3], unrestricted narrowing (even with the normalized and basic refinements [17][22]) leads to unacceptable overcomputation. For example, consider the following parsing program (adapted from [23]):

noun(X,man(X),L) = [man|L].
noun(X,man(X),L) = [man|L].
propernoun(john,L) = [john|L].
tv(X,Y,loves(X,Y),L) = [loves|L].
intv(X,lives(X),L) = [lives|L].
det(X,P1,P2,"all X/P1 => P2",L) = [every|L].
det(X,P1,P2,"exists X/P1 & P2",L) = [a|L].
np(X,P1,P,L) = det(X,P1,P2,P,noun(X,P,L)).
np(X,P,P,L) = propernoun(X,L).
vp(X,P,L) = tv(X,Y,P1,np(Y,P1,L)).
vp(X,P,L) = intv(X,P,L).
s(P,L) = np(X,P1,P,vp(X,P1,L)).

When solving the goal

s(P,[]) = [every,man,loves,a,woman] ,

the normalized (basic) narrowing stategy gives the answer

P = "all X/ man(X) => exists Y/ woman(Y) & loves(X,Y)"

one hundred and twenty times. Indeed, *s(P,[])* is reduced to

det(X,P1,P2,P,noun(X,P2,tv(X,Y,P3,det(Y,P3,P4,P1,noun(Y,P4,L))))) , and there is then a list of 5 possible subterm positions for narrowing, which leads to *factorial*(5) possible derivations of the solution.

For the sake of efficiency, it is necessary to consider strategies where only one position (rather than 5) is to be selected for narrowing in the example above. For example, one might select the innermost position (viz., that of *noun(Y,P4,L)*) or the outermost one (viz., that of *det(X,P1,P2,...)*). The **innermost** strategy and the **lazy/outermost** strategy have both been studied in the special framework of *theories with constructors*. It is very natural to limit ourselves to such theories when using equations as a programming language, because functions can generally be defined over special primitive symbols, – the data **constructors**. The signature Σ, i.e. the set of functional symbols, is then partitioned as the set *F* of **defined functions** symbols and the set *C* of constructors symbols. (Without loss of understanding, we will use the term "functions" instead of "defined functions"). The innermost strategy consists in applying narrowing at a selected (e.g., the leftmost) position of an *innermost* term, that is a term which contains no function symbol except the one located at the root [10][11]. It corresponds to call-by-value semantics. On the opposite, the lazy strategy selects, roughly speaking, an outermost term position for narrowing [25]. It corresponds to call-by-name semantics. The innermost strategy is easier to be efficiently implemented but neither avoids unnecessary computation as much as the second one does, nor allows for the manipulation of partially defined functions and infinite objects.

The innermost strategy is complete (w.r.t. normalized solutions) if
1) each left-hand side of the program is innermost,
2) the program forms a confluent term-rewriting system, and
3) all the functions are totally defined over constructors.

Points 1) and 2) suffice to ensure completeness for the lazy strategy[3].
Note that points 1) and 2) imply that there is *no equation between data constructors* in the equational theory of the program. Note also that, a program satisfying points 1)-2)-3) necessarily forms a terminating rewrite system. For such a program, completeness is still preserved if innermost narrowing is combined with normalization [11].

3. Simulation of narrowing by SLD-resolution

In order to take advantage of the efficiency of Prolog technology, it may be interesting to attempt to compile functional programs (with constructors) as Prolog programs, then to simulate narrowing through a form of **SLD-resolution**.

3.1. Normalized innermost narrowing

Innermost narrowing can be very simply simulated by decomposing nested functional subterms into conjunctions of equations, and replacing equations involving n-ary function symbols by atoms involving $(n+1)$-ary predicate symbols. (the $(n+1)$-th argument represents the value of the function applied to the n first arguments.) If the atoms corresponding to the inner subterms are put on the left of the ones corresponding to outer subterms, then the Prolog mechanism of leftmost SLD-resolution exactly simulates the process of leftmost-innermost narrowing [7][3] (cf [20][31][30]). Normalization can also be efficiently simulated on flattened programs by performing an additional mechanism, called **simplification**, which has priority over Prolog resolution, and which is performed on all the atoms of the goal (not only the leftmost one) [12]. For every $(n+1)$-ary predicate in a goal, simplification attempts to "match" the arguments w.r.t. to those appearing in the head of a program definite clause, by performing one-way unification for the n first

[3] Actually, the concept of "confluence" should be replaced by that of "i-confluence" (see [13]).

arguments and (two-way) unification for the *(n+1)*-th argument. When simplification of an atom succeeds, it yields the same result than resolution upon this atom, so Prolog augmented with simplification is still SLD-resolution but with a different atom selection function. As reduction, simplification is backtracking-free and allows for the detection of contradictions, thus precipitating backtracking on failure.

Example

Consider the functional program

append([],X) = X.

append([A|X],Y) = [A|append(X,Y)].

The goal

append(append([1|X],Y),Z) = [2|W].

reduces first to

append([1|append(X,Y)],Z) = [2|W].

then to

[1|append(append(X,Y),Z)] = [2|W].

which an equation between constructors, so is unsolvable. On the other hand, the translation of the functional program under a Prolog form gives

append([],X,X).

append([A|X],Y,[A|V]) ← append(X,Y,V).

The goal is translated as

← append([1|X],Y,U), append(U,Z,[2|W]).

The left literal is simplifiable by the second append clause since the two first arguments *[1/X]* and *Y* are one-way unifiable with their counterparts in the clause (using a substitution which binds *A* to *1*), and since the last argument *U* is unifiable with its counterpart (using a substitution which binds *U* to *[1/Us]*). This yields

← append([1|Us],Z,[2|W]).

A contradiction is then detected because the input arguments *[1/Us]* and *Z* are one-way unifiable (using a substitution which bounds *A* to *1*) but the output argument *[2/W]* cannot then be unified with its counterpart *[1/V]*. Note that Prolog resolution would have failed to terminate on this goal, whatever the order of subgoals is.

However, simplification in the Prolog framework does not exactly simulate functional reduction. Simplification indeed leaves intact any atom that corresponds, at the functional level, to a subterm which matches with a variable appearing in the rule left-hand side but not in the right-hand side. By contrast, reduction at the functional level makes the subterm disappear.

Example

Consider the rule

$*(0,X) = 0.$

where the variable X appears in the left-hand side but not in the right-hand side. The rule reduces the expression $*(0,f(X)) = Y$ to $0 = Y$, thus making the subterm $f(X)$ disappear. On the other hand, the clause corresponding to the rule is

$*(0,X,0).$

It simplifies the corresponding goal $\leftarrow f(X,Z), *(0,Z,Y),$ to $\leftarrow f(X,Z)$ (using the substitution which binds Y to 0).

The correspondence between reduction and simplification, to be exact, requires the incorporation of an auxiliary rule (see [5]), which corresponds to the **atom elimination** rule of [1], or to the **termination** rule of [15]).

3.2. Lazy narrowing

As the innermost strategy, the lazy strategy can also be simulated by decomposing nested functional terms and applying SLD-resolution on the flat equations. However, the equation resolved upon is dynamically selected (it is no longer the leftmost one, as in Prolog). Furthermore the equations *a priori* cannot be transformed into atoms (by transforming n-ary function symbols into $(n+1)$-ary predicates), because lazy evaluation requires to perform the resolution with the axiom X=X (see, e.g., [29]). Unfortunately, the resolution with X=X is source of inefficiency, when applied without control. Actually, it is shown in [4] that resolution with X=X can be restricted to equations between terms containing no function symbol, the resolution of the other equations being simulated by the rule of atom elimination. Therefore, the presence of functions is purely formal, and, here again, equations could be replaced by atoms. The interpreter in [4] then simulates in an efficient way Reddy's lazy narrowing [25].

4. Conclusion

Thus, it turns out that the control information implicitly contained in each functional program with constructors can still be exploited after compilation under a Prolog form, by means of:

(1) modification of the (leftmost atom) selection strategy used in Prolog SLD-resolution,

(2) addition of the rule of atom elimination.

For both innermost and lazy strategies, it leads to an implementation equivalent as regards to the computation process, but much more efficient than a direct implementation of narrowing at the functional level.

References

[1] Barbuti, R., Bellia, M., Levi, G., Martelli, M., "LEAF: A Language which Integrates Logic, Equations and Functions", in *Logic Programming: Functions, Relations and Equations*, D. DeGroot and G. Linstrom, eds., Prentice-Hall, 1986, pp. 201-238.

[2] Bellia, M., Levi G., "The Relation between Logic and Functional Languages: A Survey", J. Logic Programming 3:3, Oct. 1986, pp. 217-236.

[3] Bosco, P.G., Giovannetti, E., Moiso, C., "Refined Strategies for Semantic Unification", Proc. TAPSOFT, LNCS 250, Springer-Verlag, 1987, pp. 89-95.

[4] Bosco, P.G., Giovannetti, E., Levi, G., Moiso, C., Palamidessi, C., "A Complete Characterization of K-LEAF, a Logic Language with Partial Functions", Proc. Intl. Symp. on Logic Programming, 1987, pp. 89-95.

[5] Cheong, P.H., "Techniques de mise en oeuvre de langages logico-fonctionnels", D.E.A. Report, Laboratoire d'Informatique de l'Ecole Normale Superieure, Paris,1988.

[6] Colmerauer, A., "Metamorphosis Grammars", in *Natural Language Communication with Computers*, LNCS 63, Springer-Verlag, New York, 1978, pp. 133-189.

[7] Deransart, P., "An Operational Algebraic Semantics of PROLOG Programs" Proc. Programmation en Logique, Perros-Guirrec, CNET-Lannion, Mar. 1983.

[8] Dershowitz, N., Plaisted, D., "Logic Programming cum Applicative Programming", Proc. IEEE Intl. Symp. on Logic Programming, Boston, MA, 1985, pp. 54-66.

[9] Fay, M., "First-order unification in an equational theory", Proc. Fourth Workshop on Automated Deduction, Austin, TX, 1979, pp. 161-167.

[10] Fribourg, L., "A Narrowing Procedure for Theories with Constructors", Proc. Intl. Conf. on Automated Deduction, LNCS 170, 1984, pp. 259-301.

[11] Fribourg, L., "SLOG: a Logic Programming Language Interpreter Based on Clausal Superposition and Rewriting", Proc. IEEE Intl. Symp. on Logic Programming, Boston, MA, 1985, pp. 172-184.

[12] Fribourg, L., "PROLOG with Simplification", 1st France-Japan Symp. on Artifical Intelligence and Computer Science, Fuchi and Nivat, eds., Elsevier Science Publishers

B.V. (North-Holland), 1988, pp. 161-183.

[13] Giovanetti, E., Moiso, C., "A Completeness Result for E-unification Algorithms Based on Conditional Narrowing", Fundations of Logic and Functional Programming, LNCS 306, Springer-Verlag, 1988, pp.157-167.

[14] Goguen. J.A., Meseguer, J., "Equality, Types, Modules and (why not?) Generics for Logic Programming", J. Logic Programming, Vol. 1, NO 2, 1984, pp. 179-210.

[15] Hansson, A., Haridi, S., Tarnlund, S.-A., "Properties of a Logic Programming Language", in *Logic Programming*, K. Clark and S.-A. Tarnlund, eds., Academic Press, 1982, pp. 267-280.

[16] Heilbrunner, S., Holldobler, S., "The Undecidability of the Unification and Matching Problem for Canonical Theories", Acta Informatica, Vol. 24, No. 2, 1987, pp. 157-171.

[17] Hullot, J.M., "Canonical Forms and Unification", Proc. Fifth Conf. on Automated Deduction, Les Arcs, France, 1980, pp. 318-334.

[18] Hussmann H., "Unification in Conditional-Equational Theories", Proc. EUROCAL 85 Conf., Linz, 1985.

[19] Jaffar, J. , Lassez, J.L., Maher, M.J., "A Theory of Complete Logic Programs with Equality", J. Logic Programming, Vol. 1, No. 3, 1984, pp. 211-223.

[20] Kaplan, K., "Fair Conditional Term Rewriting Systems: Unification, Termination and Confluency", Technical Report, Laboratoires de Recherche en Informatique, U. Paris-Sud, Orsay, France, 1984.

[21] Kowalski, R., "Logic Programming", Information Processing, R.E.A. Mason, ed., Elsevier Science Publishers B.V. (North-Holland), 1983, pp. 133-145.

[22] Nutt, W., Rety, P., Smolka, G., "Basic Narrowing Revisited", Technical Report SR-87-07, U. Kaiserslautern, 1987.

[23] Pereira, F.C.N., Warren, H.D., "Definite Clause Grammars for Language Analysis: a Survey of the Formalism and a Comparison with the Augmented Transition Networks", Artificial Intelligence 13, 1980, pp. 231-278.

[24] Plotkin, G.D., "Building-in Equational Theories", in Machine Intelligence, Vol. 7, B. Meltzer and D. Michie (eds.), Halsted, Wiley, NY, 1972, pp. 73-90.

[25] Reddy, U.S., "Narrowing as the Operational Semantics of Functional Languages, Proc. IEEE Intl. Symp. on Logic Programming, Boston, MA, 1985, pp. 138-151.

[26] Robinson, J.A., "A Machine-Oriented Logic Based on the Resolution Principle", J. Association for Computing Machinery, Vol. 20, 1965, pp. 23-41.

[27] Slagle, J.R., "Automated Theorem-Proving for Theories with Simplifiers, Commutativity, and Associativity", J. Association for Computing Machinery, Vol. 21, No 4, 1974, pp. 622-642.

[28] Subrahmanyam, P.A., You, J.H., "Conceptual Basis and Evaluation Strategies for Integrating Functional and Logic Programming", Proc. IEEE Intl. Symp. on Logic Programming, Atlantic City, NJ, 1984, pp. 144-153.

[29] Tamaki, H., "Semantics of a Logic Programming Language with a Reducibility Predicate", Proc. IEEE Symposium on Logic Programming, Atlantic City, Feb. 1984, pp. 259-264.

[30] Togashi, A., Noguchi, S., "A Program Transformation from Equational Programs into Logic Programs", J. Logic Programming 4, 1987, pp. 85-103.

[31] Van Emden, M.H., Yukawa, K., "Logic Programming with Equations", J. Logic Programming 4, 1987, pp. 265-288.

[32] You, J.-H., Subrahmanyam, P.A., "E-Unification Algorithms for a Class of Confluent Term Rewriting Systems", Intl. Colloquium on Automata, Languages and Programming Languages, LNCS 226, Springer-Verlag, 1986, pp. 454-463.

NARROWING AND RESOLUTION IN LOGIC-FUNCTIONAL PROGRAMMING LANGUAGES

Elio Giovannetti

Dipartimento di Informatica
Università degli Studi di Torino
Corso Svizzera 185
10149 Torino, Italy

(Abstract)

The lecture will deal with the rôle of narrowing and/or resolution in the framework of integration between logic and functional programming. Narrowing and flattening plus resolution are basically equivalent methods for performing semantical unification. The conditions under whoich this equivalence holds are carefully explained, and various "relative completeness" properties for narrowing-based and resolution-based semantic unification algorithms are derived, along with their application to the design and implementation of the logic-functional programming language K-LEAF. In addition, a sketch of the latest development of the K-LEAF/IDEAL logic-functional integration will be traced, especially for as concerns the implementation and the semantics of the higher-order component.

Nondeterministic Algebraic Specifications and Nonconfluent Term Rewriting

Heinrich Hussmann

Universität Passau
Postfach 2540
D-8390 Passau

Abstract
Algebraic specifications are generalized to the case of nondeterministic operations by admitting models with set-valued functions (multi-algebras). General (in particular, non-confluent) term rewriting systems are considered as a specification language for this semantic framework. Correctness, completeness and initiality results are given.

1. Introduction

This paper is concerned with a generalization of the algebraic specification approach to nondeterministic operations. It is quite obvious why this is an interesting goal: Nondeterminism has been introduced in order to obtain a more abstract view of a class of complex problems. Thus a rigorous abstract approach to software specification either has to integrate this phenomenon or to show that it is superfluous.

In the early days of algebraic specifications there were some approaches which aimed at the simulation of nondeterminism within the framework of classical algebraic specifications. Typical work in this direction was [Subrahmanyam 81] and [Broy, Wirsing 81]. Unfortunately, a number of paradigms of algebraic specifications were lost within these approaches, in particular, the direct relation between a syntactic term and its semantic counterpart (interpretation). More recent work has tried the other way, to change the basic concepts of algebraic specifications in order to cope with nondeterminism. [Kaplan 88] still tries to refine the classical theory, but he assumes a particular built-in "choice" operator with special syntactical and semantical properties, whereas [Nipkow 86] and [Hesselink 88] switch over to a drastically different semantical framework. Both papers take as the semantical basis for nondeterministic specifications the so-called *multi-algebras* which assume operations to be set-valued. But both papers definitely exclude the question of an appropriate specification language for nondeterministic specifications. The paper at hands is based on multi-algebras and tries to fill this gap by considering a nondeterministic algebraic specification language.

In order to keep with the basic principles of algebraic specifcations, it is important to design a calculus which is easy to handle in the sense that it can be treated mechanically. The success of equational algebraic specifications is based to a great deal on the existence of interpreters which use term rewriting techniques for assigning an operational semantics to the axiom set ("algebraic programming"). As is well known, the main prerequisite for such a treatment of the equations is the *confluence* condition.

The basic idea of this paper is to use *term rewriting systems* as a specification language for nondeterministic algebraic specifications. Corresponding to the nature of nondeterministic computations, a *set* of possible outcomes of a reduction process is appropriate. Hence we drop the confluence condition and treat *general term rewriting systems without a confluence condition.* This approach is interesting from two different points of view:

- On the one hand, the ready-made theory for the term-rewriting calculus together with a number of software tools can be used to analyse nondeterministic specifications.
- On the other hand, the new approach gives a clean semantics for applications of term rewriting in algebraic programming where the confluence condition is violated.

The following text is a sketch of the basic ideas which are worked out in detail in [Hussmann 88]. Proofs are omitted due to space limitations.

2. Basic Notions

Familiarity of the reader with the basic theory of algebraic specifications and term rewriting is presupposed. We use a notation which is similar to [Wirsing et al. 83] for algebraic specifications and similar to [Huet, Oppen 81] for term rewriting systems. In particular:

The symbol Σ always means a *signature* (i.e. a set of sort symbols S and a set of function symbols F, together with functionalities for them). $X = (X_s)_{s\in S}$ is a family of countably infinite sets of *variable symbols* for every sort s. $W(\Sigma,X)_s$ denotes the set of *terms* (of sort s) built out of the function symbols in F and the variable symbols in X. The set $W(\Sigma,\emptyset)_s$ of *ground terms* is abbreviated by $W(\Sigma)_s$. The sort indices ore omitted where they are obvious from the context. A *substitution* is a family of mappings $\sigma = (\sigma_s)_{s\in S}$ where σ_s: $X_s \to W(\Sigma,X)_s$ and $|\{ x\in X \mid \sigma(x) \neq x \}| \in \mathbb{N}$. A substitution σ is extended to an endomorphism on $W(\Sigma,X)$ by $\sigma(f(t_1,\ldots,t_n)) = f(\sigma t_1,\ldots,\sigma t_n)$.

Given a *term rewriting system* R (i.e. a set of rules of the form $l \to r$, where l and r are terms of the same sort) we denote by $\to_R{}^*$ the *term rewriting relation* generated by R (i.e. the smallest reflexive and transitive relation on terms which contains all instances of rules in R and which is a congruence wrt. to the term-building operations).

For a proper treatment of set-valued functions we need the following notions for *power sets*:

$$\wp(M) = \{N \mid N \subseteq M\} \qquad \wp^+(M) = \{N \mid N \subseteq M \wedge N \neq \emptyset\} .$$

3. Nondeterministic Algebraic Specifications

An algorithm is called *nondeterministic* if it allows for states of computation where the continuation state is ambiguous, i. e. there is a free, non-predictable choice of the algorithm between a number of alternatives. The most important design decision for the new theory is how to combine this nondeterminicity with the classical concept of an algebra. Since we want to keep close to the standard concepts, we still think of an operation f^A: $s^A \to s^A$ within an algebra as an operational unit

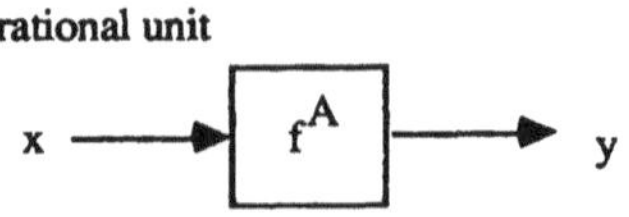

where the input to f^A is a *single data object* out of the carrier set s^A, and the output of f^A again is a single data object of s^A. The only difference to the standard case is that f^A, given a concrete input value x, can choose freely among a number of possibilities for its output value y. If we want to describe this input-output relation mathematically, it is quite natural to use a set-valued function

$$f^A: s^A \to \wp^+(s^A)$$

which gives the set of possible outcomes for each concrete input value. The basic paradigm that the input to f^A is a data object (and not a set of data objects!), has important consequences for an appropriate specification language. In particular, logical variables should range over basic (deterministic) objects, not over nondeterministic expressions.

The ideas above can be summarized in the following definition which is consistent with the notion of a multi-algebra found e.g. in [Pickett 67] and [Hansoul 83]:

A *Σ-multi-algebra* A is a tuple $A = (S^A , F^A)$, where
S^A is a family of non-empty carrier sets: $S^A = (s^A)_{s \in S}$, $s^A \neq \emptyset$ for $s \in S$
F^A is a family of set-valued functions: $F^A = (f^A)_{f \in F}$
where $f^A: s_1{}^A \times \ldots \times s_n{}^A \to P^+(s^A)$ for $[f: s_1 \times \ldots \times s_n \to s] \in F$.
The class of all Σ-multi-algebras is called MAlg(Σ).

The interpretation of terms within a Σ-multi-algebra is defined by an *additive* extension of the semantics for the function symbols. As it was mentioned above, variables within a term are to be interpreted by objects of the resp. carrier set:

Let $A = (S^A , F^A)$ be a Σ-multi-algebra.
A *valuation* β from X into A is a family $\beta = (\beta_s)_{s \in S}$ of mappings $\beta_s: X_s \to s^A$.
The *interpretation* I^A_β is a family of mappings

$$I^A_\beta = (I^A_{\beta,s})_{s \in S}, \qquad I^A_{\beta,s}: W(\Sigma,X)_s \to P^+(s^A) \qquad \text{for } s \in S.$$

$I^A_{\beta,s}$ is defined inductively as follows:

(1) If $t = x$ and $x \in X_s$, then $I^A_{\beta,s} [t] = \{ \beta(x) \}$

(2) If $t = f(t_1, \ldots, t_n)$ where $[f: s_1 \times \ldots \times s_n \to s] \in F$, then
$I^A_{\beta,s} [t] = \{ y \in f^A(x_1,\ldots,x_n) \mid x_i \in I^A_{\beta,s_i} [t_i] \}$.

If $t \in W(\Sigma)_s$, we write I^A instead of $I^A_{\beta,s}$. Again the indices for sorts are omitted frequently.

As the atomic constituent of axioms for specifying classes of multi-algebras we use *inclusion rules* as an oriented equivalent to equations. We denote an inclusion rule between two terms of the same sort by

$$t1 \to t2$$

which means informally: "t2 describes some of the possibilities for the values of t1". More formally:

Let A be a Σ-multi-algebra. An inclusion rule $t1 \to t2$ is *valid* in A ($A \models t1 \to t2$)
iff for all valuations β holds:

$$I^A_\beta [t1] \supseteq I^A_\beta [t2].$$

A *nondeterministic algebraic specification* is a tuple $T = (\Sigma, R)$, where Σ is a signature and R is a finite set of inclusion rules. A Σ-multi-algebra A is a *model* of T, iff for all inclusion rules $\Phi \in R$ holds: $A \models \Phi$. Mod(T) denotes the class of all models of T.

The following example is a specification of natural numbers, augmented by some nondeterministic operations:

spec NAT

sort Nat
func zero: → Nat, succ: Nat → Nat, add: Nat × Nat → Nat,
double: Nat → Nat, or: Nat × Nat → Nat, some: → Nat
rules
add(zero,x) → x, add(succ(x),y) → succ(add(x,y)), double(x) → add(x,x),
or(x,y) → x, or(x,y) → y,
some → zero, some → succ(some)
end

A model N of NAT takes the natural numbers $\mathbb{N}$ as the carrier set Nat^N and defines the operations as follows:

$zero^N = \{ 0 \}$, $succ^N(n) = \{ n+1 \}$, $add^N(n,m) = \{ n+m \}$,
$double^N(n) = \{ 2n \}$, $or^N(n,m) = \{ n, m \}$, $some^N = \mathbb{N}$.

4. Nonconfluent term rewriting

Term rewriting techniques can be used for deciding the equivalence of terms within an equational specification if a *confluent* set of rewriting rules can be derived from the axiom set. We want to use term rewriting also for nondeterministic specifications but in this case it is natural to drop the confluence condition and to try to establish an oriented equivalent of the classical Birkhoff theorem (correctness and completeness):

$$t1 \rightarrow_R^* t2 \Leftrightarrow Mod(R) \models (t1 \rightarrow t2) \qquad (*)$$

Unfortunately, the correctness part of (*) does not hold without a serious restriction to the syntactical form of specifications. As an example consider the specification NAT and the model N as introduced above. Within the classical term rewriting calculus, we can derive from the axioms:

$$double(or(zero,succ(zero))) \rightarrow_R^* add(or(zero,succ(zero)),or(zero,succ(zero))) \rightarrow_R^*$$
$$add(zero,or(zero,succ(zero))) \rightarrow_R^* add(zero,succ(zero)) \rightarrow_R^* succ(zero)$$

But within the model N we get the interpretations:

$$I_\beta^N[double(or(zero,succ(zero)))] = \{ 0, 2 \}, \qquad I_\beta^N[succ(zero)] = \{ 1 \}.$$

The problem sketched above is well known from the theory of nondeterministic programming languages ([Benson 79], [Hennessy 80]). Our semantic framework contains functions with a so-called "call-time-choice" parameter passing mechanism (since a function takes only objects as its parameters), but the term rewriting calculus uses a "copy-rule" semantics which is called "run-time-choice" (a function takes nondeterministic expressions as its parameters). A first, naive idea for solving the problem is to exclude the problematic cases:

A term $t \in W(S,X)$ is called *linear*, iff it does not contain multiple occurrences of variables. An inclusion rule $l \rightarrow r$ is *right-linear*, iff r is linear. A set R of inclusion rules is right-linear iff all rules in R are right linear.

Theorem 1

Let $T = (\Sigma, R)$ be a nondeterministic algebraic specification where R is right-linear and $l \notin X$ for all $l \rightarrow r \in R$. Then for $t1, t2 \in W(\Sigma,X)$:

$$Mod(T) \models t1 \rightarrow t2 \quad \Leftrightarrow \quad t1 \rightarrow_R^* t2.$$

Unfortunately, the restriction to right linear specifications is completely inacceptable from a practical point of view. For instance, the axiom for double in NAT above violates the restriction, and so do the classical rules for multiplication.

The folowing section gives an alternative solution to the problem by reconsidering the relationship of deterministic and nondeterministic values.

5. Nondeterministic Specifications wrt. a Deterministic Basis

In this section, an extension of the specification language is given which allows to designate syntactically the deterministic "basis" part of a specification. The deterministic terms correspond to the data objects in the model-theoretic semantics.

Our approach is inspired by an analogy to problems during the extension of algebraic specifications to partial functions ([Broy, Wirsing 82]). There the equational calculus turns out to be incorrect as soon as the instantiation of variables with "undefined" terms is admitted. The remedy chosen there is to introduce a *definedness predicate* which allows to restrict the instances of rules to defined terms. Analogously we introduce now a *determinicity predicate*.

A *(Σ,X-)DET-rule* is of the form

DET(t) where $t \in W(\Sigma,X)$.

This rule is valid in a Σ-multi-algebra A ($A \models DET(t)$) iff for all valuations β holds:

$$| I^{A,\beta}[t] | = 1.$$

In the following, the terms "algebraic specification" and "model" are understood to include DET-rules, too.

For instance, the specification NAT above should be enriched (at least) by the axioms

DET(zero), DET(succ(n)) .

The term rewriting relation has to be changed analoguously. The most important change is that an instance of a rule is only admitted if for all the instantiated terms the determinicity has been proved. We give below a definition of the new calculus (which now differs from the classical term rewriting definitions):

Let $T = (\Sigma, R)$ be a nondeterministic algebraic specification (including DET-rules). A formula $t1 \rightarrow t2$ or DET(t) is *derivable* within T ($T \vdash t1 \rightarrow t2$ or $T \vdash DET(t)$) iff there is a derivation for it using the following logical deduction rules:

(REFL)
$$\frac{}{t \to t} \qquad \text{if } t \in W(\Sigma,X)$$

(TRANS)
$$\frac{t1 \to t2,\ t2 \to t3}{t1 \to t3} \qquad \text{if } t1, t2, t3 \in W(\Sigma,X)$$

(CONG)
$$\frac{t_i \to t_i'}{f(t_1, \ldots, t_{i-1}, t_i, t_{i+1}, \ldots, t_n) \to f(t_1, \ldots, t_{i-1}, t_i', t_{i+1}, \ldots, t_n)}$$

$$\text{if } [f: s_1 \times \ldots \times s_n \to s] \in F,\quad t_j \in W(\Sigma,X)_{s_j} \text{ for } j \in \{1, \ldots, n\},\ t_i' \in W(\Sigma,X)_{s_i}$$

(RULE-1)
$$\frac{DET(\sigma x_1), \ldots, DET(\sigma x_n)}{\sigma l \to \sigma r} \qquad \text{if } \langle l \to r \rangle \in R,\ \sigma: X \to W(\Sigma,X),\ \{x_1, \ldots, x_n\} = Vars(l) \cup Vars(r)$$

(RULE-2)
$$\frac{DET(\sigma x_1), \ldots, DET(\sigma x_n)}{DET(\sigma t)} \qquad \text{if } \langle DET(t) \rangle \in R,\ \sigma: X \to W(\Sigma,X),\ \{x_1, \ldots, x_n\} = Vars(t)$$

(DET-D)
$$\frac{DET(t1),\ t1 \to t2}{DET(t2)} \qquad \text{if } t1, t2 \in W(\Sigma,X)$$

(DET-R)
$$\frac{DET(t1),\ t1 \to t2}{t2 \to t1} \qquad \text{if } t1, t2 \in W(\Sigma,X)$$

Now we can prove a correctness result without the right-linearity condition:

Theorem 2

Let $T=(\Sigma,R)$ be a nondeterministic algebraic specification. For $t1, t2 \in W(\Sigma,X)$ holds:

$$T \vdash t1 \to t2 \quad \Rightarrow \quad Mod(T) \models t1 \to t2$$
$$T \vdash DET(t1) \quad \Rightarrow \quad Mod(T) \models DET(t1)$$

It may be considered unsatisfactory that our new correct calculus is different to the term rewriting calculus. But it is quite easy to simulate the calculus for a large class of specifications by a standard term rewriting machine which is able to handle *conditional rewriting rules*. This is possible whenever no application of the (DET-R) rule is necessary. Then we just have to add a predicate simulating DET, to add the DET-rules and to prefix each axiom by determinicity conditions for its variables. Moreover, it turns out that some term rewriting machines use implementation tricks which automatically circumvent the difficulties of run-time-choice. The system RAP ([Hussmann 85/87]), for instance, can be used correctly for the specification NAT above without any change, because it uses DAGs for the representation of terms. Terms instantiated for a variable are *shared* automatically within this data structure, similar to [Astesiano, Costa 79].

In order to get similar results to those in the classical theory, we would like to prove a completeness result by the construction of a term model which should be initial within the model class. Unfortunately, there are specifications where this construction fails.

6. Additive Specifications

Obviously, the carrier sets of a simple (initial) term model should be formed by the terms for which determinicity can be derived. The following specification represents a typical kind of a "non-local" specification style which inhibits such a construction:

spec NI

sort s

func a: → s, b: → s, g: → s, f: s → s

rules

g → a, f(a) → b, f(b) → a, f(g) → a,

DET(a), DET(b)

end

This specification does not describe exactly the behaviour of *f* and *g* on deterministic terms, but gives some information about how *f* and *g* work together . If we interpret *g* by containing *a* only, we automatically have to assume *f(a) → a*, and conversely, if we want *f(a)* to contain *b* only, we have to assume *g → b*.

The property which is missed by the specification NI above can be formalised as follows:

A specification T=(Σ,R) is called *DET-additive* iff:

$\forall [f: s_1 \times \ldots \times s_n \rightarrow s] \in F$:

$\forall\ t_1 \in W(\Sigma)_{s_1}, \ldots, t_n \in W(\Sigma)_{s_n}, t \in W(\Sigma)_s$:

$T \vdash f(t_1,\ldots,t_n) \rightarrow t \ \wedge\ T \vdash DET(t) \ \Rightarrow$

$\exists\ t_1' \in W(\Sigma)_{s_1}, \ldots, t_n' \in W(\Sigma)_{s_n}$:

$T \vdash f(t_1',\ldots,t_n') \rightarrow t \ \wedge$

$T \vdash t_1 \rightarrow t_1' \wedge \ldots \wedge T \vdash t_n \rightarrow t_n' \wedge\ T \vdash DET(t_1') \wedge \ldots \wedge T \vdash DET(t_n')$

Another property is useful in order to construct term models:

A specification T=(Σ,R) is called *DET-complete* iff:

$\forall t \in W(\Sigma): \exists t' \in W(\Sigma):\ T \vdash t \rightarrow t' \ \wedge\ T \vdash DET(t')$.

DET-completeness is very similar to ***sufficient completeness*** w. r. t. to a base part of a specification. DET-additivity ensures another kind of completeness of the specification. This means that the whole behaviour of the specified operations can be simulated by the additive extension of operations on deterministic terms.

The DET-addditivity of a specification is not easy to be checked in general. But it holds for an important class of specifications:

A specification is called *constructor-based* iff there is a subset $C \subseteq F$ of the function symbols which are assumed to be determinate in all models and if all axioms are of the form

$f(c_1,\ldots,c_n) \to r$ where $f \in F \setminus C$, all the c_i are built of variables and C-functions only.

Constructor-based specifications are automatically DET-additive.

For a DET-complete and DET-additive specification we are able to construct a term model D which takes as its carrier sets the ground terms t for which DET(t) is deducible (more precisely, congruence classes of such terms) and which interprets terms as follows:

$I^D_\sigma[t] = \{ [t'] \mid \vdash DET(t') \wedge \vdash \sigma t \to t' \}$.

Using this model, we get a (weak) completeness result:

Theorem 3

Let $T=(\Sigma,R)$ be a DET-complete and DET-additive specification, $A \in Mod(T)$. Then for $t1, t2 \in W(\Sigma,X)$ holds:

$T \vdash DET(t2) \;\wedge\; Mod(T) \models t1 \to t2 \;\Rightarrow\; T \vdash t1 \to t2.$

In order to get a classical initiality result, we have to restrict the model class, too, as it is shown below.

7. Maximally Deterministic Algebras

Before stating an initiality result we should clarify the notion of a homomorphism:

Let $\Sigma = (S, F)$ be a signature, $A, B \in MAlg(\Sigma)$. A *(tight)Σ-homomorphism* φ from A to B is a family of mappings $\varphi = (\varphi_s)_{s \in S}$

$\varphi_s\colon s^A \to \wp^+(s^B)$,

where for all $[f\colon s_1 \times \ldots \times s_n \to s] \in F$ and all $e_1 \in s_1^A, \ldots, e_n \in s_n^A$ holds:

(*) $\{ e' \in \varphi_s(e) \mid e \in f^A(e_1, \ldots, e_n) \} = \{ e' \in f^B(e_1', \ldots, e_n') \mid e_1' \in \varphi_{s_1}(e_1), \ldots, e_n' \in \varphi_{s_n}(e_n) \}$

φ is called a *loose* Σ-homomorphism, iff in (*) only the inclusion "$\subseteq$" holds instead of equality.

The definition above is a bit more general than hitherto in research ([Pickett 67], [Hansoul 83], [Nipkow 86], [Hesselink 88]), where homomorphisms for multi-algebras are defined pointwise (i.e. $\varphi_s\colon s^A \to s^B$). Our definition covers the pointwise definition as a special case.

We can designate now a class of Σ-multi-algebras which is particularly well-adapted to specifications with a deterministic basis. We get the nwe notion by generalization of the "no-junk" principle. In a classical algebra "no junk" means that every object of a carrier set is the interpretation of some term. Here we want that every object of a carrier set is the interpretation of a deterministic term and, moreover, we would like to force the models to avoid superfluous objects also within the interpretation of nondeterministic terms.

A' is called a *refinement* of A, iff there is a loose Σ-homomorphism $\varphi\colon A' \to A$.

A' is called *more deterministic* than A, iff $\forall\, t \in W(\Sigma)\colon\ |I^A[t]| \geq |I^{A'}[t]|$.

A is called *maximally deterministic*, iff A is more deterministic than all refinements of A.

A model A of a specification $T=(\Sigma,R)$ is called *term-generated* iff

$\forall s \in S: \forall e \in s^A: \exists t \in W(\Sigma)_s: I^A[t] = \{ e \}.$

The class of all term-generated and maximally deterministic models of T is called MGen(T).

An alternative syntactical characterization may be easier to understand:

Lemma

Let $T = (\Sigma, R)$ be a DET-additive and DET-complete specification, $A \in$ Gen(T). Then A is maximally deterministic iff

$$\forall t \in W(\Sigma): \forall x \in I^A[t]: \exists t' \in W(\Sigma): T \vdash t \to t' \wedge T \vdash DET(t') \wedge I^A[t'] = \{ x \}.$$

For the term model D which has been sketched above, we now have the initiality result:

Theorem 4

Let $T=(\Sigma,R)$ be a DET-complete and DET-additive specification. Then there is a unique loose homomorphism from D to any model in Mod(T), and there is a unique tight homomorphism from D to any model in MGen(T).

It turns out that this initiality result is compatible to the pointwise notion of homomorphism, i.e. the unique homomorphism from D to the models always has a singleton set as its result.

Now we can state that the computations within the term rewriting model D are correct and complete for the intended model class. This means:

$$D \models t1 \to t2 \quad \Leftrightarrow \quad MGen(T) \models t1 \to t2$$

8. Concluding Remarks

We have shown that term rewriting without a confluence condition is an appropriate specification language for the nondeterministic algebras as used in [Nipkow 86] and [Hesselink 88]. We had to impose restrictions on the specifications and on the term rewriting calculus in order to get significant results. But a large class of practically relevant specifications fulfills the restriction on specifications; and the restricted calculus can be simulated for this class easily using existing software tools for term rewriting.

It is interesting that [Kaplan 88], although using a different approach by specifying the set union instead of the set intersection, comes up with a restriction to a very similar kind of specifications (regular specifications). The question which one of both approaches is more powerful for the treatment of practical problems, can be answered only from practical experiments.

As an application for our approach, it is possible to write down a specification of the transitions for communicationg sequential processes (CSP) using nonconfluent rewrite rules and to prototype this specification using the RAP system. Although this specification is inconsistent within the classical semantics, RAP correctly and completely enumerates traces and refusals for example processes. RAP is particularly well-suited for these enumerations because its narrowing algorithm already contains the features for handling a search tree of terms. (Even the narrowing process seems to work, an interesting question of further research.)

Another important question to be treated is to find out, under which conditions the efficient technique of confluent rewriting can be combined with the general (nonconfluent) algorithm, to keep the search tree as small as possible. Using functions rather than predicates, uch a general term rewriting algorithm, extended to narrowing, could become an attractive alternative to PROLOG.

References:

[Astesiano, Costa 79]
E. Astesiano, G. Costa, Sharing in nondeterminism, in: H. A. Maurer (ed.), *6th Internat. Coll on Automata, Languages and Programming*, Lecture Notes in Computer Science 71 (Springer, Berlin, 1979), 1-13.

[Benson 79]
D. B. Benson, Parameter passing in nondeterministic recursive programs, *J. Comput. System Sci.* **19** (1979) 50-62.

[Broy, Wirsing 81]
M. Broy, M. Wirsing, On the algebraic specification of nondeterministic programming languages, in: E. Astesiano, C. Böhm (eds.), *6th Coll. on Trees in Algebra and Programming*, Lecture Notes in Computer Science **112** (Springer, Berlin, 1981) 162-179.

[Broy, Wirsing 82]
M. Broy, M. Wirsing, Partial abstract types, *Acta Informat.* **18** (1982) 47-64.

[Hansoul 83]
G. E. Hansoul, A subdirect decomposition theorem for multialgebras, *Algebra Universalis*, **16** (1983) 275-281.

[Hennessy 80]
M. C. B. Hennessy, The semantics of call-by-value and call-by-name in a nondeterministic environment. *SIAM J. Comput.* **9** (1980) 67-84.

[Hesselink 88]
W. H. Hesselink, A mathematical approach to nondeterminism in data types. *ACM Trans. on Progr. Languages and System* **10** (1988) 87-117

[Huet, Oppen 81]
G. Huet, D. C. Oppen, Equations and rewrite rules: a survey, in: R. V. Book (ed.), *Formal Language Theory: Perspectives and Open Problems*, (Academic Press,New York,1980).

[Hussmann 85/87]
H. Hussmann, Rapid prototyping for algebraic specifications - RAP system user´s manual. Report MIP-8504, Universität Passau, Passau, 1985, 2nd, extended edition 1987.

[Hussmann 88]
H. Hussmann, Nichtdeterministische Algebraische Spezifikationen (in German), Ph. D. thesis, University of Passau, 1988.

[Kaplan 88]
S. Kaplan, Rewriting with a nondeterministic choice operator, *Theoret. Comput. Sci.* **56** (1988) 37-57.

[Nipkow 86]
T. Nipkow, Nondeterministic data types: Models and implementations, *Acta Informat.* **22**(1986) 629-661.

[Pickett 67]
H. E. Pickett, Homomorphisms and subalgebras of multialgebras, *Pacific J. of Math.* **21** (1967) 327-342.

[Subrahmanyam 81]
P. A. Subrahmanyam, Nondeterminism in abstract data types, in: S. Even, O. Kariv (eds.), *8th Internat. Coll. on Algorithms, Languages and Programming*, Lecture Notes in Computer Science **115** (Springer, Berlin, 1981) 148-164.

[Wirsing et al. 83]
M. Wirsing, P. Pepper, H. Partsch, W. Dosch, M. Broy, On hierarchies of abstract data types, *Acta Informat.* **20** (1983) 1-33.

OBJ: Programming with Equalities, Subsorts, Overloading and Parameterization*

Jean-Pierre Jouannaud
LRI†

Claude Kirchner
LORIA & CRIN‡

Hélène Kirchner
LORIA & CRIN

Aristide Mégrelis
LORIA & CRIN

Abstract

OBJ is a declarative language, with mathematical semantics given by order-sorted equational logic and an efficient operational semantics based on order-sorted term-rewriting. In addition, OBJ has a uniquely powerful generic module mechanism, including non-executable "theories" as well as executable "objects," plus "module expressions" that describe and construct whole subsystems. OBJ also has user-definable abstract data types with user-definable mixfix syntax and a powerful and flexible type system that supports overloading and subtypes. OBJ is thus a wide-spectrum language that elegantly integrates coding, specification and design into a single framework.

1 Introduction

The OBJ programming language was developed first at UCLA [20,9,19], then at SRI International [18,8,7], and now at several sites around the world [5,1,38,3].

OBJ is a logical programming language whose latest versions (OBJ-2 [4] and OBJ-3 [10]) are based on order-sorted equational logic. Programs are order-sorted equational specifications and computation is an efficient form of equational deduction by rewriting.

Type structure supports conceptual clarity and detection of many errors at program entry time. However, implementing strong typing by many-sorted logic can be too rigid and lacks the expressive power needed for handling errors and partiality [22]. Many of these problems are overcome by an order-sorted type structure that brings inheritance, operator overloading, and error handling within the realm of equational logic, and makes many seemingly partial or problematic functions total and well-defined on the right subsort [16,17]. An order-sorted algebra is like a many-sorted one, except that there is a partial ordering on the set of sorts that is interpreted as set inclusion. For instance, a subsort relation `Nat < Int` is interpreted as the inclusion $N \subseteq Z$ of the naturals into the integers in the standard model. In addition, function symbols such as `_+_` may be overloaded as `_+_: Nat Nat -> Nat`, `_+_: Int Int -> Int`, and are required to agree in their results when restricted to arguments in the same subsorts. All the basic results of equational logic generalize to the many-sorted case [16]. However, order-sorted deduction is more subtle, and its relationship to concepts such as replacement of equals for equals and term rewriting requires a careful analysis that was initiated in [21] and is further developed in [31].

We briefly summarize in Section 2 current features of the OBJ language, and in Section 3, discuss some specifics of the implementations. Then Section 4 proposes some future developments.

2 OBJ features

Modularity and strong typing are two important features of a programming language, supported by OBJ in a powerful way.

*Supported by CNRS (Greco Programmation) and INRIA (France).
†Université de Paris Sud. Centre d'Orsay. Bat 490. 91405 Orsay Cedex, France
‡Campus Scientifique. BP 239. 54506 Vandœuvre-les-Nancy Cedex, France

2.1 Modules and generics

OBJ has three basic kinds of building block:

Objects declare new sorts of data, and define new operations by equations, which become executable code when interpreted as rewrite rules.

Theories also declare sorts, operations and equations, but these are used to define properties of modules, including requirements for generic modules interfaces; thus, their equations are not considered executable. Objects and theories are both **modules**.

Views express bindings of actual modules to requirements, and are used to combine modules into larger program units.

OBJ modules can be parameterized (generic), or can also import other modules directly and then use their capabilities; this leads to the module hierarchy. Interface declarations for parameterized modules are not purely syntactic, as for example in Ada, but are given by theories that may contain semantic requirements. Module expressions can be seen as generalizing the UNIX `make` command to allow generic modules. All this makes code much more reusable, since it can be tuned for different applications by modifying module expressions and parameter values. Moreover, debugging, maintenance, readibility and portability are all improved.

2.2 Subsorts and polymorphism

One sort of data is often contained in another. For example, the natural numbers are contained in the integers; thus the sort `Nat` is a subsort of `Int`. Moreover an operation like `+` can be defined at several levels of the sort hierarchy, and have the same values on common domains. The OBJ subsort mechanism provides, within the framework of an initial algebra semantics, the following capabilities:

- Polymorphism with overloaded operations, as in the example above, but also within parameterized modules, since their operations are available to all their instances.
- Multiple inheritance in the sense of object-oriented programming, allowing a sort to be a subsort of two or more others.
- The common difficulties for abstract data types with partial operations (such as `tail` for lists) disappear by viewing the operations as total on the right subsorts.
- Errors can be treated in several styles, without the need for any special error handling mechanisms, either syntactic or semantic.

3 OBJ specifics

3.1 Operational semantics

The OBJ-2 and OBJ-3 systems implement the same mathematical semantics based on order-sorted equational logic, but differ in their operational semantics. For OBJ-2[1], the operational semantics [21] used a translation of order-sorted algebra into many-sorted algebra that reduced computation to standard term rewriting but could generate many rules with an associated loss in efficiency. By contrast, OBJ-3[2] achieves a much more simple and efficient solution through the direct performance of *order-sorted rewriting*.

Simple examples of integer and rational arithmetic may illustrate some of the peculiarities involved in the operational semantics. Consider for instance the axiom

$$\mathtt{N} + \mathtt{s}(\mathtt{M}) = \mathtt{s}(\mathtt{N} + \mathtt{M}) \qquad (1)$$

[1] OBJ-2 has been implemented in 1983-84 at SRI International

[2] OBJ-3 has been implemented in 1985-87 at SRI International in Kyoto Common Lisp [39]

where N and M are variables of sort Nat and s : Nat -> Nat is the successor function. If we apply such axiom as a rewrite rule to the integer expression

$$(-4) + \mathtt{s}(0) \tag{2}$$

we would obtain the term s((-4) + 0), an ill-formed term if we assume that the successor function only exists for the naturals and not for the integers. Therefore, it becomes crucial to check the sorts of the variables when matching a lefthand side. Ascertaining the sort of a subterm would in principle require parsing. However, such sort information can be precomputed at parse time and kept stored in the term to be reduced, in order to be used when needed or to be entirely disregarded otherwise. Our expression then would look as follows:

```
                     +(Int Int, Int)
                    /   \
(NzInt, NzInt )-         s(Nat,NzNat)
               |         |
         (NzNat)4        0(Nat)
```

where NzNat (resp. NzInt) is the subsort of nonzero natural numbers (resp. nonzero integers) and we assume that -_ is overloaded as -_: Int -> Int and -_: NzInt -> NzInt. Such a form is called the *lowest parse* of the expression, and we have placed the ranks of the operators in parenthesis to suggest that the rank information can be forgotten and the tree can then be considered identical to the original expression; i.e. we can "turn" that information "on and off" depending on the need. For instance, we can match our axiom in the standard way, forgetting rank information, but when reading a variable, we then inspect the sort of the term matched to it to ensure correctness. In this case, the variable N matched a term of sort NzInt resulting in failure. Even better, we may associate the axiom (1) with the top operator +(Nat Nat, Nat) obtained by lowest-parsing its left-hand side. In that case, we would not even attempt matching (1) to (2), since only terms with top operator +(Nat Nat, Nat) would be candidates. Also, in that context, there is no need whatsoever to check the sorts of the variables N and M, since they are as general as possible for + of rank (Nat Nat, Nat). By contrast, the sort of the integer variable I in the rule

$$\mathtt{I} * (\mathtt{R}/\mathtt{R}') = (\mathtt{I} * \mathtt{R})/\mathtt{R}' \tag{3}$$

(where R has sort Rat, R' sort NzRat, and _/_ denotes rational division) is not as general as possible and has to be checked, even after associating (3) to the operator _*_(Rat Rat, Rat). Note that, assuming that we also had the operator _*_(NzRat NzRat, NzRat) and _/_(NzRat NzRat, NzRat), we should also associate (3) to the operator _*_(NzRat NzRat, NzRat) in order to rewrite the expression 7*(3/4), i.e. we have to "specialize" axiom (3) to smaller top operators that are possible for instances of the left-hand side.

After matching succeeds, the instance of the left-hand side has to be replaced by the corresponding instance of the right-hand side. This can be done in such a way that the appropriate lowest parse of the right-hand side is compiled in advance and we obtain a right-hand side instance in lowest-parsed form in a straightforward way. In summary, the following approach is taken: At compile time:

1. localize rules by associating them to disambiguated top function symbols,
2. specialize rules to smaller overloaded top function symbols,
3. precompute lowest parses of right-hand side for different cases of left-hand side instances.

At run time:

1. look at the rank of the top function symbol and try only the rules associated to that operator,
2. match left-hand sides with complete disregard of the sort information, except for checking the sort of those variables that are not fully general,
3. replace the instance of the left-hand side by the precomputed instance of the right-hand side in lowest parsed form,

4. if necessary, update the rank information above the occurrence where the replacement has taken place whenever the sort has decreased at that occurrence.

The efficiency of such a process is entirely similar to standard term rewriting; the only minor differences are the possible need to check sorts, that can be done in constant time, and the occasional performance of step (4) that is done only by need. The gain in expressive power is considerable. For example, arithmetic computation can take place in the entire number hierarchy, with a syntax that entirely agrees with that of standard mathematical practice.

The necessary foundations for the operational semantics involve the following points:

1. Relating order-sorted equational deduction to order-sorted rewriting, giving conditions for the soundness and completeness of OBJ's operational semantics.

2. Generalizing ordinary rewriting to order-sorted rewriting modulo axioms, since OBJ supports rewriting modulo axioms such as commutativity, associativity and commutativity.

3. Studying in detail the process of specialization of rules, providing criteria and algorithms for such a process and for the associated aspects of generality of variables and precomputation of right-hand sides.

4. Developing correctness criteria for the operational semantics of modules that are structured in a hierarchical fashion with different semantic relations between modules in such a hierarchy, such as `using`, `extending` or `protecting`. In particular, developing correctness criteria for separate "compilation" of modules.

3.2 The mixfix parser

For the parser, a term is a well-formed string made of operator symbols, variables, parentheses and commas, such as '`f␣(␣pop␣(␣x␣)␣,␣y␣)`' (9 tokens) or '`i␣+␣j`' (3 tokens). In these examples, successive symbols are separated by spaces (marked '␣'). In general, these spaces are not necessary; they are used here mostly for readability and to avoid possible misunderstanding.

The notation used in programming languages for terms (and other constructs) can be cumbersome, since the user is limited by a syntax fixed by the language designer. OBJ avoids this difficulty by allowing user-definable syntax. The notation for operators may be functional, prefix, infix, postfix and even "mixfix" (where the operator is written in several parts around the operands). In this way, many parentheses can be avoided. For example, it is possible to declare "modulus" and "factorial" operators and write: `| z |`, `| z | ! !`, or even `[ 1 / | z | ] ! + 3 * 2 i`, where `[ r ]` designates the truncated part of a real number `r`.

To achieve this, the mechanism which is implemented takes two arguments into account. First, it uses information about sorts. Second, it uses precedence and associativity rules, which are given independently for each operator. To describe these features, it is useful at this point to introduce an example of a complete declaration of an operator, which looks like this:

```
op _ + _ :  Int Int -> Int [ prec 10 gather ( e E ) ] .
```

The attribute `prec` refers to the precedence of the operator `+`; `gather` refers to its gathering rule (our synonym of "associativity rule").

Sort information

Some parses are unacceptable simply because the arguments of some operator are of the wrong sort. We detect this early to avoid unnecessary processing. This bare rule is slighty modified to comply with two features of *OBJ*. First, the argument may be of a subsort of the expected sort (a natural number is right were an integer is expected); second, to allow the parsing of terms such as `( 8 / 4 ) !` (factorial of the integer 2), the argument may also be of a supersort of the expected sort. The handling of such "ill-formed" terms is described in the next section.

In this way, it is possible to discriminate the correct parse when processing a string such as: `length l !` (the factorial of the length of the list `l`, and not the list of the factorial of `l` which is meaningless).

Precedence and gathering rule

These two purely syntactic attributes work hand in hand. In designing them, we were inspired by some features available in some *PROLOG* implementations [2].

Precedence information is used to compute, for each substring likely to be a subterm, a quantity that we call ***weight*** (a better word for *precedence* could be ***density***). Then, comes the question whether that substring is acceptable as an argument to some higher-level operator. The weight of the substring is matched against the gathering rule of the higher-level operator and the question is decided.

To be clearer, we shall use the following example. We have declared:

```
op _ !  :  Int -> Int [ prec 2 gather ( E ) ] .
op _ ^ _ :  Int Int -> Int [ prec 5 gaher ( e E ) ] .
op _ * _ :  Int Int -> Int [ prec 10 gather ( e E ) ] .
op _ + _ :  Int Int -> Int [ prec 20 gather ( e E ) ] .
```

and we intend to parse `5 ^ 5 ^ 3 !  + 3 * 7 ^ 4`.

How to start computing the weights? — Substrings made of a constant, a variable or a function written functionnally, and parenthesized substrings have, by definition, a 0 weight. In the example above, `5`, as a substring, weighs 0. Other examples of substrings weighing 0 (in another context): `f ( x ), ( x + y )`.

How to combine subterms to build a larger subterm? — If not made of a constant, a variable or a function written functionnally, or if not parenthesized, strings are made using operators written in a ***mixfix*** notation. The rule in this case is that such an operator accepts as arguments marked `E` substrings whose weights are less than or equal to its precedence; and accepts as arguments marked `e` substrings whose weights are (strictly) less than its precedence. The weight of the total string is then the precedence of its "top" operator.

For instance, the substring `3 !` is made of the postfix operator `!` and a left argument which is the substring `3`: `3` of weight 0 is acceptable to `!`, for $0 \leq 2$; therefore `3 !` is parsable as a *subterm*. The weight of `3 !` is then the precedence of its "top" operator, in this case 2.

In the same way, `7 ^ 4` is parsable as a subterm and weighs 5. So that `3 * 7 ^ 4` is itself parsable as a subterm and weighs 10. The same rule applies recursively until the string is completely parsed.

It is easy to check that the string given above parses as a term, and as expected, i.e. as:

```
( 5 ^ ( 5 ^ ( 3 !  ) ) ) + ( 3 * ( 7 ^ 4 ) )
```

3.3 Type-checking in OBJ

The OBJ user declares sorts, subsorts, and functions with the only restriction that type-checking is decidable: an OBJ term must have exactly one type. Unfortunately, it may be the case that one term has several types. Let us be more precise in order to illustrate this problem.

Let OBJSPEC be an OBJ specification. To every term t, we associate a set of sorts, written *sort(t)* as follows:

1. If $(x : s) \in OBJSPEC$ then $s \in sort(x)$

2. If $f : s_1 ... s_n \rightarrow s \in OBJSPEC$ and $s_1 \in sort(t_1), \ldots, s_n \in sort(t_n)$ then $s \in sort(f(t_1 ... t_n))$

3. If $s < s' \in OBJSPEC$ and $s \in sort(t)$ then $s' \in sort(t)$

The type of t is now defined as its least sort, provided *sort(t)* has a least element. Signatures which provide a type for every term whose set of sorts is not empty are called regular. As a consequence, only regular signatures are allowed in OBJ. In case the user writes a non-regular signature, a warning is issued by the system. Here is an example, written in the OBJ syntax:

```
obj DUMMY is
sorts s s' s'' .
subsorts s<s', s<s'' .
op f : s' -> s' .
op f : s'' -> s'' .
op a : -> s .
jbo
```

The previous signature is not regular, because f(a) has the two sorts s' and s'', but no-one is smaller than the other. We can see here that problems may arise only because of overloading. In this case, a very natural rule is that overloading is "closed" with respect to subsorts. Here, for example, the signature can be made regular by adding the following new operator declaration: op f : s -> s.

So, we can clearly check at parsing time whether an OBJ term is well-formed, and if it has a type. The second property can actually be checked on the signature itself, which may improve the computation of types.

However, computing types of well-formed terms is not enough for practice: there are non-well-formed terms which we would like to assign a meaning to. Here is a simple famous example:

```
obj STACK [X :: TRIV] is
sorts stack non-empty-stack .
subsorts non-empty-stack < stack .
op : nil -> stack .
op push : elem stack -> non-empty-stack .
op pop_ : non-empty-stack -> stack .
op top_ : non-empty-stack -> elem .
var e : elem, s : stack .
eq : top push(e,s) == e .
eq : pop push(e,s) == s .
jbo
```

Now, instanciating X by INT yields a stack of integers:
obj STACK-INT is STACK[INT] jbo.

We may now compute STACK-INT expressions, for example: top push (1,nil) that gives the expected result 1.

But here is a more surprising example: the evaluation of pop pop push (1, push (2, push (3,nil))) gives the result parse failed.

The problem is that pop push (1, push (2, push (3,nil))) is of type stack, hence pop cannot be applied, since it requires an argument of sort non-empty-stack, which is strictly smaller than stack. Of course, subsorts would be almost useless if such expressions could not be computed, hence such terms must be parsed and computed.

Remark that the first subterm of the whole term, i.e. pop push(1, push (2, push (3,nil))) is well-formed, and results in push (2, push (3,nil)). Applying pop to the result is now possible and yields the expected stack: push (3,nil). So, we may well forget about terms whose set of sorts is empty, provided computations happen always on well-formed subterms, just as previously.

This can be formalized by constructing a conservative extension of stacks as follows [37]:

```
obj STACK [X :: TRIV] is
sorts stack non-empty-stack error-stack error-elem.
subsorts non-empty-stack < stack < error-stack .
subsorts elem < error-elem .
op : nil -> stack .
op push : elem stack -> non-empty-stack .
op pop_ : non-empty-stack -> stack .
op top_ : non-empty-stack -> elem .
op push : error-elem error-stack -> error-stack .
op pop_ : error-stack -> error-stack .
op top_ : error-stack -> error-elem .
var e : elem, s : stack .
eq : top push(e,s) == e .
eq : pop push(e,s) == s .
jbo
```

Now, `pop pop push (1, push (2, push (3,nil)))` has type `error-stack`, and its subterm `pop push (1, push (2, push (3,nil)))` is still of type `stack`. Remark that the last equation in STACK does not apply to the whole term, since sorts do not agree, but it of course applies on the subterm, yielding `pop push (2, push (3,nil))`. The type of the whole term is now `stack`, it has been lowered down, since the type of the first subterm has itself been lowered down from `stack` to `non-empty-stack`. Hence the same equation applies now on top, yielding the expected result.

This shows that type information cannot be computed only at parsing time, since types change at runtime, due to overloading and subsorts. However, typechecking can be made static by computing in a extended signature, as shown before. This signature is a conservative extension of the previous one, hence it behaves as expected.

A drawback of the previous extension is that it adds quite a lot of new sorts, and it forces the computations to be bottom-up. In some cases, for example computations on streams, bottom-up evaluation is not feasible, an other mechanism must be provided. Such a mechanism was provided in OBJ-2, and consisted of another conservative extension of signatures [21]:

```
obj STACK [X :: TRIV] is
sorts stack non-empty-stack .
subsorts non-empty-stack < stack .
op : nil -> stack .
op push : elem stack -> non-empty-stack .
op pop_ : non-empty-stack -> stack .
op top_ : non-empty-stack -> elem .
op c_ : non-empty-stack -> stack .
op r_ : stack -> non-empty-stack .
var e : elem, s : stack, ns : non-empty-stack .
eq : top push(e,s) == e .
eq : pop push(e,s) == s .
eq : r c ns == ns .
jbo
```

Injections from a subsort into a supersort are called *coercions*, while operators going the other way around are called *retracts*. The added equation says that the retract of a coerced expression yields that very same expression unchanged. But the converse is not true, coercing a retracted expression does not have any meaning in general.

Now, the previous expression is parsed with retracts and coercions and can be evaluated in the extended signature: `pop r pop push (1, c push (2, c push (3,nil)))` evaluates first to `pop r c push (2, c push (3,nil))`, then to `pop push (2, c push (3,nil))`, finally to `c push(3,nil)`, and the result is a non-empty-stack. Coercions and retracts act as type decorations, and the equation `r c ns == ns` allows checking types at runtime.

Note that the first conservative extension does not imply any overhead while the second implies some (very reasonnable) overhead. On the other hand, the second is not restricted to bottom-up strategies, hence is more general. Some work is currently being done for combining the two disciplines in order to have the advantages of both.

Sort constraints are another OBJ construct which requires systematic runtime type-checking. Since not all subsorts can be defined by constructors, OBJ allows to define a subsort of some sort s as the set of elements t of sort s which enjoy some equationally defined property. Type checking then requires of course evaluation, or as before, the definition of a conservative extension. Here is an example:

```
obj BOUNDED-STACK [X :: TRIV] is
sorts stack bounded-stack non-empty-stack .
subsorts non-empty-stack < bounded-stack < stack .
op : nil -> bounded-stack .
op push : elem bounded-stack -> stack .
op pop_ : non-empty-stack -> bounded-stack .
op top_ : non-empty-stack -> elem .
op length_ : stack -> int .
var x : elem, s : stack, bs : bounded-stack .
as non-empty-stack : push(x,s) if length s < bound .
eq : top push(e,bs) == e .
eq : pop push(e,bs) == s .
eq : length nil == 0 .
eq : length push(x,s) == S length(s) .
jbo
```

The conservative extension is defined in a similar way as before, and the sort-constraint is transformed into a conditional equation:

```
obj BOUNDED-STACK [X :: TRIV] is
sorts stack bounded-stack non-empty-stack .
subsorts non-empty-stack < bounded-stack < stack .
op : nil -> bounded-stack .
op push : elem bounded-stack -> stack .
op pushas : elem bounded-stack -> non-empty-stack .
op pop_ : non-empty-stack -> bounded-stack .
op top_ : non-empty-stack -> elem .
op length_ : stack -> int .
var x : elem, s : stack, bs : bounded-stack .
eq : push(x,s) == pushas(x,s) if length s < bound .
eq : top pushas(e,bs) == e .
eq : pop pushas(e,bs) == s .
eq : length nil = 0 .
eq : length push(x,s) == S length(s) .
jbo
```

As we can see, the **push** operator either remains a **push** when it can be parsed or becomes a **pushas** operator otherwise. Of course, there is still the need of the previous conservative extension in order to compute with terms which cannot be parsed at parsing time. This is left to the reader.

3.4 Module expression evaluation

Another feature of OBJ is its support for **parameterized programming** [8], which allows modifying and combining modules to form larger systems, with the interfaces between program units described by theories. These commands appear in **module expressions**, which describe and create complex combinations of modules. **Module expressions** are built from given modules by instantiating, summing, and renaming.

The simplest module expressions are previously defined non-parameterized modules. Some of these, such as the integers and booleans, are built in.

To **instantiate** a parameterized object means to provide actual objects satisfying each of its requirement theories. Actual objects are designated by **views**, which tell how to *bind* required sorts and operations to those actually provided; i.e., a view maps the sort and operation symbols in the formal requirement theory to those in the actual object in such a way that all axioms of the requirement theory are satisfied. More generally, a view can map an operation in the formal to an expression in the actual. The result of such an instantiation replaces each requirement theory by its corresponding actual module, using the views to bind actual names to formal names, being careful that multiple copies of shared submodules are not produced.

Note that there can be more than one view from a theory to an actual. Often there is a **default view** that "does the obvious thing," for example, in instantiating a sorting module by binding the ordering predicate to the usual order on the integers [8].

Renaming uses a sort mapping and an operation mapping, to create a new module from an old one where the names of sorts and the syntax of operations have been changed.

A **sum** denotes a new module that **adds** or **combines** all the information in its summands.

In the implementation, canonicalized module expressions are used internally as the names of modules. The goal of the canonicalization is to identify modules expressions that are trivial variations or easily seen to be equivalent. Exactly one instance of a module is created for each canonicalized name: it is *the* module referred to by the name. An important issue is that modules are used more than once should be *shared* i.e., only one copy should be made. This identification of equivalent modules is not perfect (that is not possible), but is satisfactory in practice.

The module expression canonicalization process proceeds by first canonicalizing sub-expressions and then checking to see if each expression has already been created. If so the canonicalization process uses this previously created value. Otherwise renames are pushed inward into instantiations and views, and composed with other renames, default views are created, and sums are pushed into views. This canonicalization process may require creating new sub-objects for the resulting module.

The module expression evaluation process proceeds by canonicalizing the expression, checking to see if the desired module has already been created, in which case the previously created module is returned as the value of the expression, and otherwise building the module described by the canonicalized expression. In instantiating an object, some of its sub-objects may be instantiated, in which case their names are created and evaluated as just described.

The kind of module composition supported by module expressions has several advantages:

1. This composition is more powerful than the purely functional composition of functional programming, in that a single module instantiation can perform many different function compositions at once. For example, a complex arithmetic module `CPXA` that takes real arithmetic modules as actual parameters could be instantiated with single precision reals, `CPXA[SP-REAL]`, or double precision reals, `CPXA[DP-REAL]`, or multiple precision reals, `CPXA[MP-REAL]`.

   Each instantiation involves the substitution of dozens of functions into dozens of other functions; this would be *much* more effort with just functional composition available. Thus a part of expressiveness of higher-order programming is available in a structured and flexible, but still rigorous way. (See [7] for more examples, including some hardware verification.)

2. The logic remains first-order, so that understanding and verifying code is much simpler than with higher-order logic.

3. Semantic declarations are allowed at module interfaces given by requirement theories.

4. Besides instantiation, module expressions also allow renaming module parts and "summing" modules i.e., making all their contents available at once.

4 Future

OBJ can be and should be extended in many ways. Let us hint a few of them:

Efficiency: Rewrite rule compilers have been designed in the non order-sorted framework in [26, 24,28,34,27,29]. A compiler is currently being written for OBJ [23]. This should provide a gain in efficiency of at least an order of magnitude. However, compiling equations like associativity and commutativity is still open, as well as compiling subsort information.

Type-system: Currently, OBJ handles overloading, subsorts, and polymorphic operators defined in parameterized modules. But really polymorphic operators are reduced to the `if_then_else` operator and to the built-in equality operator. An attractive framework for handling full polymorphism in algebraic specifications has been recently proposed by P. Moses [35], but has not been implemented yet.

Handling predicates: Enhancing the power of OBJ by providing a true logic programming language is surely desirable [14]. This goal was subjected to the discovery of complete linear strategies for first-order Horn clause calculus with equality. Recent work in this area is very promising and should lead to interesting applications [25,32], carrying over the semantically clean features of OBJ (sorts, subsorts, parameterized modules) to the logic programming world.

Object-oriented programming: The ground theory for OBJ has been extended to cope with states as they appear in object-oriented programming [15]. This work should lead to a solid basis for integrating the object-oriented programming approach with the logic programming approach.

Verification tools: OBJ-3 implements order-sorted rewriting rather than many-sorted rewriting via a functorial transformation like in OBJ-2. As a consequence, new tools will be developped for checking properties in order-sorted theories, for example order-sorted completion [6], order-sorted unification [30], and order-sorted inductive completion. This work is currently under way.

Architecture: Finally, a last area of research is hardware design for executing OBJ programs. Among such attempts, let us point out the concurrent term rewriting model of computation [12] for the Rewrite Rule Machine developed at SRI [13,40,11,33], and the pattern matching hardware at Stony Brook [36].

Acknowledgements: We thank Joseph Goguen and José Meseguer for giving us the opportunity of working on OBJ. Parts of this paper are borrowed from joined papers [31,10].

References

[1] C. Cavenaghi, M. De Zanet, and G. Mauri. *MC-OBJ: a C interpreter for OBJ.* Technical Report, Dipmentarto Scienze dell'Informazione, Universita di Milano, 1987.

[2] W. Clocksin and C. Mellish. *Programming in Prolog.* Springer-Verlag, 1981.

[3] Derek Coleman, Robin Gallimore, and Victoria Stavridou. The design of a rewrite rule interpreter from algebraic specifications. *IEEE Software Engineering Journal*, July:95–104, 1987.

[4] K. Futatsugi, J. Goguen, J-P. Jouannaud, and J. Meseguer. Principles of OBJ-2. In B. Reid, editor, *Proceedings of 12th ACM Symposium on Principles of Programming Languages*, pages 52–66, Association for Computing Machinery, 1985.

[5] Kokichi Futatsugi, J. Goguen, J. Meseguer, and Koji Okada. Parameterized programming in OBJ-2. In Robert Balzer, editor, *Proceedings, Ninth International Conference on Software Engineering*, pages 51–60, IEEE Computer Society Press, March 1987.

[6] I. Gnaedig, C. Kirchner, and H. Kirchner. Equational completion in order-sorted algebras. In M. Dauchet and M. Nivat, editors, *Proceedings of the 13th Colloquium on Trees in Algebra and Programming*, pages 165–184, Springer-Verlag, Nancy (France), 1988.

[7] J. Goguen. Higher-order functions considered unnecessary for higher-order programming. In *Proceedings, University of Texas, Year of Programming, Institute on Declarative Programming*, Addison-Wesley, 1988. To appear.

[8] J. Goguen. Parameterized programming. *Transactions on Software Engineering*, SE-10(5):528–543, September 1984.

[9] J. Goguen. Some design principles and theory for OBJ-0, a language for expressing and executing algebraic specifications of programs. In *Proceedings, Mathematical Studies of Information Processing*, pages 425–473, Springer-Verlag, 1979. Lecture Notes in Computer Science, Volume 75; Proceedings of a Workshop held August 1978.

[10] J. Goguen, C. Kirchner, H. Kirchner, A. Megrelis, J. Meseguer, and T. Winkler. An introduction to OBJ-3. In J-P. Jouannaud and S. Kaplan, editors, *Proceedings of the First International Workshop on Conditional Term Rewriting Systems*, Springer-Verlag, Orsay (France), June 1988. Also as internal report CRIN: 88-R-001.

[11] J. Goguen, C. Kirchner, S. Leinwand, J. Meseguer, and T. Winkler. Progress report on the rewrite rule machine. *IEEE Computer Architecture Technical Commitee Newsletter*, 7–21, march 1986.

[12] J. Goguen, C. Kirchner, and J. Meseguer. Concurrent term rewriting as a model of computation. In R. Keller and J. Fasel, editors, *Proceedings, Graph Reduction Workshop*, pages 53–93, Springer-Verlag, 1987.

[13] J. Goguen, C. Kirchner, J. Meseguer, and T. Winkler. OBJ as a language for concurrent programming. In *Proceedings, Second International Supercomputing Conference, Volume I*, pages 195–198, International Supercomputing Institute, Inc., 1987.

[14] J. Goguen and J. Meseguer. Eqlog: equality, types, and generic modules for logic programming. In Douglas DeGroot and Gary Lindstrom, editors, *Functional and Logic Programming*, pages 295–363, Prentice-Hall. 1986. An earlier version appears in *Journal of Logic Programming*, Volume 1, Number 2. pages 179-210, September 1984.

[15] J. Goguen and J. Meseguer. Extensions and foundations for object-oriented programming. In Bruce Shriver and Peter Wegner, editors, *Research Directions in Object-Oriented Programming*, pages 417–477, MIT Press, 1987. Preliminary version in *SIGPLAN N*otices, Volume 21, Number 10, pages 153-162, October 1986; also, Technical Report CSLI-87-93, Center for the Study of Language and Information, Stanford University, March 1987.

[16] J. Goguen and J. Meseguer. *Order-Sorted Algebra I: Partial and Overloaded Operations, Errors and Inheritance*. Technical Report To appear, SRI International, Computer Science Lab, 1988. Given as lecture at Seminar on Types, Carnegie-Mellon University, June 1983.

[17] J. Goguen and J. Meseguer. Order-sorted algebra solves the constructor-selector, multiple representation and coercion problem. In *Proceeding of the second symposium on Logic In Computer Science*, pages 18–29, IEEE Computer Society Press, 1987.

[18] J. Goguen, J. Meseguer, and David Plaisted. Programming with parameterized abstract objects in OBJ. In *Theory and Practice of Software Technology*, pages 163–193, North-Holland, 1983.

[19] J. Goguen and J. Tardo. An introduction to OBJ: a language for writing and testing software specifications. In Marvin K. Zelkowitz, editor, *Specification of Reliable Software*, pages 170–189, IEEE Press, 1979. Reprinted in *Software Specification Techniques*, Nehan Gehani and Andrew McGettrick, Eds., Addison-Wesley, 1985, pages 391-420.

[20] J. Goguen and J. Tardo. *OBJ-0 Preliminary Users Manual*. Semantics and Theory of Computation Report 10, UCLA, 1977.

[21] J.A. Goguen, J.P. Jouannaud, and J. Meseguer. Operational semantics for order-sorted algebra. In *Proceeding of the 12th ICALP, Nafplion (Greece)*, pages 221–231, 1985.

[22] J.A. Goguen and J. Meseguer. Remarks on remarks on many-sorted equational logic. *Bulletin of EATCS*, 1(30):66–73, October 1986. Also in *SIGPLAN Notices*, Volume 22, Number 4, pages 41-48, April 1987.

[23] T. Heuillard and J.P. Jouannaud. Compilation of term rewriting systems. In preparation, 1987.

[24] Christoph M. Hoffmann and Michael O'Donnell. Programming with equations. *Transactions on Programming Languages and Systems*, 1(4):83–112, 1982.

[25] J. Hsiang and J-P. Jouannaud. General e-unification revisited. In *Proceedings of 2nd Workshop on Unification*, 1988.

[26] Gérard Huet and Jean-Jacques Levy. *Computations in Non-ambiguous Linear Term Rewriting Systems*. Technical Report, INRIA Laboria, 1979.

[27] Thomas Johnsson. Target code generation from G-machine code. In J. H. Fasel and Robert M. Keller, editors, *Graph Reduction*, pages 119–159, Springer-Verlag, 1987. Lecture Notes in Computer Science, Volume 279.

[28] S. Kaplan. A compiler for term rewriting system. In *Proceedings Second Conference on Rewriting Techniques and Applications*, Springer Verlag, Bordeaux (France), May 1987.

[29] Richard B. Kieburtz. The G-machine: a fast, graph-reduction evaluator. In J-P. Jouannaud, editor, *Proceedings. Conference on Functional Programming Languages and Computer Architecture*, Springer-Verlag, 1985. Lecture Notes in Computer Science, Volume 201.

[30] C. Kirchner. Order-sorted equational unification. In *Proceeding of the Fifth International Conference on Logic Programming*, 1988.

[31] C. Kirchner, H. Kirchner, and J. Meseguer. Operational semantics of OBJ-3. In *Proceedings of ICALP'88*, Springer-Verlag, 1988.

[32] E. Kounalis and M. Rusinowitch. On word problems in horn theories. In E. Lusk and R. Overbeek, editors, *Proceedings 9th International Conference on Automated Deduction*, pages 527–537, Springer-Verlag, 1988.

[33] S. Leinwand and J. Goguen. Architectural options for the rewrite rule machine. In S. Kartashev and S. Kartashev, editors, *Proceedings, Second International Supercomputing Conference, Volume I*, pages 63–70, International Supercomputing Institute, Inc., 1987.

[34] U. Montanari and J. Goguen. *An Abstract Machine for Fast Parallel Matching of Linear Patterns.* Technical Report SRI-CSL-87-3, Computer Science Lab, SRI International, May 1987.

[35] P. Moses. *Abstract data types as lattices.* Technical Report DAIMI IR-78, Aarhus University, Computer Science Department, 1988.

[36] I. Ramakrishnan. R2m: a reconfigurable rewrite machine. In *Proceedings of 2nd Workshop on Unification*, 1988.

[37] G. Smolka, W. Nutt, J.A. Goguen, and J. Meseguer. Order sorted equational computation. In *Proceedings of the Colloquium on Resolution of Equations in Algebraic Structures*, Austin (Texas), May 1987.

[38] S. Sridhar. An implementation of OBJ-2: an object-oriented language for abstract program specification. In *Proceedings, Sixth Conference on Foundations of Software Technology and Theoretical Computer Science*, pages 81–95, Springer-Verlag, 1986.

[39] G.L. Steele. *Common Lisp: The Language.* Digital Press, 1984.

[40] T. Winkler, S. Leinwand, and J. Goguen. Simulation of concurrent term rewriting. In S. Kartashev and S. Kartashev, editors, *Proceedings, Second International Supercomputing Conference, Volume I*, pages 199–208, International Supercomputing Institute, Inc., 1987.

Logic Programming with Polymorphically Order-Sorted Types

Gert Smolka
WT LILOG, IBM Deutschland
7000 Stuttgart 80, West Germany
smolka@ds0lilog.bitnet

1 Introduction

This paper presents the foundations for relational logic programming with polymorphically order-sorted data types. This type discipline combines the notion of parametric polymorphism [Milner 78], which has been developed for higher-order functional programming [Harper et al. 86], with the notion of order-sorted typing [Goguen 78, Smolka et al. 87], which has been developed for equational first-order specification and programming [Futatsugi et al. 85]. Both notions are important for practical reasons. With parametric polymorphism one avoids the need for redefining lists and other parametric data types for every type they are used with. Subsorts not only provide for more natural type specifications, but also yield more computational power: variables can be constrained to sorts rather than to single values and typed unification computes directly with sort constraints, thus reducing the need for expensive backtracking.

Figure 1.1 shows some examples of sort and relation definitions in our language. Sort constants and sort functions are defined by equations, and relations are defined by a declaration and a collection of definite clauses. The sort `bool` has two elements, which are given by the value constants `true` and `false`. The sort `int` is defined as the union of its subsorts `inat` and `nat`. The elements of the sort `posint` are obtained by applying the value function `s` to elements of `nat`. Since `nat` is defined as the union of its subsorts `zero` and `posint`, the elements of `nat` are `o`, `s(o)`, `s(s(o))` and so forth. The relation `le` is a less or equal test on the elements of `int`. Some of the variables occurring in the clauses defining `le` are explicitly constrained to sorts. For variables that aren't explicitly constrained, a most general sort exists, which is automatically derived by a type inference algorithm.

Sort functions are defined analogously to sort constants, except that the defining equation is parameterized with respect to sort variables (one for every argument of the sort function to be defined), which range over the set of all sorts. A sort is a ground term built from sort functions. Sorts are partially ordered by a so-called inclusion order, which is defined as the rewriting relation generated by the sort rewriting rules $\sigma \to \tau$, where σ is the left-hand side of a sort equation and τ is one of the sort terms given as subsort at the right hand side. Sort functions are monotone with respect to the inclusion order; for instance, `list(nat)` is a subsort of `list(int)`, since `list(int)` rewrites to `list(nat)` using the rule `int` $\to$ `nat`. The inclusion order is compatible with the containment relation, that is, if σ is a subsort of τ, then every value contained in σ is contained in τ.

The list concatenation relation `append` is defined with exactly the same clauses one would use in untyped logic programming. The declaration for `append` is used for type checking and type inference. The type inference algorithm determines a sort term for every value variable and adds the missing sort arguments of the polymorphic relations. For instance, the second clause of `append` is completed to

```
append(T, cons(H,R), L, cons(H,RL)) <--
    H:T & R:list(T) & L:list(T) & RL:list(T) & append(T,R,L,RL).
```

Two well-known advantages of typed programming languages, which apply to typed logic programming in particular, are:

- The data structures used by a program can be defined explicitly. This leads to clearer, much easier to understand programs. The explicit definition of data structures is particularly beneficial if they are complex, as it is typically the case in Artificial Intelligence.

The research reported here has been funded by the Bundesminister für Forschung und Technologie under grant ITR8501A and also by the EUREKA Project Protos (EU 56).

```
bool := {true, false}.

int := inat ++ nat.
inat := zero ++ negint.
negint := {p: inat}.
zero := {o}.
nat := zero ++ posint.
posint :={s: nat}.

le: int x int x bool.
    le(p(I), p(J), B) <-- le(I,J,B).
    le(s(I), s(J), B) <-- le(I,J,B).
    le(o, I, true) <-- I:nat.
    le(o, I, false) <-- I:negint.
    le(I, o, true) <-- I:inat.
    le(I, o, false) <-- I:posint.

list(T) := elist ++ nelist(T).
elist := {nil}.
nelist(T) := {cons: T x list(T)}.
pair(S,T) := {cp: S x T}.
difflist(T) := pair(list(T), list(T)).

error_or_list(E,T) := errormsg(E) ++ list(T).
errormsg(E) := {error: nat x list(pair(nat, E))}.

append: list(T) x list(T) x list(T).
    append(nil, L, L).
    append(cons(H,R), L, cons(H,RL)) <-- append(R,L,RL).
```

Figure 1.1. Some examples of sort and relation definitions.

- Type checking detects many programming errors at compile time, a feature whose importance is proportional to the size of the program under development.

Well-known disadvantages of typed programming languages, whose weight has been significantly reduced by the invention of type inference and parametric polymorphism [Milner 78], are that the programmer is burdened with specifying redundant type information and that typed programs tend to be unnecessarily complicated since the programmer is sometimes forced to program around the type discipline. For pure logic programming, however, the introduction of a type discipline actually amounts to a generalization rather than a restriction. By introducing only one sort and declaring every function as a constructor of this sort, every untyped logic program becomes a well-typed program. Of course, for logic programming to be practical, one needs extra-logical features. The programming language TEL [Smolka 88a], which embodies the logical language presented in this paper, demonstrates that the necessary nonlogical features can be integrated such that they are type-safe and still practical.

As mentioned before, typed unification adds expressive power by exploiting the inclusion order on sorts. In untyped logic programming, one could express a sort as a unary predicate holding for the elements of the sort. To express, say, that the variable `X` is in the sort `negint`, one could write the atom `negint(X)`. Now suppose that during the course of a computation the additional constraint `posint(X)` is imposed. While typed unification would immediately recognize that there is no value for `X` left, untyped logic programming cannot recognize this conflict. All it can do is bind `X` successively to elements of either `posint` and `negint` and find out each time anew that the other constraint is violated.

The existing work on types for logic programming can be classified into a syntactic and a semantic approach.

The syntactic approach sees types as a syntactic discipline that does not need to be accounted for semantically and thus must not change the computational mechanisms. Mycroft and O'Keefe [84] show how Milner's [78] polymorphic type discipline can be adapted to Pure Prolog. Their system relies on type declarations given by the programmer but does not require type declarations for the variables occurring in a clause. Dietrich [88] extends Mycroft and O'Keefe's work to include subtypes. Since his approach is purely syntactic, he has to restrict the class of well-typed programs severely by imposing a mode discipline. Moreover, it is an open problem whether his type checking discipline is decidable. Another direction of the syntactic approach investigates type inference for logic programs in the complete absence of programmer-provided type declarations [Mishra 84, Zobel 87].

The semantic approach, to which this paper belongs, bases logic programming with types on logics that account for sorts. Consequently, the computational mechanisms may change, which typically shows up in the unification procedure. One direction, which appeared with Eqlog [Goguen/Meseguer 86], takes order-sorted logic as base and employs order-sorted unification [Goguen 78, Goguen/Meseguer 87, Huber/Varsek 87, Meseguer et al. 87, Smolka et al. 87, Walther 83, 87, 88]. The present paper generalizes this line of research by incorporating parametric polymorphism.

A second direction of the semantic approach was initiated by Aït-Kaci and Nasr's [86] language LOGIN, which replaces ordinary first-order terms with record structures and employs a typed unification called ψ-unification. Mukai's [87] language CIL is similar to LOGIN but has no subsorts. Smolka and Aït-Kaci [87] show how LOGIN can be captured in order-sorted logic and device a framework that combines order-sorted types with LOGIN's feature types. Recent research [Smolka 88b, Höhfeld/Smolka 88] devices a framework for logic programming with feature types, where intersections, unions and complements of sorts are available.

The paper is organized as follows. Section 2 presents the underlying predicate logic, which has sort functions, inclusion of sorts and containment of values in sorts. Section 3 discusses the properties of a class of sort rewriting systems that will be used for specifying inclusion orders and whose properties are crucial for unification and type inference. Section 4 makes precise which type specifications are possible in our language and discusses their basic properties. Section 5 presents the typed unification algorithm used for computing in our language. Section 6 defines well-typed relational programs, gives their semantics, and presents a sound and complete interpreter. The final section presents the type checking and type inference algorithms for our language. The theorems of the paper come without proofs, which can be found in my thesis [Smolka 88c].

2 Polymorphically Order-Sorted Logic

Polymorhically order-sorted logic (POS-logic for short) is a typed predicate logic having values and sorts as objects. There is a containment relation defining which values are contained in which sorts. The sorts are partially ordered by an inclusion relation that is compatible with the containment relation, that is, if A is a subsort of B, then every value contained in A is contained in B. There are two kinds of functions: value functions mapping values to values and sort functions mapping sorts to sorts. While value functions are typically partial, sort functions are required to be total. Furthermore, sort functions are required to be monotone with respect to the inclusion order. In addition to functions, there are relations taking both values and sorts as arguments.

Major differences of POS-logic with many-sorted or order-sorted predicate logics [Goguen 78, Goguen/Meseguer 86, Smolka et al. 87] are that sort functions are not limited to sort constants, that sorts are not necessarily interpreted as sets, that containment and inclusion are present at the object level rather than the meta-level, that variables are not intrinsically typed, and that no notion of well-typedness is built-in.

We start by assuming three disjoint sets of symbols: **value function symbols** (f, g, h), **sort function symbols** (ξ, η, ζ), and **relation symbols** (p, q, r). Every function symbol comes with an arity (saying how many arguments it takes), and every relation symbol comes with a sort arity (saying how many sort arguments it takes), and a value arity (saying how many value arguments it takes). The letters given above for each symbol class will always denote symbols of the corresponding class. The letter Σ will always denote a set of function and relation symbols.

A **POS-structure** $\mathcal{A}$ is a tuple

$$(\Sigma^{\mathcal{A}},\ \mathbf{V}^{\mathcal{A}},\ \mathbf{S}^{\mathcal{A}},\ :^{\mathcal{A}},\ \leq^{\mathcal{A}},\ (D_f^{\mathcal{A}})_{f\in\Sigma^{\mathcal{A}}},\ (f^{\mathcal{A}})_{f\in\Sigma^{\mathcal{A}}},\ (\xi^{\mathcal{A}})_{\xi\in\Sigma^{\mathcal{A}}},\ (r^{\mathcal{A}})_{r\in\Sigma^{\mathcal{A}}})$$

such that

- $\Sigma^{\mathcal{A}}$ is a set of function and relation symbols
- $(\mathbf{S}^{\mathcal{A}}, \leq^{\mathcal{A}})$ is a partially ordered set (the set of **sorts** ordered by **inclusion**)
- $\mathbf{V}^{\mathcal{A}}$ is a set that is disjoint from $\mathbf{S}^{\mathcal{A}}$ (the set of **values**)
- $:^{\mathcal{A}}$ (the **containment relation**) is a subset of $\mathbf{V}^{\mathcal{A}} \times \mathbf{S}^{\mathcal{A}}$ such that
  - for every $a \in \mathbf{V}^{\mathcal{A}}$ there exists an $A \in \mathbf{S}^{\mathcal{A}}$ such that $a :^{\mathcal{A}} A$
  - if $a :^{\mathcal{A}} A$ and $A \leq^{\mathcal{A}} B$, then $a :^{\mathcal{A}} B$
- if $f \in \Sigma^{\mathcal{A}}$ is a value function symbol taking n arguments, then $D_f^{\mathcal{A}}$ (the **domain of f in $\mathcal{A}$**) is a subset of the cartesian product $(\mathbf{V}^{\mathcal{A}})^n$ and $f^{\mathcal{A}}$ is a function $D_f^{\mathcal{A}} \to \mathbf{V}^{\mathcal{A}}$
- if $\xi \in \Sigma^{\mathcal{A}}$ is a sort function symbol taking n arguments, then $\xi^{\mathcal{A}}$ is a function $(\mathbf{S}^{\mathcal{A}})^n \to \mathbf{S}^{\mathcal{A}}$ that is monotone with respect to $\leq^{\mathcal{A}}$
- if $r \in \Sigma^{\mathcal{A}}$ is a relation symbol, then $r^{\mathcal{A}}$ is a subset of the cartesian product

  $$\mathbf{S}^{\mathcal{A}} \times \cdots \times \mathbf{S}^{\mathcal{A}} \times \mathbf{V}^{\mathcal{A}} \times \cdots \times \mathbf{V}^{\mathcal{A}}$$

  having a factor $\mathbf{S}^{\mathcal{A}}$ for every sort argument of r and a factor $\mathbf{V}^{\mathcal{A}}$ for every value argument of r.

A structure $\mathcal{A}$ is called a Σ-structure if $\Sigma \subseteq \Sigma^{\mathcal{A}}$.

A structure $\mathcal{A}$ is called **separated** if

$$A \leq^{\mathcal{A}} B \quad\Longleftrightarrow\quad \{a \in \mathbf{V}^{\mathcal{A}} \mid a :^{\mathcal{A}} A\} \subseteq \{b \in \mathbf{V}^{\mathcal{A}} \mid b :^{\mathcal{A}} B\}$$

for every two $A, B \in \mathbf{S}^{\mathcal{A}}$. In a separated structure a sort can be identified with the set of all values it contains. Then the containment relation $:^{\mathcal{A}}$ is exactly set containment and the inclusion relation $\leq^{\mathcal{A}}$ is exactly set inclusion. In a nonseparated structure there exist distinct sorts that have exactly the same elements. In Section 4 we will point out that for practical applications it is useful to admit nonseparated structures.

We assume that two disjoint infinite sets of symbols are given whose elements are called **value variables** and **sort variables**, respectively. The letters x, y, z will always denote value variables and the letters α, β, γ will always denote sort variables. Terms and formulas are defined as follows:

value terms:	$s, t, u, v \longrightarrow$	x	
		$f(\vec{s})$	
sort terms:	$\sigma, \tau, \mu, \nu \longrightarrow$	α	
		$\xi(\vec{\sigma})$	
formulas:	$F, G \longrightarrow$	$s \doteq t$	**equation**
		$s{:}\sigma$	**containment**
		$\sigma \mathrel{\dot{\leq}} \tau$	**inclusion**
		$r(\vec{\sigma}\,\vec{s})$	**atom**
		$\emptyset$	**empty conjunction**
		$F\&G$	**nonempty conjunction**
		$F \to G$	**implication.**

A value term $f(\vec{s})$ must respect the arity of f, that is, if f takes n arguments, $\vec{s}$ must be a tuple of n value terms. Analogously, a sort term $\xi(\vec{\sigma})$ must respect the arity of ξ, and an atom $r(\vec{\sigma}\,\vec{s})$ must repect both the sort and the value arity of r.

For notational convenience, we identify a conjunction $F_1 \& \ldots \& F_n$ of $n \geq 0$ nonconjunctive formulas with the set $\{F_1, \ldots, F_n\}$. Furthermore, we may write an inclusion $\sigma \dot{\leq} \tau$ as $\tau \dot{\geq} \sigma$ and an implication $F \rightarrow G$ as $F \leftarrow G$.

Terms and formulas are called **syntactic objects**. We use $\mathcal{V}(O)$ to denote the set of all variables occurring in a syntactic object O. A syntactic object is called **ground** if it contains no variables, **monomorphic** if it contains no sort variables, and **polymorphic** if it contains at least one sort variable.

In the following V will always denote a set of variables such that there are infinitely many value variables and infinitely many sort variables not contained in V. A syntactical object is called a ΣV-object if it contains only symbols in $\Sigma \cup V$.

Let $\mathcal{A}$ be a structure. A **$(V, \mathcal{A})$-assignment** is a function that maps every value variable in V to an element of $\mathbf{V}^{\mathcal{A}}$ and every sort variable in V to an element of $\mathbf{S}^{\mathcal{A}}$.

The **denotation of ΣV-terms** in a Σ-structure $\mathcal{A}$ under a $(V, \mathcal{A})$-assignment δ is defined inductively as follows:

- $[\![\alpha]\!]_\delta^{\mathcal{A}} = \delta(\alpha)$
- $[\![\xi(\sigma_1, \ldots, \sigma_n)]\!]_\delta^{\mathcal{A}} = \xi^{\mathcal{A}}([\![\sigma_1]\!]_\delta^{\mathcal{A}}, \ldots, [\![\sigma_n]\!]_\delta^{\mathcal{A}})$
- $[\![x]\!]_\delta^{\mathcal{A}} = \delta(x)$
- $[\![f(s_1, \ldots, s_n)]\!]_\delta^{\mathcal{A}} = f^{\mathcal{A}}([\![s_1]\!]_\delta^{\mathcal{A}}, \ldots, [\![s_n]\!]_\delta^{\mathcal{A}})$ if the tuple $([\![s_1]\!]_\delta^{\mathcal{A}}, \ldots, [\![s_n]\!]_\delta^{\mathcal{A}})$ is defined and is in $D_f^{\mathcal{A}}$.

While sort terms always denote elements of $\mathbf{S}^{\mathcal{A}}$, value terms may fail to denote an element of $\mathbf{V}^{\mathcal{A}}$. This stems from the fact that value functions are interpreted as partial functions. Denotation is defined to be strict, that is, a term denotes if and only if every subterm denotes. Also note that the denotation of a ground term does not depend on the employed assignment.

Validity of ΣV-formulas in a Σ-structure $\mathcal{A}$ under a $(V, \mathcal{A})$-assignment δ is defined as follows:

- $\mathcal{A}, \delta \models s = t$ if $[\![s]\!]_\delta^{\mathcal{A}}$ and $[\![t]\!]_\delta^{\mathcal{A}}$ are defined and equal
- $\mathcal{A}, \delta \models s : \sigma$ if $[\![s]\!]_\delta^{\mathcal{A}}$ is defined and $[\![s]\!]_\delta^{\mathcal{A}} :^{\mathcal{A}} [\![\sigma]\!]_\delta^{\mathcal{A}}$
- $\mathcal{A}, \delta \models \sigma \dot{\leq} \tau$ if $[\![\sigma]\!]_\delta^{\mathcal{A}} \leq^{\mathcal{A}} [\![\tau]\!]_\delta^{\mathcal{A}}$
- $\mathcal{A}, \delta \models r(\vec{\sigma}\,\vec{s})$ if $([\![\vec{\sigma}]\!]_\delta^{\mathcal{A}}[\![\vec{s}]\!]_\delta^{\mathcal{A}}) \in r^{\mathcal{A}}$
- $\mathcal{A}, \delta \models \emptyset$ always holds
- $\mathcal{A}, \delta \models F \& G$ if $\mathcal{A}, \delta \models F$ and $\mathcal{A}, \delta \models G$
- $\mathcal{A}, \delta \models F \rightarrow G$ if not $\mathcal{A}, \delta \models F$ or $\mathcal{A}, \delta \models G$.

We say that a structure $\mathcal{A}$ **satisfies** a formula F (or that F is **valid in** $\mathcal{A}$) if every nonvariable symbol occurring in F is in $\Sigma^{\mathcal{A}}$ and $\mathcal{A}, \delta \models F$ for every $(\mathcal{V}(F), \mathcal{A})$-assignment δ.

A structure $\mathcal{A}$ is a **model** of a set $\mathcal{F}$ of formulas if $\mathcal{A}$ satisfies every formula in $\mathcal{F}$.

A **substitution** is a function θ mapping every sort term to a sort term and every value term to a value term such that

- $\theta\xi(\sigma_1, \ldots, \sigma_n) = \xi(\theta\sigma_1, \ldots, \theta\sigma_n)$
- $\theta f(s_1, \ldots, s_n) = f(\theta s_1, \ldots, \theta s_n)$.

Following the usual abuse of notation, we call

$$\mathcal{D}\theta := \{\alpha \mid \theta\alpha \neq \alpha\} \cup \{x \mid \theta x \neq x\}$$

the **domain** of θ. The letters θ, ψ, and ϕ will always range over substitutions. A substitution is called **ground** if it maps every variable in its domain to a ground term. A substitution is called **finite** if its domain is finite.

The composition of two substitutions is again a substitution. Two substitutions are equal if and only if they have the same domain and they agree on every variable in their domain. Thus we can identify a substitution θ with the set

$$\{x/\theta x \mid x \in \mathcal{D}\theta\} \cup \{\alpha/\theta\alpha \mid \alpha \in \mathcal{D}\theta\}$$

consisting of term pairs x/s and α/σ.

The **restriction** $\theta|_V$ **of a substitution** θ to a set V of variables is defined as follows: $\theta|_V(x) := \theta x$ if $x \in V$ and $\theta|_V(x) := x$ otherwise.

Substitutions are extended to formulas as one would expect. A syntactic object O is called an **instance** of a syntactic object O' if there exists a substitution θ such that $O = \theta O'$.

3 Sort Rewriting Systems

Inclusion orders on sorts will be specified by means of inclusional axioms of the form $\sigma \dot{\geq} \tau$. If every inclusional axiom $\sigma \dot{\geq} \tau$ is taken as a term rewriting rule $\sigma \rightarrow \tau$, then the thus defined rewriting relation coincides with the inclusion order specified by the axioms. The sort rewriting rules belonging to the sort definitions in Figure 1.1 are:

```
int → inat,      int → nat,
inat → zero,      inat → negint,
nat → zero,      nat → posint,
list(T) → elist,      list(T) → nelist(T),
difflist(T) → pair(list(T),list(T)),
error_or_list(T) → errormsg(E),      error_or_list(T) → list(T).
```

The feasibility of unification and type inference depends crucially on the properties of the sort rewriting system given by the inclusional axioms. In this section we will define a class of rewriting systems having the necessary properties.

To be compatible with the standard notation for rewriting systems, we use in this section our notation for value terms although in the rest of the paper sort rewriting systems will only involve sort terms.

A **rewrite rule** is an ordered pair $s \rightarrow t$ consisting of two terms s and t such that $\mathcal{V}(t) \subseteq \mathcal{V}(s)$. A **rewriting system** is a set of rewrite rules. If $\rightarrow$ is a binary relation, we use $\rightarrow^*$ to denote its reflexive and transitive closure.

Let $\mathcal{R}$ be a rewriting system. Then we write

- $s \Rightarrow_{\mathcal{R}} t$ if $s \rightarrow t$ is an instance of a rule of $\mathcal{R}$
- $f \Rightarrow_{\mathcal{R}} g$ if there exist terms such that $f(s_1, \ldots, s_n) \Rightarrow_{\mathcal{R}} g(t_1, \ldots, t_m)$
- $s \rightarrow_{\mathcal{R}} t$ if t can be obtained from s by replacing a subterm u of s with v, where $u \Rightarrow_{\mathcal{R}} v$
- $\theta \rightarrow^*_{\mathcal{R}} \psi$ if $\theta x \rightarrow^*_{\mathcal{R}} \psi x$ for every variable x.

A partially ordered set $(M, \leq)$ is a **lower quasi-lattice** if the greatest lower bound of a and b (called the **infimum** of a and b and denoted by $a \sqcap b$) exists whenever a and b have a common lower bound. A partially ordered set $(M, \leq)$ is an **upper quasi-lattice** if the least upper bound of a and b (called the **supremum** of a and b and denoted by $a \sqcup b$) exists whenever a and b have a common upper bound. A partially ordered set $(M, \leq)$ is a **quasi-lattice** if it is an upper and a lower quasi-lattice. It is easy to show that a finite partially ordered set is a lower quasi-lattice if and only if it is an upper quasi-lattice.

In the following we assume that $\bot$ is a nullary function symbol.

A **sort rewriting system** is a finite rewriting system $\mathcal{R}$ such that

- *(Shallowness)* $\mathcal{R}$ contains a rule $x \to \bot$, no other rule of $\mathcal{R}$ contains $\bot$, and every other rule of $\mathcal{R}$ has the form $f(\vec{x}) \to g(\vec{s})$, where $\vec{x}$ is a tuple of pairwise distinct variables
- *(Termination)* there are no infinite chains $s_1 \to_{\mathcal{R}} s_2 \to_{\mathcal{R}} \cdots$
- *(Coherence)* for every term s and every function symbol f there exists at most one tuple $\vec{t}$ such that $s \Rightarrow^*_{\mathcal{R}} f(\vec{t})$
- *(Completeness)* the set of function symbols occurring in $\mathcal{R}$ is a quasi-lattice under the partial order "$f \Rightarrow^*_{\mathcal{R}} g$".

Having the rule $x \to \bot$ in sort rewriting systems is crucial for the following theorems to hold. The symbol $\bot$ is interpreted as the empty sort that is a subsort of every sort and contains no values.

Theorem 3.1. *It is decidable whether a finite rewriting system is a sort rewriting system.*

Theorem 3.2. *Let $\mathcal{R}$ be a sort rewriting system. Then the set of terms equipped with the decidable partial order $s \to^*_{\mathcal{R}} t$ is a quasi-lattice having $\bot$ as its least element. In this quasi-lattice every two terms have a computable infimum. Furthermore, for two terms it is decidable whether they have a supremum and their supremum is computable if it exists.*

Let $\mathcal{R}$ be a sort rewriting system. We say that the terms $s_1, \ldots, s_n$ have a **stable infimum in** $\mathcal{R}$ if

$$\theta(\sqcap_{\mathcal{R}}\{s_1, \ldots, s_n\}) = \sqcap_{\mathcal{R}}\{\theta s_1, \ldots, \theta s_n\}$$

for every substitution θ ($\sqcap_{\mathcal{R}}$ denotes the infimum with respect to the partial order $\to^*_{\mathcal{R}}$).

Theorem 3.3. (Stable Infima) *Let $\mathcal{R}$ be a sort rewriting system. Then the equations*

(1) $\mathrm{SINF}_{\mathcal{R}}\{f_1(\vec{s}_1), \ldots, f_n(\vec{s}_n)\} = f(\mathrm{SINF}_{\mathcal{R}}\{\vec{t}_1, \ldots, \vec{t}_n\})$

$\qquad$ *if* $f = f_1 \sqcap_{\mathcal{R}} \cdots \sqcap_{\mathcal{R}} f_n$ *and* $f_i(\vec{s}_i) \Rightarrow^*_{\mathcal{R}} f(\vec{t}_i)$ *for* $i = 1, \ldots, n$

(2) $\mathrm{SINF}_{\mathcal{R}}\{x\} = x$

define a computable partial function $\mathrm{SINF}_{\mathcal{R}}\{s_1, \ldots, s_n\}$ *that computes the stable infimum of the terms* $s_1, \ldots, s_n$ *in* $\mathcal{R}$ *if it exists.*

An **inclusion system** is a possibly empty conjunction of inclusions. The **upper matchers** of an inclusion system I in a sort rewriting system $\mathcal{R}$ are defined as follows:

$$\mathrm{UM}_{\mathcal{R}}[I] := \{\theta \mid \forall\, (s \mathrel{\dot{\geq}} t) \in I.\ \theta s \to^*_{\mathcal{R}} t\}.$$

The upper matchers of an inclusion system I are partially ordered by "$\theta \to^*_{\mathcal{R}} \psi$". The set $\mathrm{UM}_{\mathcal{R}}[I]$ is upward closed, that is, if θ is an upper matcher of I and $\psi \to^*_{\mathcal{R}} \theta$, then ψ is an upper matcher of I.

An inclusion system is **upward solved** if it has the form

$$s_1 \mathrel{\dot{\leq}} x_1\ \&\ \cdots\ \&\ s_n \mathrel{\dot{\leq}} x_n,$$

where $x_1, \ldots, x_n$ are pairwise distinct variables. For an upward solved inclusion system I we define the substitution θ^I as follows: $\theta^I(x) := s$ if $(s \mathrel{\dot{\leq}} x) \in I$ and $\theta^I(x) := \bot$ otherwise.

Proposition 3.4. *If I is an upward solved inclusion system, then θ^I is the least upper matcher of I in every sort rewriting system $\mathcal{R}$.*

The following reduction rules for inclusion systems constitute an algorithm for computing least upper matchers:

(1) $f(\vec{s}) \dot{\leq} g(\vec{t})\ \&\ I \xrightarrow{\mathrm{m}}_{\mathcal{R}} \vec{s} \dot{\leq} \vec{u}\ \&\ I$ if $g(\vec{t}) \Rightarrow^*_{\mathcal{R}} f(\vec{u})$

(2) $\bot \dot{\leq} s\ \&\ I \xrightarrow{\mathrm{m}}_{\mathcal{R}} I$

(3) $f(\vec{s}) \dot{\leq} x\ \&\ g(\vec{t}) \dot{\leq} x\ \&\ I \xrightarrow{\mathrm{m}}_{\mathcal{R}} \theta^J f(\vec{x}) \dot{\leq} x\ \&\ I$

if $h := f \sqcup_{\mathcal{R}} g$, $\vec{x}$ is a tuple of pairwise distinct variables,
$f(\vec{s}) \dot{\leq} f(\vec{x})\ \&\ g(\vec{t}) \dot{\leq} f(\vec{x}) \xrightarrow{\mathrm{m}}^*_{\mathcal{R}} J$, and J is an upward solved inclusion system.

Theorem 3.5. (Upper Matchers) *Let $\mathcal{R}$ be a sort rewriting system. Then:*

1. *(Termination) there are no infinite chains $I_1 \xrightarrow{\mathrm{m}}_{\mathcal{R}} I_2 \xrightarrow{\mathrm{m}}_{\mathcal{R}} \cdots$ and there is no infinite recursion through rule (3)*
2. *(Invariance) if $I \xrightarrow{\mathrm{m}}_{\mathcal{R}} J$, then $\mathrm{UM}_{\mathcal{R}}[I] = \mathrm{UM}_{\mathcal{R}}[J]$*
3. *(Completeness) if I has an upper matcher, then there exists an upwards solved inclusion system J such that $I \xrightarrow{\mathrm{m}}^*_{\mathcal{R}} J$.*

Corollary 3.6. *Let $\mathcal{R}$ be a sort rewriting system. Then an inclusion system I has a least upper matcher in $\mathcal{R}$ if and only if I has an upper matcher in $\mathcal{R}$. Furthermore, it is decidable whether an inclusion system has an upper matcher in $\mathcal{R}$ and the least upper matcher can be computed if it exists.*

An interesting open problem for sort rewriting systems, on whose solution I've been working unsuccessfully for some time, is:

Open Problem 3.7. (Satisfiability) *Given a sort rewriting system $\mathcal{R}$, is it decidable for two terms s and t whether there exists a substitution θ such that $\theta s \rightarrow^*_{\mathcal{R}} \theta t$?*

A sort rewriting system is **homogenous** if all rules, except $x \rightarrow \bot$, have the form $f(x_1, \ldots, x_n) \rightarrow g(x_1, \ldots, x_n)$. In my thesis [Smolka 88c] I show that the satisfiability problem for homogenous sort rewriting systems is decidable.

4 Polymorphically Order-Sorted Types

We have now set up enough machinery to formalize our idea of polymorphically order-sorted types. A collection of polymorphically order-sorted types will be formalized as a POS-structure specified in a canonical way by a set of formulas in POS-logic. As with algebraic specifications based on many-sorted or order-sorted logic [Goguen et al. 78, Ehrig/Mahr 85, Smolka et al. 87], the structure specified by a type specification will be the initial model of the specification, which is unique up to isomorphism. We will only consider free types, that is, types than can be specified without equational axioms.

From now on we assume that the symbol $\bot$, called the **empty sort**, is a nullary sort function.

A **rank** for a value function symbol f is an implication

$$x_1{:}\sigma_1\ \&\ \ldots\ \&\ x_n{:}\sigma_n \ \rightarrow\ f(x_1, \ldots, x_n){:}\sigma$$

such that $x_1, \ldots, x_n$ are pairwise distinct variables. Since the validity of a rank in a structure does not depend on the particular value variables used, the abbreviated notation $f{:}\vec{\sigma} \rightarrow \sigma$ can be used; $\vec{\sigma}$ is called the **domain** of the rank and σ is called the **codomain** of the rank.

A **type specification** $\mathcal{T}$ is a finite set of inclusions and ranks such that

- the inclusions in $\mathcal{T}$ form a sort rewriting system if an inclusion $\sigma \stackrel{\cdot}{\geq} \tau$ is taken as a rewrite rule $\sigma \rightarrow \tau$
- every rank in $\mathcal{T}$ has the form $f\colon\vec{\sigma} \rightarrow \xi(\vec{\alpha})$, where $\vec{\alpha}$ is a tuple of pairwise distinct variables and $\mathcal{V}(\vec{\sigma}) \subseteq \mathcal{V}(\vec{\alpha})$
- no rank in $\mathcal{T}$ contains the empty sort $\perp$.

Sort equations as used in Figure 1.1 are a convenient syntax for writing down type specifications. Note that the sort equations in Figure 1.1 yield in fact a type specification as defined here.

For notational convenience, we assume in the following that $\mathcal{T}$ is a type specification and that all syntactic objects employ only function symbols occurring in $\mathcal{T}$.

We write $\sigma \leq_{\mathcal{T}} \tau$ (read: "σ below τ in $\mathcal{T}$") or $\tau \geq_{\mathcal{T}} \sigma$ if τ contains only sort function symbols occurring in $\mathcal{T}$ and τ rewrites to σ in the sort rewriting system given by the inclusions of $\mathcal{T}$.

A **prefix** is a conjunction $x_1{:}\sigma_1 \& \ldots \& x_n{:}\sigma_n$ such that $x_1, \ldots, x_n$ are pairwise distinct variables. The letters P and Q will always denote prefixes.

Let P be a prefix. The relation "$s :^P_{\mathcal{T}} \sigma$" (read: "$s$ in σ under P in $\mathcal{T}$") is defined inductively as follows:

- $x :^P_{\mathcal{T}} \sigma$ if $(x{:}\tau) \in P$ and $\tau \leq_{\mathcal{T}} \sigma$
- $f(s_1, \ldots, s_n) :^P_{\mathcal{T}} \sigma$ if there exists an instance $f\colon\tau_1 \cdots \tau_n \rightarrow \tau$ of a rank in $\mathcal{T}$ such that $s_i :^P_{\mathcal{T}} \tau_i$ for $i = 1, \ldots, n$ and $\tau \leq_{\mathcal{T}} \sigma$.

We say that a value term s is **well-typed** in $\mathcal{T}$ under a prefix P if there exists a sort term σ such that $s :^P_{\mathcal{T}} \sigma$.

A $\mathcal{T}$**-sort** is a ground sort term containing only sort function symbols occurring in $\mathcal{T}$. A $\mathcal{T}$**-value** is a ground value term s that is well-typed in $\mathcal{T}$ under the empty prefix.

A type specification $\mathcal{T}$ defines a POS-structure $\mathcal{A}(\mathcal{T})$ as follows:

- $\Sigma^{\mathcal{A}(\mathcal{T})}$ is the set of all function symbols occurring in $\mathcal{T}$
- $\mathbf{V}^{\mathcal{A}(\mathcal{T})}$ is the set of all $\mathcal{T}$-values
- $\mathbf{S}^{\mathcal{A}(\mathcal{T})}$ is the set of all $\mathcal{T}$-sorts
- $s :^{\mathcal{A}(\mathcal{T})} \sigma$ if and only if $s :^{\emptyset}_{\mathcal{T}} \sigma$
- $\sigma \leq^{\mathcal{A}(\mathcal{T})} \tau$ if and only if $\sigma \leq_{\mathcal{T}} \tau$
- $D_f^{\mathcal{A}(\mathcal{T})}$ is the set of all tuples $\vec{s}$ such that $f(\vec{s})$ is a $\mathcal{T}$-value.
- $f^{\mathcal{A}(\mathcal{T})}(\vec{s}) := f(\vec{s})$
- $\xi^{\mathcal{A}(\mathcal{T})}(\vec{\sigma}) := \xi(\vec{\sigma})$.

We call $\mathcal{A}(\mathcal{T})$ the **structure specified by** $\mathcal{T}$. It is easy to verify that $\mathcal{A}(\mathcal{T})$ is a model of $\mathcal{T}$. Furthermore, one can show that $\mathcal{A}(\mathcal{T})$ is an initial model of $\mathcal{T}$, provided the right notion of homomorphism is employed.

The structure specified by the sort definitions in Figure 1.1 is not separated. For instance, the sorts `elist`, `list(⊥)`, `list(pair(⊥,⊥))` all have `nil` as their only element. Furthermore, `difflist(nat)` and `pair(list(nat),list(nat))` are sorts having the same elements.

We use $\mathrm{LUM}_{\mathcal{T}}[\vec{\sigma} \stackrel{\cdot}{\leq} \vec{\tau}]$ to denote the least upper matcher of an inclusion system $\vec{\sigma} \stackrel{\cdot}{\leq} \vec{\tau}$ in the sort rewriting system defined by the inclusions of $\mathcal{T}$.

Let P be a prefix. We define a computable partial function $\sigma^P_{\mathcal{T}}$ from value terms to sort terms inductively by the following two equations:

- $\sigma_{\mathcal{T}}^{P}[x] := \sigma$ if $(x{:}\sigma) \in P$
- $\sigma_{\mathcal{T}}^{P}[f(\vec{s})] := \theta\tau$ if $\vec{\sigma} := \sigma_{\mathcal{T}}^{P}[\vec{s}]$ exists, $(f{:}\vec{\tau} \rightarrow \tau) \in \mathcal{T}$, and $\theta := \text{LUM}_{\mathcal{T}}[\vec{\sigma} \stackrel{.}{\leq} \vec{\tau}]$.

If $\sigma_{\mathcal{T}}^{P}[s]$ is defined, we say that $\sigma_{\mathcal{T}}^{P}[s]$ is the **least sort term** of s under P in $\mathcal{T}$.

Theorem 4.1. (**Least Sorts**) *Let P be a prefix. Then $s :_{\mathcal{T}}^{P} \sigma_{\mathcal{T}}^{P}[s]$ if $\sigma_{\mathcal{T}}^{P}[s]$ is defined. Furthermore, if $s :_{\mathcal{T}}^{P} \sigma$, then $\sigma_{\mathcal{T}}^{P}[s]$ is defined and $\sigma_{\mathcal{T}}^{P}[s] \leq_{\mathcal{T}} \sigma$.*

The existence of least sort terms is the most important requirement for the feasibility of unification (see Section 5) and type inference (see Section 7).

We use $\sigma \sqcap_{\mathcal{T}} \tau$ to denote the infimum of two sort terms σ and τ in the sort rewriting system defined by the inclusions of $\mathcal{T}$.

Corollary 4.2. *Let σ and τ be $\mathcal{T}$-sorts. Then $\sigma \sqcap_{\mathcal{T}} \tau$ is a $\mathcal{T}$-sort and*

$$s :_{\mathcal{T}}^{P} \sigma \ \wedge \ s :_{\mathcal{T}}^{P} \tau \quad \Longleftrightarrow \quad s :_{\mathcal{T}}^{P} (\sigma \sqcap_{\mathcal{T}} \tau).$$

A $\mathcal{T}$-sort σ is **inhabited in** $\mathcal{T}$ if there exists at least one $\mathcal{T}$-value s such that $s :_{\mathcal{T}}^{\emptyset} \sigma$. A sort term σ is **inhabited in** $\mathcal{T}$ if at least one ground instance of σ is inhabited in $\mathcal{T}$. A prefix P is **inhabited in** $\mathcal{T}$ if σ is inhabited in $\mathcal{T}$ for every containment $(x{:}\sigma) \in P$. A type specification $\mathcal{T}$ is **fully inhabited** if every $\mathcal{T}$-sort not containing $\perp$ is inhabited in $\mathcal{T}$.

Theorem 4.3. *It is decidable whether a sort term is inhabited in a type specification. Furthermore, it is decidable whether a type specification is fully inhabited.*

5 Polymorphically Order-Sorted Unification

Ordinary term unification solves equation systems in the algebra of all terms. In our setting, unification should solve systems consisting of equations and containments in the structure $\mathcal{A}(\mathcal{T})$ specified by a type specification $\mathcal{T}$. In general, solving such systems is not straightforward: the satisfiability problem for systems of equations and containments reduces in polynomial time to the satisfiability problem for inclusion systems, whose decidability is open (see Open Problem 3.7). However, for systems containing no sort variables, a practical unification procedure exists; and, fortunately, this restricted unification suffices for relational programming in the canonical structure of a type specification.

A **solution** of a formula F in a structure $\mathcal{A}$ is a $(\mathcal{V}(F), \mathcal{A})$-assignment γ such that $\mathcal{A}, \gamma \models P$. For technical reasons, it is convenient to parameterize solutions with respect to a set V of variables we want to solve for. Thus we define the V-**solutions** of a formula F in a structure $\mathcal{A}$ as follows:

$$\text{SOL}_{V}^{\mathcal{A}}[F] := \{\gamma|_V \mid \gamma \text{ is a } (V \cup \mathcal{V}(F), \mathcal{A})\text{-assignment such that } \mathcal{A}, \gamma \models F\},$$

where $\gamma|_V$ is the restriction of γ to V. Note that this definition amounts to implicitly quantifying the variables in $\mathcal{V}(F) - V$ existentially.

In this section, we assume that $\mathcal{T}$ is a type specification and that all syntactic objects contain only function symbols occurring in $\mathcal{T}$.

A **well-typed ground substitution** is a ground substitution mapping every variable in its domain to a $\mathcal{T}$-value or a $\mathcal{T}$-sort. Since $\mathcal{A}(\mathcal{T})$ contains only $\mathcal{T}$-values and $\mathcal{T}$-sorts, $(V, \mathcal{A}(\mathcal{T}))$-assignments can be identified with well-typed ground substitutions whose domain is V.

Proposition 5.1. *Let θ be a well-typed ground substitution and F, $s \doteq t$, $s{:}\sigma$ be formulas containing only variables in $\mathcal{D}\theta$. Then:*

- $\mathcal{A}(\mathcal{T}), \theta \models F \quad \Longleftrightarrow \quad \mathcal{A}(\mathcal{T}) \models \theta F$

Decomposition of Equations

$f(\vec{s}) \doteq f(\vec{t})$ & $C \xrightarrow{u}_{\mathcal{T}} \vec{s} \doteq \vec{t}$ & C

Isolation of Variables

$x \doteq s$ & $C \xrightarrow{u}_{\mathcal{T}} x \doteq s$ & $\{x/s\}C$ if x occurs in C but not in s

Anteposition of Equations

$s \doteq x$ & $C \xrightarrow{u}_{\mathcal{T}} x \doteq s$ & C if s is not a variable

Elimination of Equations

$x \doteq x$ & $C \xrightarrow{u}_{\mathcal{T}} C$

Merging of Containments

$x{:}\sigma$ & $x{:}\tau$ & $C \xrightarrow{u}_{\mathcal{T}} x{:}(\sigma \sqcap_{\mathcal{T}} \tau)$ & C

Decomposition of Containments

$f(\vec{s}){:}\sigma$ & $C \xrightarrow{u}_{\mathcal{T}} \vec{s}{:}(\{\vec{\alpha}/\vec{\sigma}\}\vec{\tau})$ & C if $(f{:}\vec{\tau} \to \eta(\vec{\alpha})) \in \mathcal{T}$ and $\sigma \Rightarrow^*_{\mathcal{T}} \eta(\vec{\sigma})$

Figure 5.1. The reduction rules for polymorphically order-sorted unification. With $\{x/s\}C$ we denote the constraint system obtained from C by applying the substitution $\{x/s\}$. With "$\sigma \Rightarrow^*_{\mathcal{T}} \tau$" we denote the relation "$\sigma \Rightarrow^*_{\mathcal{R}} \tau$", where $\mathcal{R}$ is the sort rewriting system defined by the inclusions of $\mathcal{T}$. In order for the rules to terminate, we assume that the constraint system C appearing in the left-hand side of a rule does not contain the equation or containment(s) being reduced by the rule.

- $\mathcal{A}(\mathcal{T}) \models \theta s \doteq \theta t \iff \theta s = \theta t$ *and* θs *is a* $\mathcal{T}$*-value*
- $\mathcal{A}(\mathcal{T}) \models \theta s{:}\theta\sigma \iff \theta s :^{\emptyset}_{\mathcal{T}} \theta\sigma$
- $\mathcal{A}(\mathcal{T}) \models \theta\sigma \dot{\leq} \theta\tau \iff \theta\sigma \leq_{\mathcal{T}} \theta\tau.$

A **constraint system** is a conjunction of equations and monomorphic containments. The letter C will always denote a constraint system.

We will now present a method for solving constraint systems in $\mathcal{A}(\mathcal{T})$. The primary task of our constraint solver is to decide whether the constraint system to be solved has a solution. This is done by applying the reduction rules given in Figure 5.1 as long as they are applicable. Since the reduction rules leave the solutions of the constraint system being reduced invariant, the reduced system has exactly the same solutions as the initial system. Furthermore, a reduced system has a solution if and only if it is in a particular form called "solved form", which can be checked for easily. The first four reduction rules apply to equations and are just the standard unification rules given in [Martelli/Montanari 82].

A constraint system is **quasi-solved** if it has the form

$$x_1 \doteq s_1 \ \& \ \cdots \ \& \ x_m \doteq s_m \ \& \ y_1{:}\sigma_1 \ \& \ \cdots \ \& \ y_n{:}\sigma_n,$$

where

- $x_1, \ldots, x_m$ are pairwise distinct variables not occurring in $s_1, \ldots, s_m$
- $y_1, \ldots, y_n$ are pairwise distinct variables different from $x_1, \ldots, x_m$
- $\sigma_1, \ldots, \sigma_n$ are inhabited $\mathcal{T}$-sorts.

A constraint system C is **well-typed** if for every equation $s \doteq t$ in C the following two conditions are satisfied:

- s is either a variable or there exists a prefix $P \subseteq C$ such that s is well-typed under P
- there exists a prefix $Q \subseteq C$ such that t is well-typed under Q.

A constraint system is **solved** if it is quasi-solved and well-typed. A constraint system is **reduced** if none of the reduction rules in Figure 5.1 applies to it.

Theorem 5.2. (Unification)

1. *(Termination) There are no infinite chains $C_1 \xrightarrow{u}_{\mathcal{T}} C_2 \xrightarrow{u}_{\mathcal{T}} \cdots$.*
2. *(Invariance) If $C \xrightarrow{u}_{\mathcal{T}} C'$, then $\mathrm{SOL}_V^{\mathcal{A}(\mathcal{T})}[C] = \mathrm{SOL}_V^{\mathcal{A}(\mathcal{T})}[C']$ for every V.*
3. *(Completeness) If C is well-typed, $C \xrightarrow{u}{}^{*}_{\mathcal{T}} C'$ and C' is reduced, then:*

$$C \text{ has a solution} \iff C' \text{ is quasi-solved} \iff C' \text{ is solved.}$$

Proposition 5.3. (Garbage Collection)

1. *If $x \doteq s$ & C is a solved constraint system and $x \notin V$, then C is a solved constraint system and $\mathrm{SOL}_V^{\mathcal{A}(\mathcal{T})}[C] = \mathrm{SOL}_V^{\mathcal{A}(\mathcal{T})}[x \doteq s \;\&\; C]$.*
2. *If $x{:}\sigma$ & C is a solved constraint system such that $x \notin \mathcal{V}(C) \cup V$, then C is a solved constraint system and $\mathrm{SOL}_V^{\mathcal{A}(\mathcal{T})}[C] = \mathrm{SOL}_V^{\mathcal{A}(\mathcal{T})}[x{:}\sigma \;\&\; C]$.*

6 Relational Programs

In this section we define relational programs computing on polymorphically order-sorted types. We show that our programs have a minimal model semantics and present a sound and complete interpreter for them. Our language fits nicely in with the modern view of logic programming as "constraint logic programming", which initiated with Colmerauer's [83, 84] PROLOG II and was elaborated by Jaffar and Lassez [87].

We will first consider programs in which most of the type information is explicit. In the next section we will give type inference algorithms supporting a much less tedious syntax.

A **relation declaration** for a relation symbol r is an implication

$$r(\alpha_1, \ldots, \alpha_m, x_1, \ldots, x_n) \;\rightarrow\; x_1{:}\sigma_1 \;\&\; \ldots \;\&\; x_n{:}\sigma_n,$$

where $\alpha_1, \ldots, \alpha_m$ and $x_1, \ldots, x_n$ are pairwise distinct variables and $\sigma_1, \ldots, \sigma_n$ do not contain sort variables other than $\alpha_1, \ldots, \alpha_m$. Since the validity of a relation declaration in a structure does not depend on the particular value variables used, the abbreviated notation $r{:}\vec{\alpha}\vec{\sigma}$ can be used.

A **relation scheme** for a type specification $\mathcal{T}$ is a finite set $\mathcal{R}$ of relation declarations, where no relation is declared more than once and every occurring sort function symbol occurs in $\mathcal{T}$.

In the sequel, $\mathcal{T}$ is a type specification and $\mathcal{R}$ is a relation scheme for $\mathcal{T}$. Furthermore, we assume that all syntactic objects employ only function and relation symbols occurring in $\mathcal{T} \cup \mathcal{R}$.

An atom $r(\vec{\sigma}\vec{s})$ is **well-typed** in $\mathcal{T} \cup \mathcal{R}$ under a prefix P if $\vec{\sigma}$ is a tuple of inhabited sort terms and $\mathcal{R}$ contains a declaration $r{:}\vec{\alpha}\vec{\tau}$ such that $\vec{s}{:}^{P}_{\mathcal{T}}\{\vec{\alpha}/\vec{\sigma}\}\vec{\tau}$.

A **clause** is an implication

$$r(\vec{\alpha}\vec{s}) \leftarrow P \,\&\, \vec{R},$$

where P is a prefix, $\vec{R}$ is a possibly empty conjunction of atoms, $\vec{\alpha}$ is a tuple of pairwise distinct sort variables, and P and $\vec{R}$ contain only sort variables occurring in $\vec{\alpha}$. The symbol $\vec{R}$ will always denote a possibly empty conjunction of atoms. A clause $R \leftarrow P\&\vec{R}$ is **well-typed** in $\mathcal{T} \cup \mathcal{R}$ if P is an inhabited prefix and R and every atom in $\vec{R}$ is well-typed in $\mathcal{T} \cup \mathcal{R}$ under P.

Proposition 6.1. *It is decidable whether a clause is well-typed.*

Proposition 6.2. *Let $\theta(R \leftarrow P\&\vec{R})$ be a ground instance of a well-typed clause $R \leftarrow P\&\vec{R}$. Then $\theta(R \leftarrow P\&\vec{R})$ is well-typed if θP is valid in $\mathcal{A}(\mathcal{T})$.*

Given the sort definitions and the relation declaration for append in Figure 1.1, the clauses

```
append(T, nil, L, L) ← L:T
append(T, cons(H,R), L, cons(H,RL)) ←
    H:T & R:list(T) & L:list(T) & RL:list(T) & append(T,R,L,RL)
```

are well-typed. The type inference algorithm given in the next section can derive these clauses automatically from the abbreviated syntax in Figure 1.1.

A **program** is a finite set $\mathcal{P} = \mathcal{T} \cup \mathcal{R} \cup \mathcal{C}$ of formulas such that $\mathcal{T}$ is a type specification, $\mathcal{R}$ is a relation scheme for $\mathcal{T}$ and C is a set of clauses that are well-typed in $\mathcal{T} \cup \mathcal{R}$.

In the following $\mathcal{P}$ will denote a program $\mathcal{P} = \mathcal{T} \cup \mathcal{R} \cup \mathcal{C}$.

The **base of a structure** is obtained by forgetting the denotations of all relation symbols. A **standard model of** $\mathcal{P}$ is a model of $\mathcal{P}$ whose base is $\mathcal{A}(\mathcal{T})$.

We will define the **structure** $\mathcal{A}(\mathcal{P})$ specified by a program $\mathcal{P}$ to be the least standard model of $\mathcal{P}$. To construct $\mathcal{A}(\mathcal{P})$, we define inductively a chain $T_0 \subseteq T_1 \subseteq T_2 \subseteq \cdots$ of sets of well-typed ground atoms by $T_0 := \emptyset$ and

$$\begin{aligned} T_{n+1} := T_n \cup \{R \mid\ & R \leftarrow P \,\&\, \vec{R} \text{ is a ground instance of a clause of } \mathcal{P}, \\ & \vec{R} \subseteq T_n \text{ and } P \text{ is valid in } \mathcal{A}(\mathcal{T})\ \}. \end{aligned}$$

Now we define $\mathcal{A}(\mathcal{P})$ to be the structure whose base is $\mathcal{A}(\mathcal{T})$ and whose relational denotations are given by

$$(\vec{\sigma}\vec{s}) \in r^{\mathcal{A}(\mathcal{T})} \quad \Longleftrightarrow \quad \exists\, i.\ r(\vec{\sigma}\vec{s}) \in T_i.$$

Theorem 6.3. *$\mathcal{A}(\mathcal{P})$ is the least standard model of $\mathcal{P}$.*

From the definition of $\mathcal{A}(\mathcal{P})$ it is clear that the relation declarations of $\mathcal{P}$ are semantically redundant. However, they are needed to establish a notion of well-typedness; and, since only well-typed clauses are admitted in $\mathcal{P}$, they are valid in $\mathcal{A}(\mathcal{P})$ as a consequence of Proposition 6.2.

Proposition 6.4. *If F is a formula containing no relation symbols, then $\mathrm{SOL}_V^{\mathcal{A}(\mathcal{P})}[F] = \mathrm{SOL}_V^{\mathcal{A}(\mathcal{T})}[F]$.*

Given a tuple $\vec{s}$ of $\mathcal{T}$-values and a tuple $\vec{\sigma}$ of $\mathcal{T}$-sorts such that $r(\vec{\sigma}\vec{s})$ is a well-typed atom, how does the validity of $r(\vec{\sigma}\vec{s})$ in $\mathcal{A}(\mathcal{P})$ depend on the particular choice of the sorts $\vec{\sigma}$? This question is answered by the following theorem, which justifies that we speak of "parametric" polymorphism.

Theorem 6.5. **(Independence)** *Let $r(\vec{\sigma}\,\vec{s})$ and $r(\vec{\tau}\,\vec{s})$ be two well-typed ground atoms. Then $r(\vec{\sigma}\,\vec{s})$ is valid in $\mathcal{A}(\mathcal{P})$ if and only if $r(\vec{\tau}\,\vec{s})$ is valid in $\mathcal{A}(\mathcal{P})$.*

A **simple formula** is a conjunction C & $\vec{R}$ consisting of a constraint system C and a conjunction $\vec{R}$ of atoms. A simple formula C & $\vec{R}$ is **well-typed** if C is a well-typed constraint system and for every atom R in $\vec{R}$ there exists a prefix $P \subseteq C$ such that R is well-typed under P.

Given a program $\mathcal{P}$, a set of variables V, and a well-typed simple formula F, an interpreter for our programming language should solve F in $\mathcal{A}(\mathcal{P})$ with respect to the variables in V, that is, it should enumerate solved constraint systems $C_1, C_2, \ldots$ such that

$$\mathrm{SOL}_V^{\mathcal{A}(\mathcal{P})}[F] = \mathrm{SOL}_V^{\mathcal{A}(\mathcal{T})}[C_1] \cup \mathrm{SOL}_V^{\mathcal{A}(\mathcal{T})}[C_2] \cup \cdots .$$

We will now outline such an interpreter. The interpreter employs two loosely coupled operations, called backward chaining and constraint solving, whose combination generalizes the well-known SLD-resolution [Lloyd 84] used in classical logic programming. Constraint solving takes the rôle of term unification and has already been discussed in the previous section.

A formula G is called a **variant** of a formula F if G can be obtained from F by consistent variable renaming. Note that the variant relation is an equivalence relation and that a structure satisfies a formula if and only if it satisfies each of its variants.

We define $(\mathcal{P}, V)$-**backward chaining** as the binary relation $\xrightarrow{\text{b}}_{\mathcal{P},V}$ on the set of formulas given by the rule:

$$r(\vec{\sigma}\,\vec{s}) \;\&\; F \;\xrightarrow{\text{b}}_{\mathcal{P},V}\; \vec{s} \doteq \vec{t} \;\&\; \{\vec{\alpha}/\vec{\sigma}\}G \;\&\; F$$

if $r(\vec{\alpha}\,\vec{t}) \leftarrow G$ is a variant of a clause in $\mathcal{P}$
containing no value variable in $V \cup \mathcal{V}(\vec{s}) \cup \mathcal{V}(F)$.

Theorem 6.6. **(Soundness)** *If $F \xrightarrow{\text{b}}{}^{*}_{\mathcal{P},V} G$, then $\mathrm{SOL}_V^{\mathcal{A}}[G] \subseteq \mathrm{SOL}_V^{\mathcal{A}}[F]$ for every model $\mathcal{A}$ of $\mathcal{P}$.*

Proposition 6.7. **(Well-Typedness)** *If F is a well-typed simple formula and $F \xrightarrow{\text{b}}{}^{*}_{\mathcal{P},V} G$, then G is a well-typed simple formula.*

Theorem 6.8. **(Completeness)** *If F is a simple formula and $\theta \in \mathrm{SOL}_V^{\mathcal{A}(\mathcal{P})}[F]$, then there exists a constraint system C such that $F \xrightarrow{\text{b}}{}^{*}_{\mathcal{P},V} C$ and $\theta \in \mathrm{SOL}_V^{\mathcal{A}(\mathcal{T})}[C]$.*

These three results give us already a sound and complete interpreter. Given a well-typed simple formula, the interpreter first applies backward chaining until no atoms are left and then solves the obtained constraint system. Of course, this strategy is extremely inefficient since it involves many don't know choices and inconsistencies are detected only at the very end of a computation path. To obtain a more efficient interpreter that can compete with SLD-resolution, we need to interleave backward chaining with constraint solving to detect inconsistencies early; furthermore, we need to make a careful distinction between don't know choices, for which the interpreter must explore every alternative, and don't care choices, for which it is irrelevant which alternative is explored.

We define $(\mathcal{P}, V)$-**constraint solving** as the binary relation $\xrightarrow{\text{c}}_{\mathcal{P},V}$ on the set of formulas given by the rule:

$$C \;\&\; F \;\xrightarrow{\text{c}}_{\mathcal{P},V}\; C' \;\&\; F$$

if C and C' are constraint systems such that $\mathrm{SOL}_{V\cup\mathcal{V}(F)}^{\mathcal{A}(\mathcal{T})}[C] = \mathrm{SOL}_{V\cup\mathcal{V}(F)}^{\mathcal{A}(\mathcal{T})}[C']$.

To distinguish between don't know and don't care choices, we define the following complexity measure. The $(F, V, \mathcal{P})$-**complexity** $\|\,\theta\,\|_{F,V}^{\mathcal{P}}$ of a substitution $\theta \in \mathrm{SOL}_V^{\mathcal{A}(\mathcal{P})}[F]$ is defined as the minimal n such that there exists a constraint system C such that $\theta \in \mathrm{SOL}_V^{\mathcal{A}(\mathcal{T})}[C]$ and $F \xrightarrow{\text{b}}{}^{*}_{\mathcal{P},V} C$ in n steps.

Theorem 6.9. (Strong Completeness) *Let F be a simple formula. Then:*

1. *if R is an atom in F, $D_1, \ldots, D_n$ are the clauses in $\mathcal{P}$ defining the relation symbol of R, and $F \xrightarrow{b}_{\mathcal{P},V} F_i$ by reducing R using a variant of D_i, then*

   1.1 $\mathrm{SOL}_V^{\mathcal{A}(\mathcal{P})}[F] = \mathrm{SOL}_V^{\mathcal{A}(\mathcal{P})}[F_1] \cup \cdots \cup \mathrm{SOL}_V^{\mathcal{A}(\mathcal{P})}[F_n]$

   1.2 for every $\theta \in \mathrm{SOL}_V^{\mathcal{A}(\mathcal{P})}[F]$ *there exists an i such that* $\| \theta \|_{F,V}^{\mathcal{P}} > \| \theta \|_{F_i,V}^{\mathcal{P}}$

2. *if $F \xrightarrow{c}_{\mathcal{P},V} G$, then*

   2.1 $\mathrm{SOL}_V^{\mathcal{A}(\mathcal{P})}[F] = \mathrm{SOL}_V^{\mathcal{A}(\mathcal{P})}[G]$

   1.2 $\| \theta \|_{F,V}^{\mathcal{P}} = \| \theta \|_{G,V}^{\mathcal{P}}$ *for every* $\theta \in \mathrm{SOL}_V^{\mathcal{A}(\mathcal{P})}[F]$.

Compared to SLD-resolution, our interpreter has to do some additional work: for every variable occurring in a clause employed by a backward chaining step the constraint solver has to verify a containment, and for polymorphic relations like append sort arguments have to be passed. Most of this additional work is redundant and can be optimized away. The Independence Theorem stated above already suggests that the sort arguments of relations are semantically redundant. Furthermore, the containments in the prefix of a clause can be classified into redundant and nonredundant ones. In my thesis [Smolka 88c] I show that there is no need for passing sort arguments and that redundant containments can be omitted. In particular, if there is no inclusional axiom except $x \to \bot$, every containment in a well-typed clause is redundant and can thus be omitted. Consequently, programs that don't employ subtypes will run with the same efficiency as their untyped equivalents.

7 Type Inference

Writing clauses in our typed language is tedious: one has to give a sort term for every value variable and every sort argument of an atom. Fortunately, it is possible to infer almost all of this sort information automatically. For instance, the append relation for list concatenation can in fact be written as in Figure 1.1 and the completed clauses given in the previous section can be derived automatically. The type inference algorithms compute sort information that is most general, that is, just suffices to render the clause well-typed. If the programmer wants to use more specific sorts, as in the less or equal test for integers in Figure 1.1, he can do this where necessary by writing a containment.

In this section we assume that $\mathcal{P} = \mathcal{T} \cup \mathcal{R} \cup \mathcal{C}$ is a program.

A **containment system** is a possibly empty conjunction of containments. In this section the letter C will always denote a containment system. We say that a prefix P **supports** a containment system C in $\mathcal{T}$ and write $P \vdash_{\mathcal{T}} C$ if $s :_{\mathcal{T}}^{P} \sigma$ for every containment $(s{:}\sigma) \in C$. We call a prefix P a **stable decomposition** of a containment system C in $\mathcal{T}$ if

$$Q \vdash_{\mathcal{T}} \theta C \iff Q \vdash_{\mathcal{T}} \theta P$$

for every prefix Q and every substitution θ.

Proposition 7.1. *A containment system has at most one stable decomposition.*

The following two reduction rules for containment systems, which are slight modifications of two similar rules given for polymorphically order-sorted unification in Figure 5.1, constitute an algorithm for computing stable decompositions. We use $\mathrm{SINF}_{\mathcal{T}}\{\sigma_1, \ldots, \sigma_n\}$ to denote the stable infimum of $\sigma_1, \ldots, \sigma_n$ in the sort rewriting system defined by the inclusions in $\mathcal{T}$.

(1) $f(\vec{s}){:}\sigma \;\&\; C \xrightarrow{d}_{\mathcal{T}} \vec{s}{:}(\{\vec{\alpha}/\vec{\sigma}\}\vec{\tau}) \;\&\; C$ if $(f{:}\vec{\tau} \to \eta(\vec{\alpha})) \in \mathcal{T}$ and $\sigma \Rightarrow_{\mathcal{T}}^{*} \eta(\vec{\sigma})$

(2) $x{:}\sigma_1 \;\&\; \cdots \;\&\; x{:}\sigma_n \;\&\; C \xrightarrow{d}_{\mathcal{T}} x{:}\sigma \;\&\; C$ if $x \notin \mathcal{V}(C)$ and $\sigma := \mathrm{SINF}_{\mathcal{T}}\{\sigma_1, \ldots, \sigma_n\}$ exists.

(1) $r(\vec{s}) \leftarrow P_o \,\&\, \vec{R}_o \quad \xrightarrow{t}_{\mathcal{P}} \quad r(\vec{\alpha}\vec{s}) \leftarrow P \,\&\, \vec{R}$ if

- $r{:}\,\vec{\alpha}\vec{\sigma}$ is the declaration of r in $\mathcal{P}$
- Q is the stable decomposition of $\vec{s}{:}\vec{\sigma}\ \&\ P_o$
- $Q \,\&\, \vec{R}_o \xrightarrow{t}_{\mathcal{P}} P \,\&\, \vec{R}$

(2) $P \quad \xrightarrow{t}_{\mathcal{P}} \quad P$ if P is an inhabited prefix

(3) $P_o \,\&\, r(\vec{s}) \,\&\, \vec{R}_o \quad \xrightarrow{t}_{\mathcal{P}} \quad P \,\&\, r(\vec{s}) \,\&\, \vec{R}$ if

- r has no sort arguments and $r{:}\,\vec{\sigma}$ is the declaration of r in $\mathcal{P}$
- Q is the stable decomposition of $\vec{s}{:}\vec{\sigma}\ \&\ P_o$
- $Q \,\&\, \vec{R}_o \xrightarrow{t}_{\mathcal{P}} P \,\&\, \vec{R}$

(4) $P_o \,\&\, r(\vec{s}) \,\&\, \vec{R}_o \quad \xrightarrow{t}_{\mathcal{P}} \quad P \,\&\, r(\theta\vec{\alpha}\vec{s}) \,\&\, \vec{R}$ if

- r has sort arguments and $r{:}\,\vec{\alpha}\vec{\sigma}$ is the declaration of r in $\mathcal{P}$
- $P^{\perp} := (P_o \,\&\, x_1{:}\perp \,\&\, \cdots \,\&\, x_n{:}\perp)$, where $\{x_1, \ldots, x_n\} = \mathcal{V}(\vec{s}) - \mathcal{V}(P_o)$
- $\vec{\tau} := \sigma_{\mathcal{T}}^{P^{\perp}}[\vec{s}]$ and $\theta := \mathrm{LUM}_{\mathcal{T}}[\vec{\tau} \mathrel{\dot{\leq}} \vec{\sigma}]$
- $\theta\vec{\alpha}$ is a tuple of inhabited sort terms
- Q is the stable decomposition of $\vec{s}{:}\theta\vec{\sigma}\ \&\ P_o$
- $Q \,\&\, \vec{R}_o \xrightarrow{t}_{\mathcal{P}} P \,\&\, \vec{R}$

Figure 7.1. The type inference algorithm for abbreviated clauses.

Theorem 7.2. (Stable Decompositions)

1. *(Termination) There are no infinite chains* $C_1 \xrightarrow{d}_{\mathcal{T}} C_2 \xrightarrow{d}_{\mathcal{T}} \cdots$.
2. *(Invariance) If* $C \xrightarrow{d}_{\mathcal{T}} C'$*, then, for every prefix* P *and every substitution* θ, $P \vdash_{\mathcal{T}} \theta C$ *if and only if* $P \vdash_{\mathcal{T}} \theta C'$.
3. *(Completeness) If* C *is a containment system and* P *is a prefix, then* P *is the stable decomposition of* C *if and only if* $C \xrightarrow{d}{}^{*}_{\mathcal{T}} P$.

An **abbreviated atom** has the form $r(s_1, \ldots, s_n)$, where n is the value arity of the relation symbol r. An **abbreviated clause** has the form $R \leftarrow P \& \vec{R}$, where R is an abbreviated atom, $\vec{R}$ is a possibly empty conjunction of abbreviated atoms, and P is a prefix such that every value variable occurring in P occurs in R or $\vec{R}$.

The type inference algorithm given in Figure 7.1, which also does the necessary type checking, traverses an abbreviated clause deterministically from left to right. This deterministic strategy is an important requirement for obtaining precise error messages in case the clause cannot be well-typed. An deterministic strategy also ensures that the outcome of the type inference is easily predictable for the programmer.

Theorem 7.3. *If* $R_o \leftarrow P_o \;\&\; \vec{R}_o$ *is an abbreviated clause and*

$$R_o \leftarrow P_o \;\&\; \vec{R}_o \quad \xrightarrow{t}_{\mathcal{P}} \quad R \leftarrow P \;\&\; \vec{R},$$

then $R \leftarrow P \;\&\; \vec{R}$ *is a well-typed clause.*

For abbreviated clauses containing no polymorphic relations, the type inference algorithm in Figure 7.1 is perfect: if there is a typing that renders a clause well-typed, then the algorithm computes the most general well-typing. For abbreviated clauses containing polymorphic relations, however, the well-typing computed by the type inference algorithm may depend on the order of the atoms in the body of the clause. Furthermore, in the polymorphic case the algorithm might even fail although there is a well-typing. While I find these difficulties annoying from a theoretical point of view, our experience with the programming language TEL [Smolka 88a], which implements a similar type inference algorithm, strongly suggests that failure of type inference does not occur in practical applications (if there is a well-typing, of course). Furthermore, failure of the type inference algorithm can always be prevented by furnishing the abbreviated clause with a more informative prefix.

References

H. Aït-Kaci and R. Nasr, LOGIN: A Logic Programming Language with Built-In Inheritance. The Journal of Logic Programming, 1986, 3, 185–215.

A. Colmerauer, H. Kanoui, and M. Van Caneghem, Prolog, Theoretical Principles and Current Trends. Technology and Science of Informatics 2,4, 1983, 255–292.

A. Colmerauer, Equations and Inequations on Finite and Infinite Trees. Proc. of the 2nd International Conference on Fifth Generation Computer Systems, 1984, 85–99.

R. Dietrich, A Polymorphic Type System with Subtypes for Prolog. Proc. of the 2nd European Symposium on Programming, Nancy, France, Springer LNCS 300, 1988.

H. Ehrig and B. Mahr, Fundamentals of Algebraic Specification 1, Equations and Initial Semantics. Springer Verlag, 1985.

K. Futatsugi, J.A. Goguen, J.-P. Jouannaud and J. Meseguer, Principles of OBJ2. Proc. POPL 1985, 52–66.

J.A. Goguen, Order Sorted Algebra. Semantics and Theory of Computation Report No. 14, UCLA Computer Science Department, 1978.

J.A. Goguen, J.W. Thatcher, and E.G. Wagner, An Initial Algebra Approach to the Specification, Correctness, and Implementation of Abstract Data Types. In R.T. Yeh (ed.), Current Trends in Programming Methodology, Volume IV, Data Structuring; Prentice Hall, 1978, 80–149.

J.A. Goguen and J. Meseguer, Eqlog: Equality, Types, and Generic Modules for Logic Programming. In D. DeGroot and G. Lindstrom (eds.), Logic Programming, Functions, Relations, and Equations; Prentice Hall 1986.

J.A. Goguen and J. Meseguer, Models and Equality for Logic Programming. TAPSOFT '87, Pisa, Springer LNCS 250, 1987, 1–22.

R. Harper, D. MacQueen, and R. Milner, Standard ML. Report ECS-LFCS-86-2, Department of Computer Science, University of Edinburgh, 1986.

M. Höhfeld and G. Smolka, Logic Programming with Feature Terms. Forthcoming SEKI Report, Universität Kaiserslautern, West Germany, 1988.

M. Huber and I. Varsek, Extended Prolog for Order-Sorted Resolution. Proc. of the 4th IEEE Symposium on Logic Programming, San Francisco, 1987, 34–45.

J. Jaffar and J.-L. Lassez, Constraint Logic Programming. Proc. of the 14th ACM Symposium on Principles of Programming Languages, Munich, 1987, 111–119.

J.W. Lloyd, Foundations of Logic Programming. Springer Verlag, 1984.

A. Martelli and U. Montanari, An Efficient Unification Algorithm. ACM Transactions on Programming Languages and Systems 4,2, 1982, 258–282.

J. Meseguer, J.A. Goguen, and G. Smolka, Order-Sorted Unification. Report CSLI-87-86, Center for the Study of Language and Information, Stanford University, 1987. To appear in Symbolic Computation, Special Issue on Unification.

R. Milner, A Theory of Type Polymorphism in Programming. Journal of Computer and System Sciences 17, 1978, 348–375.

P. Mishra, Towards a Theory of Types in Prolog. Proc. of the 1st IEEE Symposium on Logic Programming, 1984, 289–298.

K. Mukai, Anadic Tuples in Prolog. Technical Report TR-239, ICOT, Tokyo, 1987.

A. Mycroft and R.A. O'Keefe, A Polymorphic Type System for Prolog. Artificial Intelligence 23, 1984, 295–307.

G. Smolka, TEL (Version 0.9), Report and User Manual. SEKI Report SR-87-11, FB Informatik, Universität Kaiserslautern, West Germany, 1988a.

G. Smolka, A Feature Logic with Subsorts. LILOG Report 33, IBM Deutschland, West Germany, May 1988. Presented at the Workshop on Unification Formalisms—Syntax, Semantics and Implementation, Titisee, West Germany, April 1988b.

G. Smolka, Logic Computation with Polymorphically Order-Sorted Types. Dissertation, FB Informatik, Universität Kaiserslautern, West Germany, 1988c.

G. Smolka and H. Aït-Kaci, Inheritance Hierarchies: Semantics and Unification. MCC Report AI-057-87, MCC, Austin, Texas, 1987. To appear in Symbolic Computation, Special Issue on Unification.

G. Smolka, W. Nutt, J.A. Goguen and J. Meseguer, Order-Sorted Equational Computation. SEKI Report SR-87-14, Universität Kaiserslautern, West Germany, 1987. To appear in H. Aït-Kaci and M. Nivat, Resolution of Equations in Algebraic Structures, Academic Press.

C. Walther, A Many-Sorted Calculus Based on Resolution and Paramodulation. Proc. 8th International Joint Conference on Artificial Intelligence, 1983, W. Kaufmann, 882–891.

C. Walther, A Many-sorted Calculus Based on Resolution and Paramodulation. Pitman and Morgan Kaufman Publishers, Research Notes in Artificial Intelligence, 1987.

C. Walther, Many-Sorted Unification. Journal of the ACM 35 (1), 1988, 1–17.

J. Zobel, Derivation of Polymorphic Types for Prolog Programs. Proc. of the 4th International Conference on Logic Programming, Melbourne, Australia, 1987, 817–838.

Integrating Logic Programming and Equational Specification of Abstract Data Types

Christoph Beierle and Udo Pletat
IBM Deutschland GmbH
Science and Technology - LILOG
P.O. Box 80 08 80
7000 Stuttgart 80, W. Germany
(electronic mail on EARN/BITNET:
BEIERLE at DSØLILOG, PLETAT at DSØLILOG)

Abstract: We discuss various semantics of many-sorted logic programs with equality both from the viewpoint of algebraic specifications as well as from the viewpoint of logic programming. We define model-theoretic semantics based on initial models, least generalized Herbrand models and a least fixpoint construction, and we investigate proof-theoretic semantics based on resolution and unification modulo a set of conditional equations. Generalizing ordinary SLD derivations, we introduce so-called SLDE derivations and SLDE trees and study their correctness and completeness properties. We define a translation of logic programs LP_E with equality into equivalent logic programs $LP_\emptyset$ with empty equational part such that LP_E satisfies a goal G if and only if $LP_\emptyset$ satisfies G.

1. Introduction

Both logic programming and the specification of abstract data types have been an active field of computer science research for quite some time. While logic programming is based on the paradigm of relational programming, abstract data type specifications support the paradigm of functional programming. Since both programming styles have their advantages, it is not surprising that several approaches to combine them into a uniform framework have been suggested, see e. g. [JLM 84], [DeGL 86]. Integrating logic programming and abstract data type specifications leads to programming languages offering relations, functions, and data types. Our concern here is to study this junction of logic programming and abstract data type specifications under semantical aspects. In particular, we discuss various approaches to define model-theoretic and proof-theoretic semantics for logic programs with data types and equality and study their relationships.

On the model-theoretic side we introduce initial models, least Herbrand models, and a least fixpoint construction. The latter two approaches refer to the lattice of generalized Herbrand models (c. f. [WB 80] and [AvE 82] where also a lattice view of the class of models is taken). The least fixpoint construction generalizes the approaches of [AvE 82] and [JLM 84] to the many-sorted equational setting we investigate here. We can show the equivalence of all the three semantics definitions in the sense that (1) there is exactly one least model, that (2) the least fixpoint construction yields exactly this least model, and that (3) the model obtained as a least fixpoint is an initial one. Since the initial approach is of non-constructive nature, this latter result is convenient for several proofs referring to initial models and where it is required to know about their structure.

On the proof-theoretic side we investigate how to generalize SLD-resolution as discussed in [AvE 82] to the situation where typed equational unification (see e. g. [Si 84]) has to be used instead of ordinary Robinson unification. We propose a form of SLDE-resolution where unification with respect to a set of conditional equations is combined with SLD-resolution. The idea of performing unification with respect to a set of conditional equations is also present in the language EQLOG ([GM 86]) and is supported by the results of [Hu 85], where a complete unification procedure for a confluent set of conditional equations is presented.

The research reported here has been carried out within the international EUREKA Project PROTOS (EU 56).

Compared to [GR 86] where conditional equations may also have non-equational premises, our conditional equations only allow for equational premises. This is slightly less general than the approach of [GR 86] but it leads to a stricter separation of the unification and resolution part of the deduction process. The proposal of [GR 86] uses all equational heads of clauses for the E-unification and treats the conditions of equations as subgoals in the resolution process.

Besides this direct realization of equational Horn logic by means of SLDE-resolution we also present a transformational approach translating an equational logic program LP_E into an equivalent non-equational one LP_ϕ. Equivalence is given in the following sense: LP_E satifies a goal G iff LP_ϕ satisfies G. This is based on the observation that unification procedures can be realized as non-equational logic programs and that unifying goals with clause heads is equivalent to adding a special unification goal to the body of the corresponding clauses. This approach is also interesting from a practical point of view: since the translation is rather simple one only has to implement an E-unification algorithm in order to realize equational Horn logic on top of an existing implementation such as Prolog. The E-unification procedure could be derived from algorithms as described in e. g. [Hu 85], [RKKL 85], or [MMR 86].

Summarizing the different approaches to the semantics of many-sorted logic programs with equality this paper provides three model-theoretic semantics based on

- initial models,
- least (generalized) Herbrand models, and
- a least fixpoint construction.

Additionally, three approaches to an operational semantics are given, namely

- generating SLDE refutations,
- searching an SLDE tree, and
- translating LP_E into LP_ϕ and taking the operational semantics of LP_ϕ.

The relationship between the various model-theoretic and proof-theoretic semantics of typed logic programs with equality is established in the form of completeness and correctness theorems. Due to space limitations the proofs of these theorems had to be shortened in this paper; the complete proofs are given in [BP 88].

2. Syntax of many-sorted logic programs with equality

A signature Σ = <S, OP, PR> is a triple consisting of a set S of sorts, an S*xS-indexed family of operator symbols OP, and an S*-indexed family of predicate symbols PR. For op ε $OP_{w,s}$ (resp. pr ε PR_w) we write op: w -> s (resp. pr: w). If op ε $OP_{w,s}$ and w is the empty string of sorts, op is a constant of sort s, in which case we write op: -> s. For each sort s there is a distinguished binary predicate = in PR which is the equality symbol. It takes two arguments of sort s; thus we may write =: s s ε PR.

The family of well-formed terms over Σ and an S-indexed family of variables V is denoted by $T_\Sigma(V)$, and T_Σ = $T_\Sigma(\phi)$ is the family of ground terms over Σ. In order to avoid problems with empty sorts, we will only consider signatures Σ such that for every sort s there is at least one ground term of sort s.

For pr: $s_1 \ldots s_n$ ε PR and t_i ε $T_\Sigma(V)_{s_i}$ pr($t_1,\ldots,t_n$) is a literal, and for literals P_i, a Horn clause is of the form

$$P_0 \text{ :- } P_1,\ldots,P_n$$

where $P_1,\ldots,P_n$ must be equations if P_0 is an equation. Such a Horn clause is called **equation** if P_0 is of the form s = t and there are no conditions (i.e. n = 0), and it is called **conditional equation** if P_0 is of the form s = t and all conditions P_i are equations as well, i.e. for any i $\geq$ 0, P_i is of the form $s_i = t_i$.

A many-sorted logic program LP is a pair (Σ, C) where C is a set of Horn clauses over the signature Σ and some family of variables V. The **equational part** E $\subseteq$ C of LP contains all equations and conditional equations of C. Note that C - E may contain clauses with non-equational heads and both equational and non-equational conditions. A Σ-goal G is a conjunction of literals over the alphabet Σ, written :- $P_1,\ldots,P_n$.

3. Model-theoretic semantics

We introduce the general concepts of models for logic programs. The notion of model morphism then allows us to discuss initial models.

Given a signature $\Sigma = <S, OP, PR>$ a Σ-model consists of an S-indexed family of sets A, a function $op_A : A_w \to A_s$ for every op: w -> s ε OP, and a relation $pr_A \subseteq A_w$ for every pr: w ε PR. The equality symbols =: s s ε PR must be interpreted by the identity on the carrier sets A_s, i.e. $=_A$ is the relation $\{(a,a) \mid a \varepsilon A_s\}$.

Note that by forgetting all relations pr_A in a Σ-model A, we obtain a <S,OP>-algebra. Such <S,OP>-algebras are exactly the models of equational abstract data type specifications ([GTW 78]).

Given two Σ-models A and A', a Σ-model morphism h: A -> A' is an S-indexed family of functions $h_s : A_s \to A'_s$ preserving both the operations and predicates, i.e. for op: w -> s ε OP and pr: w ε PR we have for $w = s_1 \ldots s_n$

$(a_1, \ldots, a_n) \varepsilon A_w$

implies

$h_s(op_A(a_1, \ldots, a_n)) = op_{A'}(h_{s1}(a_1), \ldots, h_{sn}(a_n))$

and

$(a_1, \ldots, a_n) \varepsilon pr_A$ implies $(h_{s1}(a_1), \ldots, h_{sn}(a_n)) \varepsilon pr_A$

Mod(Σ) denotes the class of all Σ-models together with all Σ-model morphisms. For instance, one particular Σ-model is the free term model $T_\Sigma(V)$ over a family of variables V. The carriers of $T_\Sigma(V)$ are the terms over Σ and V, the functions are the term generating operations, and all non-equality relations are empty.

Let A be a Σ-model and α: V -> A be a sort respecting function assigning a carrier element a ε A_s to each variable v ε V_s. The unique extension of α to a morphism from T_Σ (V) to A is also denoted by α. We say that A satisfies

- a literal $P = p(t_1, \ldots, t_n)$ under the assignment α: V -> A, written $A \models_\alpha P$, iff $(\alpha(t_1), \ldots, \alpha(t_n)) \varepsilon p_A$.
  In this definition V must contain all variables occuring in P.

- a Horn clause P_0 :- $P_1, \ldots, P_n$, written
  $A \models P_0$:- $P_1, \ldots, P_n$, iff for every variable assignment α: V -> A we have:
  $A \models_\alpha P_i$ for $1 \le i \le n$ implies $A \models_\alpha P_0$

  Here, V is the set of variables occurring in $P_0, \ldots, P_n$.

- a set C of clauses, written $A \models C$, iff it satisfies every clause in C; we then say that A is a (Σ,C)-model. Mod(Σ,C) denotes the class of all (Σ,C)-models together with all model morphisms between them.

- a goal G = :- $P_1, \ldots, P_n$, written $A \models G$, iff there exists a variable assignment α: V -> A such that $A \models_\alpha P_i$ for every i.
  As above, V is the set of variables occurring in $P_1, \ldots, P_n$.

We say that a logic program LP satisfies a goal G, written LP $\models$ G, iff every A ε Mod(Σ,C) satisfies G.

The definitions above are all straightforward generalizations of the algebraic approach to the specification of abstract data types as given in e. g. [GTW 78] or [EM 85]. There, the set of predicates only contains the equality predicates. As in the case of purely equational specifications, every logic program LP = (Σ, C) with equality also denotes a (up to isomorphisms) unique model, which we denote by $T_{\Sigma,C}$. $T_{\Sigma,C}$ is the (isomorphism class of the) initial model in Mod(Σ,C), i.e. it is characterized by the property that for every C-model A there is a unique Σ-model morphism h: $T_{\Sigma,C}$ -> A. For instance, the term model $T_\Sigma(\emptyset)$ is an initial model in Mod(Σ).

The existence of $T_{\Sigma,C}$ follows from general results about logics that admit initial models (c.f. [MM 84]). It is also possible to give a definition of $T_{\Sigma,C}$ in analogy to the quotient term algebra construction of [GTW 78], using the provability relation $\vdash$ for many-sorted first order logic with equality (e.g. as in [GM 86]). The importance of initial models is based on the following theorem.

Theorem:
A logic program LP satisfies a goal G iff the initial model of LP satisfies G.

A proof of this theorem is given in e.g. [GM 86]. Note that the construction of the initial model in the case where we do not have any equations and only one sort leads to the same result as the least fixpoint construction as studied in [AvE 82]. [JLM 84] generalizes this construction to the one-sorted case where we have also equations. In Section 6 we will extend these fixpoint constructions to the many-sorted setting as introduced above.

4. Unification in equational theories

We briefly recall the basic definitions needed for dealing with unification in equational theories; for a survey we refer to [Si 84].

Given a set E of equations, we say that a substitution σ is an E-unifier of the terms s and t iff $\sigma(s) =_E \sigma(t)$ where $=_E$ is the least congruence relation on $T_\Sigma(V)$ generated by the equations in E. σ is more general than δ, written $\delta \leq_E \sigma$, iff there is a substitution β such that $\delta =_E \beta^\circ\sigma$, where for the substitution composition $\beta^\circ\sigma(x)$ is defined by $\beta(\sigma(x))$. A set of substitutions Subst is a complete set of E-unifiers for s and t iff for all $\sigma \in$ Subst we have $\sigma(s) =_E \sigma(t)$ (correctness), and for every E-unifier δ of s and t there is some $\sigma \in$ Subst so that $\delta \leq_E \sigma$ (completeness). Additionally, Subst is called **minimal** iff for all $\sigma_1, \sigma_2 \in$ Subst we have $\sigma_1 \leq_E \sigma_2$ implies $\sigma_1 = \sigma_2$.

Note that for Σ containing only one sort the given definitions correspond exactly to the unsorted case as treated in e.g. [Si 84]; if additionally E is empty we have the situation of classical Robinson unification of unsorted terms. Note that the definitions also apply to the case of conditional equations.

With the notation introduced above we can precisely state the problem of solving a goal :- s = t with respect to a logic program LP whose equational part is E: it is exactly the unification problem whether there is a substitution σ such that $\sigma(s) =_E \sigma(t)$.

5. Least models and the least fixpoint construction

In the one-sorted non-equational case a model of a logic program is completely determined by its underlying carrier set and the relations interpreting the predicates. By fixing the underlying carrier set to be the set of ground terms over the functions symbols occuring in the program, one only has to consider the relations. Such models are typically called Herbrand models and are studied in e.g. [vEK 76], [AvE 82]. Dealing with the one-sorted equational case, [JLM 84] also argues to stay in a fixed domain, namely the set of equivalence classes of terms generated by the equations. For the many-sorted non-equational case which is treated in [PB 87] the corresponding standard domain is the set T_Σ of well-sorted ground terms, leading to so-called Σ-Herbrand models. Here, we generalize these three cases to the many-sorted equational setting, covering also conditional equations.

Definition:
Let LP = (Σ, E u C) with Σ = <S, OP, PR> be a logic program with equational part E, and let A be a Σ-model. Let $=_E$ be the least congruence relation on T_Σ generated by E, let $T_{\Sigma,E}$ be the factorization of T_Σ by $=_E$, and for $t \in T_\Sigma$ let [t] be the equivalence class of t in $T_{\Sigma,E}$. A is a <Σ,E>-Herbrand model iff for every $s \in S$ and every op: $s_1 \dots s_n \to s_0 \in OP$ we have:

1. $A_s = (T_{\Sigma,E})_s$

2. for all $t_i \in (T_\Sigma)_{s_i}$ $\quad op_A([t_1],\dots,[t_n]) = [op(t_1,\dots,t_n)]$
(This definition is independent of the choice of the class representatives t_i.)

Note that for every <Σ,E>-Herbrand model the interpretation of the sort symbols, the operation symbols, and the equality predicate is fixed. Thus, two <Σ,E>-Herbrand models may differ only in the interpretation of the predicate symbols other than equality. In fact, two <Σ,E>-Herbrand models A1, A2 can be partially ordered by

$A1 \leq A2$ iff for every pr: $w \in PR$. $pr_{A1} \subseteq pr_{A2}$.

The fixpoint construction of a <Σ,E>-Herbrand model satisfying the clauses of a logic program will use this ordering. Obviously, the set of all <Σ,E>-Herbrand models with the order relation ≤ forms a complete partially ordered set (cpo). More precisely, it is even a lattice with respect to ≤. The next fact characterizes the least element in this lattice.

Definition and Fact:
The <Σ,E>-Herbrand model where all relations (except equality which is the identity relation) are empty is the least <Σ,E>-Herbrand model with respect to ≤. It will be denoted by $T_{\Sigma,E,\phi}$.

Let $(\Sigma,E)^*$ be the set of all <Σ,E>-Herbrand models. We will now define a transformation from $(\Sigma,E)^*$ to $(\Sigma,E)^*$.

Definition:
Let LP = (Σ, E u C) as above. For every A1 ε $(\Sigma,E)^*$

$TRANS_{LP}(A1) := A2$

is defined as follows: The sorts, the operation symbols and the equality symbol are interpreted as in A1. For a non-equality symbol pr: $s_1 \ldots s_n$ ε PR

$([t_1],\ldots,[t_n]) \in pr_{A2}$

iff there is a clause

$pr(t_1',\ldots,t_n') :- B_1,\ldots,B_m$

in C and an E-unifier σ of $pr(t_1,\ldots,t_n)$ and $pr(t_1',\ldots,t_n')$ such that for i ε {1,...,m} we have

$([\sigma(r_1)],\ldots,[\sigma(r_{ik})]) \in pr_{i\,A1}$

if B_i is the literal $pr_i(r_1,\ldots,r_{ik})$.

(Note that as before, this definition is independent of the choice of the class representatives t_j. Moreover, note that we can take t_j instead of $\sigma(t_j)$ since t_j ground and therefore $\sigma(t_j) = t_j$.)

We can now state some properties of $TRANS_{LP}$ which hold for any LP and which generalize the one-sorted non-equational ([AvE 82]), the one-sorted equational ([JLM 82]), and the many-sorted non-equational case ([PB 87]).

Lemma: For any A ε $(\Sigma,E)^*$, $TRANS_{LP}(A)$ is a <Σ,E>-Herbrand model. Thus, $TRANS_{LP}$ is a function $TRANS_{LP}: (\Sigma,E)^* \to (\Sigma,E)^*$.

Lemma: A <Σ,E>-Herbrand model A is a model of LP iff $TRANS_{LP}(A) \leq A$.

Lemma: $TRANS_{LP}$ is continuous.

The next lemma follows from standard fixpoint theorems for continuous functions.

Lemma: $TRANS_{LP}$ has a least fixpoint which is $\lim_{i=0}^{\infty} (TRANS_{LP}{}^i(T_{\Sigma,E,\phi}))$.
We denote this least fixpoint by $lfp(TRANS_{LP})$.

We have now introduced the least <Σ, E>-Herbrand model $T_{\Sigma,E,\phi}$ as well as the least fixpoint $lfp(TRANS_{LP})$ of $TRANS_{LP}$. Both these models are elements of the lattice of <Σ, E>-Herbrand models. Furthermore, there is also a least model among those models which additionally satisfy all the clauses of the logic program LP.

Definition and Fact:
Let LP = <Σ, E u C> be a logic program. The set of all <Σ,E>-Herbrand models satisfying all clauses in C constitutes a sublattice of <(Σ,E)*, ≤>. This sublattice has a least element which we denote by min(Mod(LP)).

The last theorem of this section finally states that the three semantics based on

- least <Σ,E>-Herbrand models,
- the least fixpoint construction, and
- initial models

are all equivalent.

Theorem:

1. $lfp(TRANS_{LP}) = min(Mod(LP))$, i. e. $lfp(TRANS_{LP})$ is the least <Σ,E>-Herbrand model which is also a model of LP.

2. $lfp(TRANS_{LP})$ is an initial model of LP.

Before moving to the subject of SLD derivations we would like to comment on the properties and advantages of the different semantics definitions. Although they are all essentially equivalent as stated in the theorem above, the initial model approach is in some sense the most abstract one among them: It is non-constructive in nature, and it is completely representation independent since it does not distinguish between isomorphic models that differ only in the representation of their carrier sets. However, for certain tasks like carrying out proofs, it is sometimes more convenient to have a specific initial model at hand. In particular, in the proofs of the next sections we will use the initial model generated by the least fixpoint construction.

6. SLDE derivations and trees

The evaluation of non-equational logic programs is based on SLD-derivations which are discussed in detail in [AvE 82]. In this section we extend the notion of SLD-derivations and SLD-trees to SLDE-derivation and SLDE-trees for many-sorted logic programs with equality. As in [JLM 84] we use E-unifiers where E is a set of conditional equations, while in [GR 86] the condition of an equation may also be a non-equational literal. Therefore, the SLDE derivation of [GR 86] take into account only the equations occuring in the heads of the clauses while the conditions of the equations must be treated explicitly as separate subgoals in the resolution part of the derivation process.

6.1. SLDE derivations

Definition:

Let LP = (Σ, E u C) be a logic program with equational part E, and let G = :- $P_1, .., P_n$ be a Σ-goal. An **SLDE-derivation** of LP and G is a sequence $N_0, N_1, ...$ of Σ-goals such that G = N_0 and the following holds for every $i \geq 0$:
Let N_i = :- $A_1, ..., A_k, ..., A_m$ with $m \geq 1$. N_{i+1} is obtained according to either of the following rules:

1. A_k is not an equation. Then there is some clause
A :- $B_1, ..., B_n$ in C (possibly with some variables renamed so that it does not contain any variables that also occur in A_k) and some E-unifier σ_i of A and A_k with
$N_{i+1} = \sigma_i$(:- $A_1, ..., A_{k-1}, B_1, ..., B_n, A_{k+1}, ..., A_m$)
where σ_i(...) stands for the list of literals with σ_i applied to every element in the list, or

2. A_k is an equation s = t. Then there is some E-unifier σ_i of s and t with
$N_{i+1} = \sigma_i$(:- $A_1, ..., A_{k-1}, A_{k+1}, ..., A_m$)

An SLDE-derivation is called an **SLDE-refutation** iff it contains the empty goal (which is then the last element, say N_{r+1}, of the derivation sequence). In this case, the composition $\sigma_r \circ ... \circ \sigma_1$ is called the **answer substitution**.

In the situation where E = ∅, SLDE-derivations are exactly the well-known SLD-derivations. Moreover, analogously to the SLD case (c.f. [AvE 82]) we can prove both correctness and completeness of SLDE-refutations.

Our correctness proof makes use of initial models as discussed in Section 3.2. In the following, let T_{LP} be the initial model of LP, and let V be the set of all variables. We start with some lemmas.

Lemma 1:

If T_{LP} satisfies a literal A for every variable assignment, then T_{LP} satisfies also every instance of A for every variable assignment, i.e.

$$\text{for all } \alpha: V \to T_{LP} \,.\, T_{LP} \models_\alpha A$$
$$\Rightarrow$$
$$\text{for all } \sigma: V \to T_\Sigma(V) \,.\, \text{ for all } \alpha: V \to T_{LP} \,.\, T_{LP} \models_\alpha \sigma(A)$$

The next lemma holds because the carrier sets of T_{LP} are equivalence classes of terms under the least congruence $=_E$ generated by E.

Lemma 2:

If σ is an E-unifier of the Σ-terms s and t, then T_{LP} satisfies $\sigma(s) = \sigma(t)$ for every variable assignment, i.e.

$$\text{for all } \sigma: V \to T_\Sigma(V) .\ (\sigma(s) =_E \sigma(t) \Rightarrow \text{for all } \alpha: V \to T_{LP} .\ T_{LP} \models_\alpha \sigma(s) = \sigma(t))$$

The third lemma follows from successive applications of lemma 2 to the arguments of two non-equational literals.

Lemma 3:

If σ is an E-unifier of the non-equational literals A and A' and if T_{LP} satisfies $\sigma(A)$ for every variable assignment, then T_{LP} also satisfies $\sigma(A')$ for every variable assignment, i.e.

$$\text{for all } \sigma: V \to T_\Sigma(V) .\ ((\sigma(A) =_E \sigma(A') \ \&\ \text{for all } \alpha: V \to T_{LP} .\ T_{LP} \models_\alpha \sigma(A)) \Rightarrow \text{for all } \alpha: V \to T_{LP} .\ T_{LP} \models_\alpha \sigma(A'))$$

Now we can prove the correctness of SLDE refutations with respect to the initial model.

Theorem:

If there is an SLDE refutation of LP and G with answer substitution σ_{ANS}, then T_{LP} satisfies every atom in $\sigma_{ANS}(G)$ for every variable assignment.

Proof:

Let $N_1, \ldots, N_{r+1}$ be the SLDE-refutation and σ_i be the substitution used in stepping from N_i to N_{i+1}. Thus, $\sigma_{ANS} = \sigma_r \circ \ldots \circ \sigma_1$.

For $i \in \{0,\ldots,r-1\}$ let ASS(i) be the following assertion:

For every atom A in $\sigma_r \circ \ldots \circ \sigma_{r-i}(N_{r-i})$, T_{LP} satisfies A for every variable assignment, i.e.

$$A \in \sigma_r \circ \ldots \circ \sigma_{r-i}(N_{r-i}) \Rightarrow \text{for all } \alpha: V \to T_{LP} .\ T_{LP} \models_\alpha A.$$

ASS(r-1) is the assertion of the theorem, and is proved by induction on i.

In order to prove the SLDE refutation completeness we use again the initial model T_{LP}.

Lemma 4:

Let A be a ground Σ-atom. If T_{LP} satisfies A, then there is an SLDE refutation of LP and :- A.

Proof:

The initiality of $lfp(TRANS_{LP})$ and the characterization of $lfp(TRANS_{LP})$ in terms of the limit of

$$TRANS_{LP}^{\,i}(T_{\Sigma,E,\phi})$$

implies that there is some finite k such that $A \in TRANS_{LP}^{\,k}(T_{\Sigma,E,\phi})$.

(Recall that $T_{\Sigma,E,\phi}$ is the least $\langle\Sigma,E\rangle$-Herbrand model, with all relations apart from the equality relation being empty.) By induction on k we can show that there is an SLDE refutation of LP and :- A.

Theorem:

If T_{LP} satisfies G then there is an SLDE refutation of LP and G.

Proof:

If LP satisfies the goal $G = \text{:- } A_1,\ldots,A_n$ there is a variable assignment $\alpha: V \to T_{LP}$ (with V the variables occuring in G) such that for every $i \in \{1,\ldots m\}$ $T_{LP} \models_\alpha A_i$.

Now for every variable v, $\alpha(v)$ is an equivalence class of ground terms over the operation symbols of LP. Choosing any term $t \in \alpha(v)$ and sending v to t yields a ground term substitution $\sigma: V \to T_\Sigma$, and obviously, T_{LP} satisfies every $\sigma(A_i)$ and therefore also the goal $\sigma(G)$. Since $\sigma(A_i)$ is ground, there is an SLDE refutation for LP and :- $\sigma(A_i)$ according to Lemma 4, and thus also an SLDE refutation, say R, for LP and $\sigma(G)$.

In order to show that there is also an SLDE refutation for LP and G let $\sigma(A_j)$ be the first selected atom in R. Furthermore, let

$A' :- B_1, \ldots, B_m.$

be the first used clause and τ the E-unifier of $\sigma(A_j)$ and A'. Thus, there exists an SLDE-refutation of LP and

$$:- \tau(\sigma(A_1), \ldots, \sigma(A_{j-1}), B_1, \ldots, B_m, \sigma(A_{j+1}), \ldots, \sigma(A_n)).$$

Without loss of generality we may assume that σ does not change any variables in LP. Thus, $\sigma(B_i) = B_i$ and we can write the goal above as

$$:- \tau^\circ\sigma(A_1, \ldots, A_{j-1}, B_1, \ldots, B_m, A_{j+1}, \ldots, A_n).$$

Furthermore, since $\sigma(A') = A'$, $\tau^\circ\sigma$ is an E-unifier of A_j and A'. Therefore, there is an SLDE refutation of LP and G using A_j as the first selcted atom, the clause with head A' as above, and $\tau^\circ\sigma$ as the used substitution, followed by R.

Theorem:

Let LP = (Σ, E u C) be a logic program with equational part E and G a Σ-goal. LP satisfies G if and only if there exists an SLDE-refutation for LP and G.

Proof:

By the theorems above, LP satisfies G iff the initial model T_{LP} satisfies G (c.f. Section 3) iff there is an SLDE refutation for LP and G.

6.2. SLDE trees

In order to check the validity of a goal G a refutation for G has to be found, i.e. we have to construct a derivation which contains the empty goal.

The standard way of finding a refutation for G is to organize the derivations in a tree-like manner. The above definition of a derivation indicates that each derivation step is determined by making three choices in order to proceed to the next state of a proof:

1. choice of a selected atom A_k

2. choice of a program clause whose head A is E-unifiable with A_k

3. choice of an E-unifier σ_i of A_k and A

Apt and van Emden [AvE 82] discuss the influence of these three choices on constructing a proof based on SLD-resolutions. They show that it does not matter which atom is selected, so the choice in (1) is of "don't care" style, whereas each clause the head of which unifies with the selected atom must be considered in (2) in order to achieve completeness. Their results carry over to our setting as far as the selection of an atom and a matching clause is concerned. Differences arise for the third choice to be made. Since SLD-refutations are based on unification in the absence of equations, it suffices to consider only the most general unifier σ of A and A_k. Therefore, we don't have a proper choice of unifiers in (3) since there is at most one such σ. SLDE-derivations, however, are based on unification in the presence of equations. Here the concept of a most general unifier has to be generalized to that of complete sets of unifiers (c.f. Section 4) to which we can restrict the search space for (3). Since complete sets of unifiers are in general not singletons, SLDE-refutations do require a proper choice under (3) leading to an enlarged possibility for subsequent proof steps.

Definition:

Let U_E be a function taking every pair (A,B) of terms resp. atoms to a complete set of E-unifiers of A and B. An SLDE-tree with respect to U_E for G and LP is a tree T with root G. A non-empty node in T has one atom which is the selected atom. A node

$N = :- A_1, \ldots, A_k, \ldots, A_m$

with selected atom A_k has the following descendants:

1. If A_k is not an equation, then for every clause
   $A :- B_1, \ldots, B_n$ in C such that A and A_k are E-unifiable there is a descendant
   $\sigma(:- A_1, \ldots, A_{k-1}, B_1, \ldots, B_n, A_{k+1}, \ldots, A_m)$
   for every $\sigma \in U_E(A, A_k)$.

2. If A_k is an equation s = t, then there is a descendant
   $\sigma(:- A_1, \ldots, A_{k-1}, A_{k+1}, \ldots, A_m)$

for every $\sigma \in U_E(s,t)$.

In both cases σ is the label of the edge from N to the new node. ANS(T) is the set of all answer substitutions that are the compositions of the substitutions labelling the edges along a path from the root of T to an empty node.

SLDE-trees may be of infinite depth just as the SLD-trees studied in [AvE 82]. If only finite logic programs are considered, SLD-trees are of finite breadth since there is at most one most general unifier between any two literals. SLDE-trees, however, may be of infinite breadth since the complete set of E-unifiers $U_E(A,B)$ may be infinite [Si 84].

We can show that searching any arbitrary SLDE-tree is correct and complete with respect to goal satisfaction.

Theorem:
Let LP = (Σ, E u C) be a logic program with equational part E and G a Σ-goal. Then for any SLDE-tree T with respect to U_E for G and LP we have

1. For every ground term substitution σ such that LP satisfies σ(G) there exists an $\alpha \in$ ANS(T) with $\sigma \leq_E \alpha$.
2. For every $\alpha \in$ ANS(T), LP satisfies α(G).
3. LP satisfies G if and only if ANS(T) $\neq \phi$.

Proof:
Assertions (1) and (2) are strong forms of correctness resp. completeness of SLDE refutations and can be proven similar as the theorems in Sections 6.2 and 6.3, using the fact that $U_E(A,A_k)$ resp. $U_E(s,t)$ always yields a correct and complete set of unifiers. Assertion (3) follows immediately from (1) and (2).

7. Translation of logic programs with equality

We intend to compute E-unifiers by means of non-equational logic programs and are thus going to define what it means that a program serves this purpose.

Starting from an equational logic program LP = (Σ, E u C) we move to the extended program LPa = (Σa, E u C) allowing us to view Σ-literals and Σ-terms uniformly as Σa-terms of the distinguished sort "atom" and to view any Σ-unification problem as a Σa-goal
:- unify(A, B).
Thus for a signature Σ = (S, OP, PR) with OP $\cap$ PR = ϕ, and "s-atom", "atom", and "unify" symbols not occuring in Σ we define
Σa = (S u {atom}, OP u OP', PR u {unify: atom atom})
as Σ with **atomic terms** where
OP' = {p: $s_1 \ldots s_n$ -> atom | p: $s_1 \ldots s_n \in$ PR} u
{s-atom: s -> atom | s $\in$ S}

Definition:
Let LP = (Σ, E u C) be any logic program with equational part E, Σa the extended signature as constructed above, and LP_{UNI} = (Σ_{UNI}, C_{UNI}) a non-equational logic program with Σa $\subseteq \Sigma_{UNI}$. Let A and B be either two Σ-atoms or two Σa-terms of the form s-atom(t) where t is a Σ-term.
LP_{UNI} **computes [complete] sets of E-unifiers** iff for any SLD-tree T for unify(A, B) and LP_{UNI}, ANS(T) is a [complete] set of E-unifiers for A and B.

Let LP_E = (Σ, C u E) be a logic program with equational part E. We define a logic program LP_ϕ = (Σ_{UNI}, TR(C) u C_{UNI}) with empty equational part so that LP_ϕ is in some sense equivalent to LP_E (see below for a precise meaning of equivalent in this context). First, we define a non-equational logic program (Σa, TR(C)) over the enlarged signature Σa with TR(C) resulting from C by translating every (non-equational) clause to a clause with explicit calls to the unification predicate as subgoals.

Definition:
Let LP_E and Σa be as above.

1. If A is the Σ-equation $t_1 =_s t_2$, the translation TR(A) is the Σa-literal
unify(s-atom(t_1),s-atom(t_2))

2. If A is a non-equational Σ-literal, TR(A) is A again.

3. For a list $A_1,\ldots,A_n$ of Σ-literals, TR($A_1,\ldots,A_n$) is the list TR(A_1),...,TR(A_n).

4. If CL is the Σ-clause A :- $A_1,\ldots,A_n$ with non-equational A, the translation TR(CL) is the Σa-clause
   A' :- unify(A, A'), TR(A_1, ..., A_n)
   where for A = p($t_1,\ldots,t_r$) we let A' = p($x_1,\ldots,x_r$) such that the variables x_i do not occur in the original clause and sort(x_i) = sort(t_i).

5. For the set C of Σ-clauses, TR(C) is the set {TR(CL) | CL ε C}.

We still need a logic program that defines the predicate unify. Thus, let LP_{UNI} = (Σ_{UNI}, C_{UNI}) be any logic program with empty equational part computing complete sets of E-unifiers. We assume that Σa ⊆ Σ_{UNI} and that C_{UNI} does not contain any clauses with predicates from Σ, i.e. LP_{UNI} only defines the predicate unify and some auxiliary predicates in Σ_{UNI} - Σ.

The non-equational logic program LP_ϕ resulting from LP_E is then the combination of TR(C) and C_{UNI}, i.e.
   LP_ϕ = (Σ_{UNI}, TR(C) u C_{UNI}).
For the rest of this section we further assume that G_E is any Σ-goal and that G_ϕ is the Σ_{UNI}-goal TR(G_E).

Before we will prove the main theorem of this section, we start with another lemma stating that for every E-unifier σ of two terms or atoms over Σ there is an SLD refutation for the corresponding 'unify' goal using LP_ϕ with answer substitution σ and vice versa.

Lemma 5:
Let t_1, t_2 be Σ-terms of sort s [resp. A_1, A_2 non-equational Σ-atoms]. Then the following statements are equivalent:

1. σ is an E-unifier of t_1 and t_2 [resp. A_1 and A_2].

2. There is an SLD refutation for
   :- unify(s-atom(t_1),s-atom(t_2))
   [resp. :- unify(A_1,A_2)]
   and LP_ϕ with answer substitution σ.

Theorem:
If there exists an SLDE refutation for G_E and LP_E, then there exists an SLD refutation for G_ϕ and LP_ϕ with the same answer substitution.

Proof:
Let $N_1,\ldots,N_{r+1}$ be an SLDE refutation for G_E and LP_E, and let σ_j be the E-unifier used in stepping from N_j to N_{j+1}. We will prove the theorem by showing that for every j ε {1,...r} there is an SLD derivation starting from TR(N_j) and leading to TR(N_{j+1}) using only clauses from LP_ϕ and producing σ_j as answer substitution. This situation is sketched in the following commuting diagram:

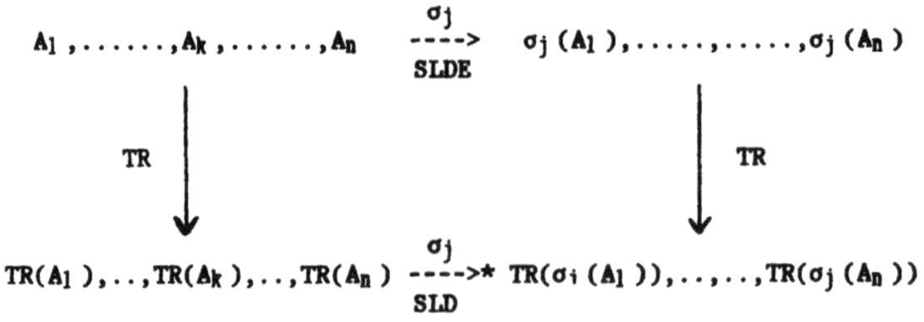

Let A be the selected atom of N_j. The complete proof now distinguishes the two cases of A being an equation or being a non-eqautional atom. In both cases the SLDE-derivation step to N_{j+1} is mirrored by a corresponding sequence of SLD-derivation steps starting with TR(A) as selected atom and using Lemma 5 in order to produce the E-unifier σ_j.

Theorem:

If there exists an SLD refutation for G_ϕ and LP_ϕ, then there exists an SLDE refutation for G_E and LP_E with the same answer substitution.

Proof:

Let $N_1, \ldots, N_{r+1}$ be an SLD refutation for G_ϕ and LP_ϕ, and for $j \in \{1, \ldots r\}$ let C_j resp. σ_j be the clause resp. substitution used in stepping from N_j to N_{j+1}. We will show how we can replace subsequences of SLD derivation steps in $N_1, \ldots, N_{r+1}$ by single SLDE derivation steps, yielding an SLDE refutation for G_E and LP_E with the same answer substitution.

The completeness of SLD refutations using SLD trees ensures that in every N_j any atom may be selected so that the refutation may still be completed. Moreover, we can rearrange the derivation steps and still get a refutation of the same length and with the same answer substitution as long as we select the same clause for every selected atom and choose the substitutions according to the original refutation. Thus, without loss of generality, we can make some assumptions about the next selected atom.

Let A_k be the selected atom of N_j = :- $A_1, \ldots, A_{k-1}, A_k, A_{k+1}, \ldots, A_n$.

If A_k is of the form unify(A,B) we assume that no other atom of N_j is selected until an SLD refutation R_j for unify(A,B) is obtained. Let π_j be the answer substitution of R_j. By Lemma 5 we conclude that R_j uses only clauses from $C_{\cup \mathbb{N} I}$ and that π_j is an E-unifier of A and B. Thus, there is a uniquely determined f(j) with j < f(j) < r+1 such that

$$N_{f(j)+1} = \text{:- } \pi_j(A_1, \ldots, A_{k-1}, A_{k+1}, \ldots, A_n)$$

and

$$\pi_j = \sigma_{f(j)} \circ \sigma_{f(j)-1} \circ \ldots \circ \sigma_j .$$

Thus, we assume that whenever the selected atom A_k in N_j is a unify goal the SLD derivation steps from N_j to $N_{f(j)+1}$ will exactly refute the unify goal and produce the E-unifier π_j as answer substitution. On the other hand, if the selected atom A_k in N_j is an atom $p(t_1, .., t_u)$ with p a predicate symbol from the original program LP_E, then the matching clause C_j must be of the form TR(C) where C is a clause in LP_E. Using these two observations we can replace the computation of an E-unifier in the SLD derivation by the selection of this E-unifier in the SLDE derivation. Similar, the application of TR(C) corresponds to the application of C in the SLDE case.

Now we state the main theorem regarding the relationship of LP_E and its translation LP_ϕ.

Theorem:

1. There exists an SLDE-refutation for G_E and LP_E
   <==>
   there exists an SLD-refutation for G_ϕ and LP_ϕ.

2. $LP_E \models G_E$ <==> $LP_\phi \models G_\phi$.

Proof:

The two implications of assertion (1) follow from the two previous theorems. Assertion (2) follows from (1) since SLD and SLDE refutations are correct and complete with respect to goal satisfaction.

8. Implementation aspects and conclusions

We have studied the integration of logic programming and equational specification of abstract data types. Various approaches to the semantics of many-sorted logic programs with equality are given, namely

1. a model-theoretic one based on initial models,
2. a semantics based on least Herbrand models, and
3. a least fixpoint semantics,

and three approaches to an operational semantics:

- Generate SLDE derivations and look for a refutation
- Search an SLDE tree
- Translate LP_E to LP_ϕ and take the operational semantics of LP_ϕ

The last approach demonstrates how to reduce the operational semantics of Horn logic with equality to that of non-equational Horn logic. These theoretical considerations form the basis for a rapid implementation of a typed Prolog-like language enhanced by equality in ordinary Prolog. Using an untyped Prolog version, our approach was implemented by Christof Schadt and Hinrich Schütze in VM/Prolog. This required only two things: A simple syntactic translation of LP_E and the implementation of an E-unification algorithm. For the latter we used the narrowing method , c.f. [RRKL 85]. However, problems arise if the process of generating a complete set of E-unifiers does not halt. In this case a strategy that allows the alternation of resolution steps and computing steps for E-unification might be more appropriate; [Ya 87] addresses such a situation for the case of unconditional equations, and [BGM 87] also studies the correspondence between narrowing and SLD-resolution.

Acknowledgements: We would like to thank our colleagues Karl Hans Bläsius and Hinrich Schütze for carefully reading an earlier draft of this paper and many helpful comments.

References

[AvE 82] Apt, K. R. and van Emden, M. H.: Contributions to The Theory of Logic Programming. Journal of the ACM, Volume 29, Number 3, 1982.

[BGM 87] Bosco, P. G., Giovannetti, E. and Moiso, C.: Refined Strategies for Semantic Unification. Proc. TAPSOFT-87, LNCS Vol 250, Springer Verlag, 1987.

[BP 88] Beierle, C. and Pletat, U.: Semantics of Logic Programs with Equational Abstract Data Type Specifications. LILOG Report 38, IBM Deutschland GmbH, Stuttgart, 1988.

[CM 84] Clocksin, W. F. and Mellish, C. S.: Programming in Prolog. Springer Verlag, Berlin 1982.

[DeGL 86] De Groot, D. and Lindstrom, G. (eds): Functional and Logic Programming, Prentice Hall, 1986.

[vEK 76] van Emden, M. H. and Kowalski, R. A.: The Semantics of Predicate Logic as a Programming Language. Journal of the ACM, Volume 23, Number 4, 1976.

[EM 85] Ehrig, H. and Mahr, B.: Foundations of Algebraic Specification 1. Springer Verlag, Berlin 1985.

[GM 86] Goguen, J. G. and Meseguer, J.: EQLOG: Equality, Types and Generic Modules for Logic Programming. In [DeGL 86].

[GR 86] Gallier, J. H. and Raatz, S.: SLD-Resolution Methods for Horn Clauses with Equality Based on E-Unification. In. Proc. 1986 Symposium on Logic Programming, Salt Lake City, IEEE Computer Society Press, 1986.

[GTW 78] Goguen, J. A. and Thatcher, J. W. and Wagner, E. An Initial Algebra Approach to The Specification, Correctness and Implementation of Abstract Data Types. In: Current Trends in Programming Methodology, R. T. Yeh, (ed), Prentice-Hall, 1978.

[Hö 87] Hölldobler, S.: Equational Logic Programming. Proc. Symposium on Logic Programming, 1987.

[Hu 85] Hussmann, H.: Unification in Conditional-Equational Theories. Proc. EUROCAL 85, LNCS, Springer Verlag, 1985.

[JLM 84] Jaffar, J., Lassez, J.-L., Maher, M.J.: A Theory of Complete Logic Programs with Equality. J. of Logic Programming, 1984, pp. 211-223.

[JLM 86] Jaffar, J., Lassez, J.-L., Maher, M.J.: A Logic Programming Language scheme. In [DeGl 86].

[Ll 84] Lloyd, J. W.: Foundations of Logic Programming. Springer Verlag, Berlin 1984.

[MM 82] Martelli, A., Montanari, U.: An Efficient Unification Algorithm. ACM TOPLAS, Vol. 4(2), 1982, pp. 258-282.

[MM 84] Mahr, B. and Makowsky, J. A.: Characterizing Specification Languages which Admit Initial Semantics. Theoretical Computer Science, Volume 31, North-Holland 1984.

[MMR 86] Martelli, A., Moiso, C. and Rossi, G.F.: An Algorithm for Unification in Equational Theories. In: Proc. 1986 Symposium on Logic Programming, Salt Lake City, IEEE Computer Society Press, 1986.

[PB 87] Pletat, U., Beierle, C.: The Semantics of Asserting and Retracting Clauses to Logic Programs. LILOG Report 7, IBM Deutschland GmbH, Stuttgart 1987.

[RKKL 85] Rety, P., Kirchner, C., Kirchner, H. and Lescanne, P.: NARROWER: a new algorithm for unification and its application to Logic Programming. In: Proc. Term Rewriting Techniques and Applications, J. P. Jouannaud (ed), LNCS Vol. 202, Springer Verlag, Berlin 1985.

[Si 84] Siekmann, J.: Universal Unification. Proc. 7th International Conference on Automated Deduction, R. E. Shostak (ed), LNCS Vol. 170, Springer Verlag, Berlin 1984.

[WB 80] Broy, M. and Wirsing, M.: Abstract Data types as Lattices of Finitely Generated Models. Proc. 9th MFCS, Lecture Notes in Computer Science, Vol. 88, Springer Verlag, Berlin 1980.

[Ya 87] Yamamoto, A.: A Theoretical Combination of SLD-Resolution and Narrowing. Proc. 4th Int. Conf. on Logic Programming, May 1987.

NARROWING WITH BUILT-IN THEORIES

Alexander Bockmayr
Sonderforschungsbereich 314 "Künstliche Intelligenz - Wissensbasierte Systeme"
Institut für Logik, Komplexität und Deduktionssysteme
Universität Karlsruhe, Postfach 6980, D-7500 Karlsruhe 1, F. R. Germany

Abstract

Rewriting and narrowing provide a nice theoretical framework for the integration of logic and functional programming. For practical applications however narrowing is still much too inefficient. In this paper we show how narrowing modulo equality theories may considerably increase the efficiency of the narrowing process.

1. Introduction

During the last years there has been an increasing interest in the integration of logic and (first order) functional programming (DeGroot/Lindstrom 86, Bellia/Levi 87). While some authors in a more pragmatic approach propose a synthesis of the existing programming languages PROLOG and LISP, others develop the idea of logic-functional programming on a more theoretical level.

It has turned out that (conditional) term rewriting systems provide a nice theoretical framework for logic-functional programming (Dershowitz/Plaisted 85, Bockmayr 86a). In the rewriting process the rewrite rules are employed for the simplification or evaluation of terms ("functional programming"), whereas in the narrowing process, which is closely related to resolution, the same rewrite rules are used for the solution of goals or equations ("logic programming").

From a theoretical point of view, narrowing provides a complete unification procedure for any equational theory that can be defined by a canonical term rewriting system (Fay 79, Hullot 80, Réty et al. 85). For practical applications however, narrowing in its original form is much too inefficient (Bockmayr 86b). Many optimizations have been proposed (Hullot 80, Fribourg 85, Nutt/Réty/Smolka 87, Padawitz 87, Bosco/Giovannetti/Moiso 87), most of which restrict the set of occurrences at which a narrowing step is performed. But, even the most sophisticated narrowing procedure would be inadequate to solve for example a system of linear equations. It is therefore necessary to incorporate special theories and their unification algorithms into the narrowing process.

Building-in equality theories into the deduction machinery is an old idea. In automated theorem proving for example it goes back to Plotkin 72. Theory resolution has been extensively discussed in (Stickel 85). To deal with equality theories in logic programming Jaffar/Lassez/Maher 86 introduced a logic programming language scheme. Prolog II and Prolog III (Colmerauer 87) are special instances of it.

In this paper we study built-in equality theories in the context of narrowing. We show how narrowing modulo equality may considerably increase the efficiency of the narrowing process.

The organization of the paper is as follows: After some preliminaries in section 2, we present in section 3 the motivation for our work. In section 4 we introduce narrowing modulo equality and present the correctness and completeness result of Jouannaud/Kirchner/Kirchner 83. In section 5 we show how narrowing modulo equality may dramatically reduce the search space of the narrowing algorithm.

We believe that a logic-functional programming language without built-in theories cannot meet the requirements of practical applications. Therefore the study of special theories and their integration into logic-functional programming languages will play an important role in future research.

2. Preliminaries

We recall briefly some basic notions that are needed in the sequel. More details can be found in the classical survey of Huet/Oppen 80.

$\Sigma = (S,F)$ denotes a signature where S is a finite set of sort symbols, and F is a finite set of function symbols together with an arity function.

A Σ-algebra A consists of a family of sets $(A_s)_{s \in S}$ and a family of functions $(f^A)_{f \in F}$ such that if $f{:}s_1 \times ... \times s_n \to s$ then $f^A: A_{s1} \times ... \times A_{sn} \to A_s$.

T_Σ denotes the algebra of ground terms over Σ.

X represents a familiy $(X_s)_{s \in S}$ of countable infinite sets X_s of variables of sort s.

$T_\Sigma(X)$ is the algebra of terms (with variables) over Σ.

O(t) denotes the set of occurences in the term t, t/ω is the subterm of t at position $\omega \in O(t)$,

$t[\omega \leftarrow s]$ the term obtained from t by replacing the subterm t/ω with s.

V(t) is the set of variables occuring in t.

A substitution is a mapping $\sigma: X \to T_\Sigma(X)$ which is different from the identity only for a finite subset $D(\sigma)$ of X. $I(\sigma) =_{def} \cup_{x \in D(\sigma)} V(\sigma(x))$ is the set of variables introduced by σ.

If a substitution maps x_i to t_i for $i=1,\ldots,n$, then it is denoted by $[x_1/t_1,\ldots,x_n/t_n]$.

We do not distinguish σ from its canonical extension to $T_\Sigma(X)$.

A precongruence is a binary relation $\to$ on $T_\Sigma(X)$ or T_Σ such that $s_1 \to t_1, \ldots, s_n \to t_n$ implies $f(s_1,\ldots,s_n) \to f(t_1,\ldots,t_n)$ for all terms s_i,t_i and all $f \in F$.

By $\to^+$, $\to^*$ and $\leftrightarrow^*$ we denote the transitive, the reflexive-transitive, and the reflexive-transitive-symmetric closure of $\to$ respectively.

A congruence is a precongruence which is also an equivalence relation.

An equation is an expression of the form $l = r$ with terms $l, r \in T_\Sigma(X)$ belonging to the same sort.

Let T be a set of equations. The equational theory $=_T$ associated with T is the smallest congruence $\approx$ on $T_\Sigma(X)$ such that $\sigma(l) \approx \sigma(r)$ for all equations $l = r$ in T and all substitutions $\sigma{:}X \to T_\Sigma(X)$.

The T-subsumption preorder $\leq_T$ on $T_\Sigma(X)$ is defined by $s \leq_T t$ iff there is a substitution $\sigma{:}X \to T_\Sigma(X)$ with $t =_T \sigma(s)$. For two substitutions $\sigma,\tau: X \to T_\Sigma(X)$ and a set of variables V we say that $\sigma \leq_T \tau$ [V] iff there is a substitution λ with $\tau(x) =_T \lambda(\sigma(x))$ for all $x \in V$. By $\equiv_T$ we denote the equivalence relation associated with $\leq_T$.

Given two terms $t, u \in T_\Sigma(X)$, a T-matcher from t to u is a substitution $\sigma: X \to T_\Sigma(X)$ with $\sigma(t) =_T u$. We are only interested in the case where V(t) and V(u) are disjoint.

A T-unifier of two terms $t, u \in T_\Sigma(X)$ is a substitution $\sigma: X \to T_\Sigma(X)$ with $\sigma(t) =_T \sigma(u)$.

Let W be a finite set of variables containing $V =_{def} V(t) \cup V(u)$.

A set $CSU_T(t,u,W)$ of substitutions is called a complete set of T-unifiers of t and u away from W iff

- every $\sigma \in CSU_T(t,u,W)$ is a T-unifier of t and u
- for any T-unifier τ of t and u there is a $\sigma \in CSU_T(t,u,W)$ such that $\sigma \leq_T \tau$ [W] .
- for all $\sigma \in CSU_T(t,u,W)$: $D(\sigma) \subseteq V$ and $I(\sigma) \cap W = \emptyset$

$CSU_T(t,u,W)$ is called <u>minimal</u> iff it satisfies the additional condition

- for all $\sigma, \sigma' \in CSU_T(t,u,W) : \sigma \leq_T \sigma'$ [W] implies $\sigma = \sigma'$.

3. Motivation

To begin our investigation we consider a problem taken from the "Vollständige Anleitung zur Algebra" by Leonhard Euler (1707 - 1783).

A father leaves his three sons an inheritance of 1600 £. According to his testament the oldest son should receive 200 £ more than the middle son and the middle son should receive 100 £ more than the youngest son. How much does each son receive?

We would like to show how the narrowing algorithm solves this problem. Here we provide only an informal description of the method. The exact definitions will follow later.

First a canonical term rewriting system R is needed.

The simplest one is presumably $R = \{ 0 + x \longrightarrow x , s(x) + y \longrightarrow s(x+y) \}$.

Then we have to solve the goal:

$$?\text{-}\ x + y + z = s^{1600}(0)\ ,\ x = y + s^{200}(0)\ ,\ y = z + s^{100}(0).$$

If we had a special unification algorithm for the equational theory given by R, i.e. an algorithm for solving systems of linear diophantine equations (Clausen/Fortenbacher 87), (Käufl 87) we would immediately get the answer $x = s^{700}(0)$, $y = s^{500}(0)$, $z = s^{400}(0)$.

But how does the narrowing algorithm proceed?

The narrowing algorithm looks for a nonvariable subterm u in the system of equations that can be unified with the left-hand side of a rule $l \rightarrow r$ in R with the most general unifier μ. Then it applies the substitution μ to the whole system and replaces the subterm $\mu(u)=\mu(l)$ with $\mu(r)$.

The algorithm explores all such derivations until the system becomes unifiable in the ordinary sense.

When we formulate our goal in the form:

$$?\text{-}\ s^{400}(z+z+z) = s^{1600}(0),\ y = s^{100}(z),\ x = s^{200}(y)$$

we need 401 narrowing steps to find the solution $z = s^{400}(0)$.

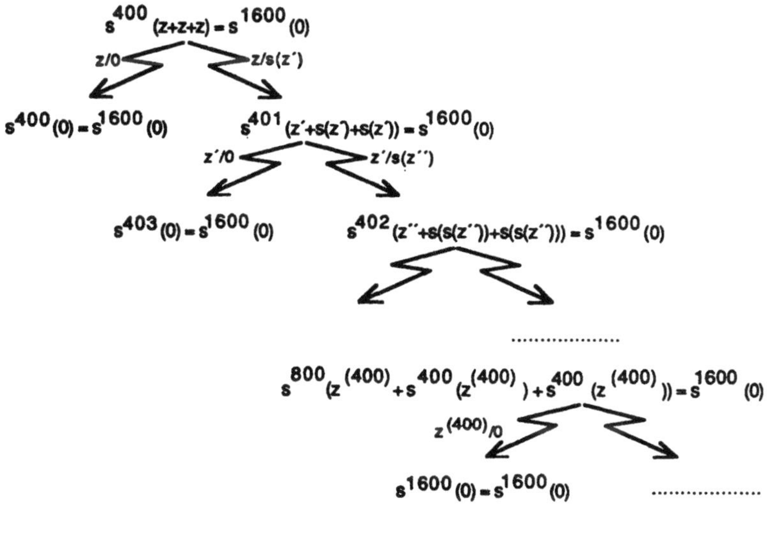

This example should convince the reader that narrowing is not an appropriate method for solving linear equations. However, such equations occur very often in practical applications. Consequently we must find a way to incorporate special unification procedures, for example an algorithm for solving linear equations, into the general narrowing machinery. This is topic of the next sections.

4. Equational term rewriting and narrowing

In this section we introduce equational term rewriting and narrowing according to Peterson/Stickel 81 and Jouannaud/Kirchner/Kirchner 83. We obtain the classical definitions when the equational theory E is empty.

Definition

A term rewriting system R is a finite set of rewrite rules of the form

$$l \rightarrow r,$$

where l and r are terms of $T_\Sigma(X)$ belonging to the same sort and $V(l) \supseteq V(r)$.

An equational term rewriting system R,E is a pair of a term rewriting system R and a set of equations E.

Definition

Given an equational TRS R,E the equational rewriting relation $\rightarrow_{R,E}$ on $T_\Sigma(X)$ is defined by

$$s \rightarrow_{R,E} t$$

iff there is a rule $l \rightarrow r$ in R, a substitution σ and an occurrence $\omega \in O(s)$,

such that $s/\omega =_E \sigma(l)$ and $t = s[\omega \leftarrow \sigma(r)]$.

A term $s \in T_\Sigma(X)$ is R,E-normalized iff there is no $t \in T_\Sigma(X)$ with $s \rightarrow_{R,E} t$.

A substitution σ: $X \longrightarrow T_\Sigma(X)$ is R,E-normalized iff $\sigma(x)$ is R,E-normalized for all $x \in X$.

It is possible to define more complex equational rewriting relations. For example, we might divide our set of rules R in a set of left-linear rules L and another set N such that E-matching is done only for the rules in N whereas, for the rules in L ordinary matching is used (Jouannaud/Kirchner 86).

Definition

An equational TRS R,E is called R,E-Church-Rosser modulo E iff for all $s, t \in T_\Sigma(X)$:

$s =_{R \cup E} t$ implies $s \downarrow_{R,E} t$

where $s \downarrow_{R,E} t$ iff there exist s' and t' such that $s \rightarrow^*_{R,E} s'$, $t \rightarrow^*_{R,E} t'$ and $s' =_E t'$.

R is called E-terminating iff there is no infinite sequence of the form

$$t_0 =_E t_0' \rightarrow_R t_1 =_E t_1' \;\ldots\ldots\; t_n =_E t_n' \rightarrow_R t_{n+1} \ldots\ldots$$

In (Jouannaud83), (Jouannaud/Munoz 84), (Jouannaud/Kirchner 86) necessary and sufficient conditions are presented to check these fundamental properties. But much work remains to be done, in particular on the topic of E-termination.

We now define equational narrowing (Jouannaud/Kirchner/Kirchner 83, Kirchner 85).

Definition

Let R,E be an equational TRS. Given two terms t, $t' \in T_\Sigma(X)$, we say that t is R,E–narrowable into t′ with narrowing substitution σ,

denoted

$$t \ \text{-}\!\!\!\!\!\vdash\!\sigma\!\!\rightarrow_{R,E} \ t',$$

iff there is a non-variable occurrence $\omega \in O(t)$ and a rule $l \mapsto r$ in R,

such that t/ω and l are E-unifiable with a E-unifier $\sigma \in CSU_E(t/\omega, l, W)$ and $t' = \sigma(t)[\omega \leftarrow \sigma(r)]$.

Here W is some set containing $V(t) \cup V(l)$ and V(t) and V(l) are disjoint.

By $\text{-}\!\!\vdash\!\sigma\!\!\rightarrow^*_{R,E}$ we denote a sequence of narrowing steps $t_0 \text{-}\!\!\vdash\!\sigma_0\!\!\rightarrow_{R,E} t_1 \text{-}\!\!\vdash\!\sigma_1\!\!\rightarrow_{R,E} \ldots\ldots \text{-}\!\!\vdash\!\sigma_{n-1}\!\!\rightarrow_{R,E} t_n$, $n \geq 0$, where $\sigma = \sigma_{n-1} \bullet \ldots\ldots \bullet \sigma_1 \bullet \sigma_0$.

The next theorem states the correctness and completeness of equational narrowing and is due to Jouannaud/Kirchner/Kirchner 83. It has been refined in (Kirchner 85).

Theorem

Let R,E be an equational TRS that is R,E-Church–Rosser modulo E and E–terminating.

Assume that E has a finite and complete E-unification algorithm.

Given two terms $t_0, t_0' \in T_\Sigma(X)$ consider the term $eq(t_0,t_0')$ where eq is a new function symbol.

Let U be the set of all substitutions τ with the following property:

There exists a R,E–narrowing derivation

$$eq(t_0,t_0') \ \text{-}\!\!\vdash\!\sigma_0\!\!\rightarrow_{R,E} eq(t_1,t_1') \text{-}\!\!\vdash\!\sigma_1\!\!\rightarrow_{R,E} \ldots\ldots \text{-}\!\!\vdash\!\sigma_{n-1}\!\!\rightarrow_{R,E} \ eq(t_n,t_n') \ , n \geq 0,$$

such that

- t_n and t_n' are E–unifiable with a E-unifier $\rho \in CSU_E(t_n,t_n', V(t_0) \cup V(t_0'))$,
- the substitutions $\tau_i = \sigma_{i-1} \bullet \ldots\ldots \bullet \sigma_1 \bullet \sigma_0$ are R,E-normalized for any $i = 1,\ldots,n$,
- $\tau = \rho \bullet \tau_n \quad [V(t_0) \cup V(t_0')]$.

Then U is a complete set of R∪E–unifiers of t_0 and t_0' away from $V(t_0) \cup V(t_0')$.

5. Increasing the Efficiency of Narrowing by Built-in Theories

After having established the theoretical framework of narrowing modulo equality we want to show how this concept may considerably increase the efficiency of the narrowing process.

Originally narrowing modulo equality was introduced to handle equations such as associativity and commutativity for which canonical term rewriting systems do not exist (Jouannaud/Kirchner/ Kirchner 83).

In this paper we take another point of view, namely the following:

Given a term rewriting system R, we may partition R in two subsystems R_0 and E. When R_0,E is R_0,E-Church-Rosser modulo E and E-terminating, then we may use equational narrowing with R_0 modulo E instead of ordinary narrowing with R. We may also add additional properties to E such as associativity and commutativity.

This may considerably prune the narrowing tree. E-unification problems are no longer solved by narrowing but by a special E-unification algorithm.

An example is as follows.

Consider the two canonical term rewriting systems

$R_0 = \{ q(0) \longrightarrow 0 , q(s(x)) \longrightarrow s((q(x) + x) + x) \}$

$E = \{ x + 0 \longrightarrow x , x + s(y) \longrightarrow s(x + y) \}$

and suppose there is the query

?- $q(x) = s(s(s(0)))$.

R_0,E is R_0,E-Church-Rosser modulo E and E-terminating.

When we do normal narrowing with the system $R = R_0 \cup E$ we need three steps to find a solution. The normal narrowing tree of depth three contains 39 nodes. Normal means that we normalize the goal after each narrowing step.

When we do equational narrowing with R_0 modulo E the rules of E need no longer to be considered in a narrowing step. Consequently the narrowing tree is much smaller. We still need three steps but the equational narrowing tree of depth three contains only 11 nodes.

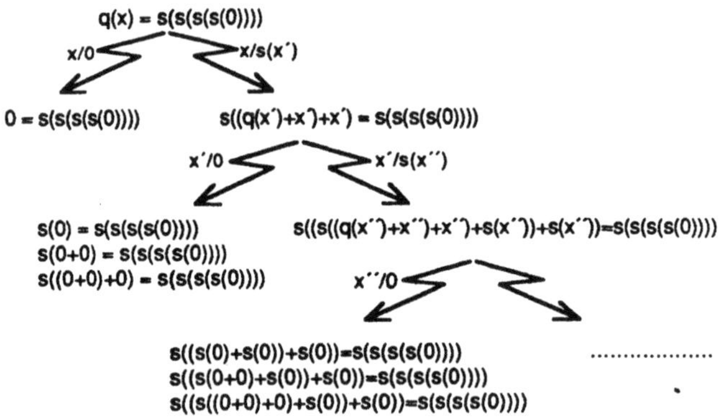

A subterm of the goal and the left-hand side of a rule that are not syntactically unifiable may be E-unifiable. This leads for example to the narrowing step

$s((q(x')+x')+x')) \rightsquigarrow_{[x'/0]} s(0)$.

E-unification of the left-hand and right-hand side of the last goal replaces the normalization step in normal narrowing.

Now consider the query

?- $q(x) + q(y) = q(z)$.

In order to find the first nontrivial solution

$x = s^4(0) , y = s^3(0) , z = s^5(0)$

the ordinary narrowing algorithm needs 15 narrowing steps.

The number of nodes in the normal narrowing tree of depth 15 is enormous. In depth 4 we have about 4000, in depth 5 about 50000 nodes.

However, the equational narrowing algorithm only needs 5 steps

$$?\text{-}\ q(x) + q(y) = q(z)$$

$\rightsquigarrow[y/s(y_1)]\rightarrow$ $?\text{-}\ q(x) + s(q(y_1) + y_1 + y_1) = q(z)$

$\rightsquigarrow[y_1/s(y_2)]\rightarrow$ $?\text{-}\ q(x) + s(\ s(q(y_2) + y_2 + y_2) + s(y_2) + s(y_2)) = q(z)$

$\rightsquigarrow[y_2/s(y_3)]\rightarrow$ $?\text{-}\ q(x) + s(\ s(\ s(q(y_3) + y_3 + y_3) + s(y_3) + s(y_3)) + s(s(y_3)) + s(s(y_3))) = q(z)$

$\rightsquigarrow[y_3/0]\rightarrow$ $?\text{-}\ q(x) + s(\ s(\ s(0 + 0 + 0) + s(0) + s(0)) + s(s(0)) + s(s(0))) = q(z)$

$\rightsquigarrow[z/s(z_1)]\rightarrow$ $?\text{-}\ q(x) + s(\ s(\ s(0 + 0 + 0) + s(0) + s(0)) + s(s(0)) + s(s(0))) = s(q(z_1) + z_1 + z_1)$

then E-unification yields at once

$$x = z_1 = s^4(0).$$

In the mathematical notation the last goal is $x^2 + 9 = z_1{}^2 + 2z_1 + 1$.

Altogether we get

$$x = s^4(0)\ ,\ y = s(y_1) = s(s(y_2)) = s(s(s(y_3))) = s^3(0)\ ,\ z = s(z_1) = s^5(0).$$

The design of a special E-unification algorithm is a hard task, but it only has to be done once. In particular the E-unification algorithm must be able to deal with free function symbols that do not occur in the equations of E. These are just the function symbols coming from R. The problem of dealing with free function symbols has been recently solved by Schmidt-Schauss 88. His algorithm however is highly indeterministic. Another solution has been announced by Jouannaud/Boudet.

The most interesting theories are the unitary unifying theories, i.e. those theories that admit a unique most general E-unifier up to $\equiv_E$. For unitary theories there is no indeterminism in the E-unification process. Examples of such theories are Boolean algebras (Büttner/Simonis 87, Martin/Nipkow 86) and primal algebras (Nipkow 88). However Boolean unification is no longer unitary when free function symbols are allowed (Schmidt-Schauss 88).

The solution of linear diophantine equations is described in (Clausen/Fortenbacher 87) and (Käufl 87).

So far the E-unifications of the left-hand side of the rule and the subterm of the goal were rather trivial. To see that this is not always the case we take an example from linear algebra that might be quite useful in practical applications.

Assume that we have a linear homomorphism

$$f\colon \text{int} \times \text{int} \longrightarrow \text{int} \times \text{int}\ .$$

and that we know the image of two basis vectors, say

$$f(2,1) = (2,3) \quad \text{and} \quad f(3,2) = (1,2).$$

From the linearity of f we obtain

$$f(m*(2,1) + n*(3,2)) = m*(2,3) + n*(1,2).$$

To express this in our formalism we consider the signature $\Sigma = (S, F)$ with

$$S = \{\ \text{int},\ \text{vector}\ \},$$

$$F = \{\ 0\colon \longrightarrow \text{int},\ s,\ p\colon \text{int} \longrightarrow \text{int},\ +\colon \text{int} \times \text{int} \longrightarrow \text{int},\ v\colon \text{int} \times \text{int} \longrightarrow \text{vector},$$
$$++\colon \text{vector} \times \text{vector} \longrightarrow \text{vector},\ f\colon \text{vector} \longrightarrow \text{vector}\ \}.$$

For the equational theory E we may use

$$E = \{\ 0 + x = x,\ s(x) + y = s(x+y),\ p(x) + y = p(x+y),$$
$$v(x,y)\ ++\ v(x',y') = v(x+x',y+y')\ \}.$$

The term rewriting system R then looks as follows:

R = { f(v(m+m,m) ++ v(n+n+n,n+n)) —> v(m+m,m+m+m) ++ v(n,n+n) }

Again R,E is R,E-Church-Rosser modulo E and E-terminating.

Hereafter we shall use the mathematical notation.

Suppose we want to compute f(1,1).

This can be done by rewriting with R modulo E.

$f(1,1) =_E f((-1)*(2,1) + 1*(3,2)) \longrightarrow_R (-1)*(2,3) + 1*(1,2) =_E (-1,-1).$

Now we want to solve the goal

?- f(x,y) = (1,1) .

In order to perform a narrowing step we E-unify

f(x,y) and f(m*(2,1) + n*(3,2))

and obtain the narrowing substitution

x = 2*m + 3*n , y = m + 2*n .

The new goal is

?- m*(2,3) + n*(1,2) = (1,1)

and E-unification gives

m = 1 , n = –1 or x = –1, y = –1.

6. Conclusion and further research

In this paper we have shown how built-in equality theories may considerably improve the efficiency of the narrowing algorithm. The main task now is to design special E-unification algorithms that are capable to deal with free function symbols. Further we should generalize equational rewriting and narrowing to include the case of conditional term rewriting systems (Kaplan 87, Hussmann 85, Giovannetti/Moiso 86). Continuing the work of Kirchner 85 the equational narrowing approach should be combined with the diverse optimizing narrowing strategies (Hullot 80, Fribourg 85 Nutt/Réty/Smolka 87, Padawitz 87, Bosco/Giovannetti/Moiso 87) that have been proposed so far. Finally we should investigate how constraint solving mechanisms which are more powerful than pure unification algorithms can be integrated into narrowing. The constraint logic programming scheme of Jaffar/Lassez 87 could serve as a model.

Acknowledgement

This work was partially supported by the Deutsche Forschungsgemeinschaft, SFB 314 – "Künstliche Intelligenz - Wissensbasierte Systeme" and was done within a joint project of the University Karlsruhe and the GMD. I would like to thank the members of the Karlsruhe Prolog Group for many fruitful discussions and in particular Laura Smith and Christoph Brzoska for their comments on a draft of this paper.

References

[Bellia/Levi 86] M. Bellia, G. Levi: The Relation between Logic and Functional languages: A Survey. J. Logic Programming 3 (1986), 217-236

[Bockmayr 86a] A. Bockmayr: Conditional rewriting and narrowing as a theoretical framework for logic-functional programming: A Survey. Int. Bericht 10/86, Fak. f. Informatik, Univ. Karlsruhe, 1986

[Bockmayr 86b] A. Bockmayr: Narrowing with Inductively Defined Functions. Int. Bericht 25/86, Fak. f. Informatik, Univ. Karlsruhe, 1986

[Bockmayr 87] A. Bockmayr: A Note on a Canonical Theory with Undecidable Unification and Matching Problem. J. Autom. Reas. 3 (1987), 379-381

[Bosco/Giovannetti/Moiso 87] P. G. Bosco, E. Giovannetti, C. Moiso: Refined Strategies for Semantic Unification. TAPSOFT 87, Pisa, 1987, Springer LNCS 250

[Büttner/Simonis 87] W. Büttner, H. Simonis: Embedding Boolean Equations into Logic Programming. J. Symb. Comput. 4 (1987), 191-205

[Colmerauer 87] A. Colmerauer: Opening the Prolog III Universe. BYTE, Aug. 1987, 177-182

[Clausen/Fortenbacher 87] M. Clausen, A.Fortenbacher: Efficient Solution of Linear Diophantine Equations. Int. Ber. 32/87, Fak. f. Informatik, Univ. Karlsruhe, 1987 ; to appear in J. Symb. Comput.

[DeGroot/Lindstrom 86] D. DeGroot, G. Lindstrom: LOGIC PROGRAMMING – Functions, Relations and Equations. Prentice Hall 1986

[Dershowitz/Plaisted 85] N. Dershowitz, D. A. Plaisted: Logic Programming cum Applicative Programming. Symp. on Logic Programming, Boston 1985, IEEE

[Fay 79] M. Fay: First Order Unification in an Equational Theory. 4th Workshop on Automated Deduction, Austin, Texas, 1979

[Fribourg 85] L. Fribourg: Handling Function Definitions Through Innermost Superposition and Rewriting. Rewriting Techniques and Applications, Dijon 1985, Springer LNCS 202

[Giovannetti/Moiso 86] E. Giovannetti, C. Moiso: A Completeness Result for E-Unification Algorithms Based on Conditional Narrowing. Foundations of Logic and Functional Programming, Trento 1986, Springer LNCS 306

[Huet/Oppen 80] G. Huet, D. C. Oppen: Equations and Rewrite Rules, A Survey. Formal Language Theory (Ed. R. V. Book), Academic Press 1980

[Hullot 80] J. M. Hullot: Canonical Forms and Unification. 5th Conference on Automated Deduction, Les Arcs 1980, Springer LNCS 87

[Hussmann 85] H. Hussmann: Unification in Conditional-Equational Theories. Technical Report MIP-8502, Univ. Passau, Jan.1985
short version: EUROCAL 85, Linz, Springer LNCS 204

[Jaffar/Lassez 87] J. Jaffar, J.-L. Lassez: Constraint Logic Programming. POPL 87, München, 1987

[Jaffar/Lassez/Maher 86] J. Jaffar, J.-L. Lassez, M. S. Maher: Logic Programming Language Scheme. In [DeGroot/Lindstrom 86]

[Jouannaud 83] J. P. Jouannaud: Church-Rosser computations with equational term rewriting systems. 8th CAAP, L'Aquila, 1983, Springer LNCS 159

[Jouannaud/Kirchner 86] J. P. Jouannaud, H. Kirchner: Completion of a set of rules modulo a set of equations. SIAM J. Comput. 15,4 (1986), 1155-1194

[Jouannaud/Kirchner/Kirchner 83] J. P. Jouannaud, C. Kirchner and H. Kirchner: Incremental construction of unification algorithms in equational theories. 10th ICALP, Barcelona, 1983, Springer LNCS 154

[Jouannaud/Munoz 84] J. P. Jouannaud, M. Munoz: Termination of a set of rules modulo a set of equations. 7th CADE, Napa Valley, 1984, Springer LNCS 170

[Kaplan 87] S.Kaplan: Simplifying Conditional Term Rewriting Systems: Unification, Termination and Confluence. J. Symb. Comput. 4 (1987), 295-334

[Käufl 87] T. Käufl: Reasoning about systems of linear inequalities. Int. Ber. 16/87, Fak. f. Informatik, Univ. Karlsruhe, 1987

[Kirchner 85] C. Kirchner: Methodes et outils de conception systématique d'algorithmes d'unification dans les théories equationnelles. Thèse d'état, Univ. Nancy, 1985

[Martin/Nipkow 86] U. Martin, T. Nipkow: Unification in Boolean Rings. CADE–8, Oxford 1986, Springer LNCS 230

[Nipkow 88] T. Nipkow: Unification in Primal Algebras. CAAP 88, Nancy 1988, Springer LNCS 299

[Nutt/Réty/Smolka 87] W. Nutt, P. Réty, G. Smolka: Basic Narrowing Revisited. SEKI-Report SR 87-07, Univ. Kaiserslautern, 1987

[Padawitz 87] P. Padawitz: Strategy-Controlled Reduction and Narrowing. Rewriting Techniques and Applications, Bordeaux 1987, Springer LNCS 256

[Peterson/Stickel 81] G. E. Peterson, M. Stickel: Complete Sets of Reductions for Some Equational Theories. J.ACM 28 (1981), 233-264

[Plotkin 72] G. D. Plotkin: Building-In Equational Theories. Machine Intelligence 7 (1972)

[Réty et al. 85] P. Réty, C. Kirchner, H. Kirchner, P. Lescanne: NARROWER: a new algorithm for unification and its application to Logic Programming. Rewriting Techniques and Applications, Dijon 1985, Springer LNCS 202

[Schmidt-Schauss 88] M. Schmidt-Schauss: Unification in a Combination of Arbitrary Disjoint Equational Theories. CADE-9, Argonne 1988, Springer LNCS 310

[Stickel 85] M. Stickel: Automated Deduction by Theory Resolution. J. Autom. Reas. 1 (1985), 333 - 355

A completion procedure for hierarchical conditional rewriting systems

Wadoud Bousdira*
CRIN-LORIA
BP 239 F54506 Vandœuvre Cedex
FRANCE
e-mail bousdira@crin.crin.fr

Abstract

The first part of this work presents two principal approaches to conditional rewriting, the "Log-algebra approach" and the "initial algebra approach". The former defines a deduction system called L which is correct and complete with respect to some class of conditional specifications. The latter provides results in the initial algebra of conditional specifications which are faithfull with respect to booleans. Under some hypotheses, both approaches share the same initial algebra. The second part of this paper presents a completion procedure for hierarchical conditional specifications. This procedure is founded on inference rules and we establish the correction of these rules with respect to the proof system L.

1 Introduction

A flexible and natural method for specifying abstract data types uses conditional specifications. The equations can be turned into rewrite rules. The rewriting method is founded on the classical Knuth-Bendix's theorem [KB70]. This theorem has been extended later to conditional theory under different approaches, including [RZ84], [Kap87], [JW86], [Gan87b].
Conditional specifications raise two problems. First one has to formalize the semantics and the models that the theory understand, second a method for proving inductive theorem has to be developed.
This paper is organized as follows : in the first section, we succintly present two principal approaches that distinguish on the one hand, by the syntax of conditional equations which they use and on the other hand, by the type of algebras which they apprehend. However, under some hypotheses, both approaches define the same initial object. In the second part, we introduce conditional rewriting systems and more particularly, hierarchical systems. In this class of systems, we define a hierarchical contextual relation and we state results of ground confluence of hierarchical systems. Form this hierarchical relation defined in this part, we define in the following section two rewriting processes, the case reasoning and the contextual rewriting. Both processes are expressed by inference rules according to the method established in [BDH86]. The fourth section concerns the correctness of the inference rules with respect to some deduction system. We precise in this part, the theory which we consider and the algebraic context in which this correctness is valid. The next section includes the completion algorithm.
Familiarity with the usual notions concerning algebraic specifications and rewriting systems is assumed (for detail, see [Hue80]). Furthermore, this paper does not contain proofs, they can be found in [Bou88].

2 Two conditional approaches

The diversity of approaches within conditional specification framework does not allow to classify totally the obtained results. However, three directions can be enlighted. The first approach apprehends conditional specifications in their whole generality. It guarantees the existence of initial algebra and in most applications the existence of free algebras. This approach has been considered by [Kap84]. The second approach defines more restricted class of conditional specifications. It allows simulating the existence of predefined logic and by this way to get a deduction system which is correct and complete with respect to the class of admissible

*This work has been partially supported by the GRECO-PRC Programmation Avancée et Outils de l'Intelligence Artificielle.

models. Finally, the last approach focus on results in the ground term algebra. It induces the consistency with respect to the boolean sort. This guarantees the existence of an initial algebra for a conditional specification which is faithfull with respect to booleans. In this paper, we are interested by the two latter approaches.

2.1 "Log-algebra" approach

This approach is algebraic and considers predefined logic. It has been proposed by Navarro [Nav87]. Conditional specifications are extensions of the specification of boolean values. The precondition of a conditional equation is any propositional formula with or without the equality symbol. More formally,

Definition 1 *Log-specification*
A conditional specification $SP = (\Sigma, E)$ is an extension of the boolean specification SP_{bool} (def 2) such that each conditional equation of E is of the form $c \Rightarrow s = t$ where s and t are of the same sort and the sort of c is boolean. Moreover, for each operator symbol $f : s_1 \times - - - \times s_n \rightarrow s$, such that s is not the boolean sort, s_i is different from the boolean sort, for each $1 \leq i \leq n$. Such conditional specification is called a Log-specification.

Definition 2 *Boolean specification*
The specification SP_{bool} of the boolean values is a specification $(\Sigma_{bool}, E_{bool})$ such that $\Sigma_{bool} = (S_{bool}, F_{bool})$ with : $S_{bool} = \{bool\}$, $F_{bool} = \{\neg, \wedge, \vee, \Rightarrow, \Leftrightarrow, true, false\}$. E_{bool} contains axioms that describe the behavior of logical operators of F_{bool} (see [Nav87] for detail).

Remark 1 *The equation $c \Rightarrow s = t$ is an abbreviation of $c = true \Rightarrow s = t$. c is the context of the equation and "$s = t$" is the conclusion.*

Definition 3 *Log-algebra*
*Let $SP = (\Sigma, E)$ be a Log-specification and Alg_{SP} be the class of algebras associated with SP. $LogAlg_{SP}$ denotes the sub-class of algebras A in Alg_{SP} such that $A_{bool} = \{$**true, false**$\}$ where A_{bool} is the boolean carrier of the algebra A and* **true, false** *represent the respective values of the constants true and false.*

Remark 2 *In [Nav87], the author proves that this approach allows to define a function symbol eq that represents the equality in any sort with the following equations :*
$eq(x, x) = true, \quad eq(x, y) \Rightarrow x = y$

In this approach, admissible models are models that satisfy the equations of the specification and whose carrier of boolean sort is exactly {**true, false**}. This approach does not guarantee the existence of initial algebra. For example,

Example 1 *[Nav87]*
Let $SP = (S, F, E)$ with $S = S_{bool}$, $F = F_{bool} \cup \{a : \rightarrow bool\}$, $E = E_{bool}$
This specification admits two models ; in the former, the interpretation of a is **true** *and in the latter, the interpretation of a is* **false**. *Therefore, the specification does not admit initial model.*

The principal contribution of this method is the definition of deduction system called L which is correct and complete with respect to the class $LogAlg_{SP}$.

Theorem 1 *Let $SP = (\Sigma, E)$ be a Log-specification and let $e : c \Rightarrow s = t$ be a conditional equation of E ; The deduction system L validates e in E and we note $E \vdash_L e$ iff for each Log-algebra A in $LogAlg_{SP}$, $A \models e$.*

Remark 3 *Deduction in the system L is semi-decidable. Indeed, the proof problems in boolean contexts are still open. There does not exist decidable method for proving for example that the disjunction of two contexts $c_1 \vee c_2$ is equivalent to true in Log-algebras.*

Study of parameterized specifications has also been made in this approach. Two aspects have been approached, namely demonstration theory and correction of parameterized specifications. Particularly, the authors define syntactic characterization of the correction of parameterized specifications from two properties, the *bool-persistency* and the *persistency* [NO84b].

2.2 "Initial algebra" approach

Conditional equations have preconditions of boolean sort. In [BR88], we consider this approach and we get a result of consistency with respect to the boolean sort which guarantees the existence of initial algebra for a conditional specification faithfull with respect to booleans.

Definition 4 *Conditional specification*

A conditional specification $SP = (\Sigma, E)$ consists of a signature $\Sigma = (S, F)$. S is a set of sorts and F is a finite set of function symbols and E is a set of conditional equations of the form $p \Rightarrow s = t$. p is of boolean sort,s and t are terms of same sort and p is the precondition of the rule. p is of the form $\&_{i=1}^{n} p_i = p'_i$ with for each i in [1..n], p'_i is either the boolean constant "true . or the boolean constant "false".
Under this form, we note p by $\vec{p}$.

We assume the boolean carrier contains only the constants true and false. We define the negation of true equal to false and conversely, the negation of false equal to true. Imposing that the rule precondition is of the aforementioned form is not too strong restriction. It allows effectively expressing the negation of any simple boolean expression p, without any logical connective under the form $p = false$; on the other hand, a conditional equation of the form $(P_1 \vee --- \vee P_n) \Rightarrow s = t$ where $\vee$ represents the disjunctive logical connective is equivalent to the following set of conditional equations : $\{P_1 \Rightarrow s = t, ---, P_n \Rightarrow s = t\}$

Definition 5 *Semantic of conditional equations*

Let $SP = (\Sigma, E)$ be a conditional specification and let A be a SP-algebra. Let e be a conditional equation of the form $\vec{p} \Rightarrow s = t$. e is true in A and we note $A \models e$ iff for each substitution $\sigma : X \rightarrow A$ such that $\sigma(\vec{p}) = true$, then $\sigma(s) = \sigma(t)$. A set E of conditional equations is true in A if each equation of E is true in A.
$\sigma(\vec{p}) = true$ means for each i in [1..n], $\sigma(p_i) = p'_i$.

Theorem 2 *Let $SP = (\Sigma, E)$ be a conditional specification. The SP-algebras, equipped with homomorphisms form the non-empty category Alg_{SP} which the initial object is $I_{SP} = T(\Sigma)/\equiv_E$ where $\equiv_E$ is the smallest congruence generated by E. $\equiv_E$ is built by induction as follows*

- *let $\equiv_0$ be the smallest congruence relation which contains the relation R_0 defined as follows : $u\ R_0\ v$ iff there exist an equation $s = t$ in E and a substitution $\sigma : X \rightarrow T(\Sigma)$ such that $\sigma(s) = u$ and $\sigma(t) = v$.*
- *let $\equiv_{i+1}$ the smallest congruence which contains the following relation R_{i+1} on $T(\Sigma)$: $u\ R_{i+1}\ v$ iff $u \equiv_i v$ or there exist a conditional equation in E let $\vec{p} \Rightarrow s = t$ and a substitution $\sigma : X \rightarrow T(\Sigma)$ such that $\sigma(s) = u$, $\sigma(t) = v$ and $\sigma(\vec{p}) \equiv_i true$*

then $\equiv_E = \cup \{\equiv_i\ /\ i \geq 0\}$
$\sigma(\vec{p}) \equiv_i true$ means for each i in [1..n], $\sigma(p_i) \equiv_i p'_i$.

This approach uses preconditions with algebraic syntax and it guarantees the existence of an initial algebra. However, using preconditions of boolean sort requires a careful treatement. Effectively, as in the following example, the initial algebra can be non-conformable according to the expected result as for its boolean part. This problem also arises in [Kap84] and [Gan87b] but the authors do not consider it in their works.

Example 2 *[KR87] Let the following specification $SP = (S, F, E)$:*
$S = \{bool\}$, $F = \{true, false, p, q : \rightarrow bool\}$, $E = \{p = false \Rightarrow q = true\}$
$\equiv_E$ is the identity and the carrier of the initial algebra of the specification SP is the set {**p, q, true, false**} *where* **p, q, true, false** *are the values of the constants p, q, true and false. Therefore, the boolean part of the initial algebra is not isomorphic to the set* {**true, false**}.

For this purpose, in [BR88], the authors define the behavior of conditional specifications with respect to the boolean sort with two properties : the *consistency* and the *completeness with respect to booleans.* For these conditional specifications which are faithfull with respect to booleans, the existence of initial algebra is guaranteed within hierarchical frame and we get sufficient condition of confluence for hierarchical contextual rewriting systems.

Definition 6 *Completeness with respect to booleans*

Let $SP = (\Sigma,\ E)$ be a conditional specification.
SP is complete with respect to booleans iff for each boolean ground term t, $t =_{SP} true$ or $t =_{SP} false$.

Definition 7 *Consistency with respect to booleans*

Let $SP = (\Sigma,\ E)$ be a conditional specification. SP is consistent with respect to booleans iff $\neg(true =_{SP} false)$. As consequence of the consistency with respect to booleans, for each boolean ground term t, if $t =_{SP} true$ then $t \neq_{SP} false$.

Within this frame, the hierarchical approach allows characterizing the consistency with respect to booleans by the ground confluence. The hierarchical frame allows to check this confluence by adopting modular process and this method is useful in practice when one deals with big specifications.
However, important problems are still to consider in this approach. They include principally use of first order formulas in order to make inductive proofs. This problem has been clearly pointed out in [Gan87b]. In this paper, the author explains the necessity of using complementary assertions which the verification is assumed in the initial algebra. Two of these assertions are fundamental :

1. $p = true$ or $p = false$ where p is variable of boolean sort
2. $\neg(true = false)$

These assertions represent the properties of completeness and consistency with respect to booleans, in order to guarantee faithfull behavior of the specification with respect to the boolean sort. [BR88] present formal definitions of these properties within the frame of conditional specifications with equations of the form $p \Rightarrow s = t\ \ with\ \ p = \&_{i=1}^{n}\ \epsilon_i\ q_i$ where for each i in [1..n], ϵ_i stands for $\neg$ or nothing and the q_i's are simple boolean expressions without any logical connective. Both forms of preconditions are equivalent. Indeed, a precondition $\&_{i=1}^{n}\ p_i = p_i'$ with $p_i' = true$ or $p_i' = false$ and where p_i is a simple boolean expression without any logical connective is equivalent to $\&_{i=1}^{n}\ \epsilon_i\ p_i$ where

$\epsilon_i = \epsilon$ if $p_i' = true$ and $\epsilon_i = \neg$ if $p_i' = false$.

The question that can be asked is the following : under which conditions the assertion $\neg p =_{SP} true \Leftrightarrow p =_{SP} false$ is valid for any boolean ground formula?

Lemma 1 *[Bou88] Let $SP = (\Sigma,\ E)$ be a conditional specification. If SP is consistent and complete with respect to booleans then, for any boolean ground formula p. $\neg p =_{SP} true \Leftrightarrow p =_{SP} false$.*

3 Hierarchical contextual rewriting

Hierarchical specification is built from a set of specifications by successive enrichments. Hierarchical approach of conditional specifications is one of the methods which are used in order to avoid recursive evaluation of preconditions. It also allows modular study of specifications.

Definition 8 *Hierarchical conditional specification*

A hierarchical conditional specification with $(m+1)$ levels. $SP = (SP_0,\ ---,\ SP_m)$ is a set of specifications such that for each i, $0 \leq i \leq m$, SP_i is a triple $(S_i,\ F_i,\ E_i)$ where S_i is a set of sorts, F_i is a set of operator symbols, E_i is a set of conditional equations of the form $\overrightarrow{p} \Rightarrow g = d$. For each k, $0 \leq k < m$,

1. *$SP_k \subset SP_{k+1}$: SP_{k+1} is an extension of SP_k, namely S_k, F_k, E_k are respectively included in S_{k+1}, F_{k+1}, E_{k+1}.*
2. *if $\overrightarrow{p} \Rightarrow g = d \in E_{k+1} - E_k$ then $\overrightarrow{p} \in T(F_k,\ X)$, $g,\ d \in T(F_{k+1},\ X)$ and either g or d is in $T(F_{k+1},\ X) - T(F_k,\ X)$.*

Our study restricts to hierarchical specifications built with successive enrichments, without any new sort. Therefore, for each k, $0 \leq k < m$, $S_{k+1} = S_k$. Of course, this restriction is strong. However, the extension to many sorts does not pose any big difficulty and it requires principally more technical details.
The set of operator symbols F of the whole specification is the set F_m.

Definition 9 *Hierarchical conditional rewriting system (HCRS)*

Intuitively, conditional rewriting system is hierarchical if the associated specification is hierarchical. More precisely, the rules of R are of the form $\overrightarrow{p} \Rightarrow g \rightarrow d$

1. $R = (R_0, ---, R_m)$ *where* $R_k \subset R_{k+1}$, *for each* k. $0 \leq k < m$;

2. *if* $\vec{p} \Rightarrow g \to d \in R_{k+1} - R_k$ *then* $g \in T(F_{k+1}, X) - T(F_k, X)$, $\vec{p} \in T(F_k, X)$ *and* $d \in T(F_{k+1}, X)$.

3. $Var(\vec{p}) \subseteq Var(g)$ *and* $Var(d) \subseteq Var(g)$.
$Var(\vec{p}) = \cup_i \{Var(p_i)\}$

Remark 4 *By convention,* $T(F_{-1}, X)$ *is the empty set, it means that the rules of* R_0 *are rules without any condition.*

We define in what follows a hierarchical relation called simple contextual relation. This relation deals with terms with variables. It is a tool which we use in order to define the inference rules of the completion procedure.

Definition 10 *Simple hierarchical contextual relation*
Let $R = (R_0, ---, R_m)$ *be an HCRS and let* s, t, c *be terms of* $T(\Sigma, X)$. $c \vdash s \to_{H,R} t$ *if there exist an index* k *in [0..m], a rule* $\vec{p} \Rightarrow g \to d$ *in* $R_k - R_{k-1}$, *an occurrence* u *of* s *and a substitution* σ *such that* $s_u = \sigma(g)$, $\sigma(\vec{p}) \in T(F_{k-1}, X)$ *then* $\vec{c} = \sigma(\vec{p})$ *and* $t = s[u \leftarrow \sigma(d)]$

The syntactical forms of the context $\vec{c}$ and of the rule preconditions are identical. $\vec{c}$ is of the form $\&_{i=1}^n c_i = c'_i$ with $c'_i = true$ or $c'_i = false$ for each i in [1..n].
The "$\_ \vdash \_ \to_{H,R} \_$" relation is not a rewriting relation. Indeed, it is a ternary relation. Instead, it is a technical tool for the expression of the inference rules.

Example 3 *Let the "infimum" specification of two terms defined on integers ;* x *and* y *are integer variables,* p *is boolean variable ;*
$SP = (S, F, E)$ *with* $S = \{bool, integer\}$,
$F = \{true, false : \to bool, \& : bool \times bool \to bool, \leq : integer \times integer \to bool,$
$inf : integer \times integer \to integer\}$,
$E = \{true \;\&\; p = p, \; false \;\&\; p = false,$
$x \leq y = true \Rightarrow inf(x,y) = x, \; x \leq y = false \Rightarrow inf(x,y) = y\}$
The rewriting system which corresponds to SP includes the equations of E oriented from left to right. There exists a hierarchical contextual simple relation on the triple
$\langle inf(x_1, inf(y_1, z_1)), inf(x_1, y_1), y_1 \leq z_1 = true \rangle$ *namely*
$y_1 \leq z_1 = true \vdash inf(x_1, inf(y_1, z_1)) \to_{H,R} inf(x_1, y_1)$ *with the substitution* $\sigma = (x \leftarrow y_1, y \leftarrow z_1)$

We establish in the following the ground confluence for hierarchical specifications under some hypotheses. This result is proved in [BR88] by using hierarchical contextual rewriting relation noted $-c \to_{H,R}$. However, this result is still valid here by the lemma 2. We firstly define the notion of contextual term which is used by the relation $-c \to_{H,R}$. Note that classical term can be seen as contextual term with by default the context "true".

Definition 11 *Contextual term*
A contextual term $(c :: t)$ *is a pair of terms such that* t *is of any sort and* c *is of boolean sort. We call* c *a context. Moreover, the syntax of* c *and the syntax of the rule preconditions are the same. We mean that* c *has the following form* $\&_{i=1}^n c_i = c'_i$ *with for each* i, $c'_i = true$ *or* $c'_i = false$.

Definition 12 *Hierarchical contextual rewriting relation*
Let $R = (R_0, ---, R_m)$ *be an HCRS. A contextual term* $(c :: t)$ *rewrites contextually to* $(c' :: t')$, *denoted* $(c :: t) - c \to_{H,R} (c' :: t')$, *if there exist an index* k *in [0..m], a rule* $p \Rightarrow g \to d$ *in* $R_k - R_{k-1}$ *and a substitution* σ *such that* $\sigma(p) \in T(F_{k-1}, X)$ *and*

- *either there exists an occurrence* u *in* t *such that* $t_u = \sigma(g)$ *then* $c' = c \;\&\; \sigma(p)$, $t' = t[u \leftarrow \sigma(d)]$; *in this case, it is term rewriting.*
- *or there exists an occurrence* w *in* c *such that* $c_w = \sigma(g)$ *then* $c' = c[w \leftarrow \sigma(d)] \;\&\; \sigma(p)$, $t' = t$; *in this case, it is context rewriting.*

In this definition, p *has the aforementioned form* $\&_{i=1}^n \epsilon_i \; q_i$

Lemma 2 *[Bou88] Let $R = (R_0, ---, R_m)$ be an HCRS, and s, t be two terms in $T(F, X)$ If $c \vdash s \rightarrow_{H,R} t$ then $(true :: s) - c \rightarrow^*_{H,R} (c :: t)$.*

We can now present the fundamental result of ground confluence for hierarchical conditional rewriting systems.

Theorem 3 *Confluence theorem [BR88]*

Let $R = (R_0, ---, R_m)$ be a c-nœtherian HCRS which is operationally well covered, operationally sufficiently complete and operationally complete with respect to booleans. Let us assume moreover that true and false are irreducible in R. If all hierarchical critical pairs in R define complete coherent sets of contextual normal forms, then R is confluent on ground terms of $T(F_m)$ and R is consistent with respect to booleans.

We explain briefly the notions used in this theorem.
R is c-nœtherian if there does not exist an infinite sequence of the form
$(c_1 :: t_1) - c \rightarrow_{H,R} (c_2 :: t_2) - c \rightarrow_{H,R} ---- c \rightarrow_{H,R} (c_n :: t_n) - c \rightarrow_{H,R} ---$
The operational well covering property means that for each maximal subset of rules of the form $\{p_i \Rightarrow g \rightarrow d_i, i \in I\}$ in R, for each ground substitution σ, there exists j in I such that $\sigma(p_j) \rightarrow^*_{H,R}$ *true*. R is operationally sufficiently complete iff for each $0 < k \leq m$, for each term $t \in T(F_k)$, there exists $t' \in T(F_{k-1})$ such that $t \rightarrow^*_{H,R} t'$. The operational definition of the completeness with respect to booleans means that for each boolean ground term t, $t \rightarrow^*_{H,R}$ *true* or $t \rightarrow^*_{H,R}$ *false*. CNFS(c :: t) means the set of contextual normal forms of $(c :: t)$. Indeed, the normal form of contextual term is not unique. Instead, it is replaced by set of contextual normal forms whose the contexts complement. In the similar way, the equality notion of two normal forms in the classical case, without preconditions, is equivalent to the coherence notion of two sets of contextual normal forms. One can find the formal definitions of these notions in [BR88] as well as the formal proof of the theorem.

The hierarchical conditional rewriting is related to the initial algebra approach. Indeed, the hierarchical approach requires the sufficient completeness property and this property is defined on ground terms. Moreover, according to [NO84a], under the sufficient completeness property, and if one restricts to ground terms, confluence of the hierarchical rewriting involves confluence of the recursive rewriting without any hierarchy hypothesis.

In the following section, we present a completion procedure for the conditional theory. This procedure is defined by inference rules according to the method which has been established by [BDH86]. The inference rules are expressed from the simple hierarchical contextual rewriting and they define rewriting process more complex which is the base of the completion procedure.

4 Conditional completion procedure

The conditional completion procedure is based on two basic principles, the case reasoning and the contextual rewriting.

4.1 Inference rules for the completion procedure

E and R are respectively a set of conditional equations and a set of conditional rewrite rules. $>$ is the comparison ordering of both conditional equation members.
The four first inference rules are these which we find in each basic completion procedure of Knuth-Bendix. The other rules are simplification rules presented within the frame of both rewriting processes.
$E \vdash_L p_1 \vee --- \vee p_n = true$ expresses the validity in the class of Log-algebras of the expression $p_1 \vee --- \vee p_n = true$. Indeed, we present later the proof system L which is correct and complete for this class of models. Similarly, $E \vdash_L c = false$ expresses the unsatisfiability of c. Therefore, we can deal with the deduction system L and prove the validity of assertions by L. However, note that the deduction system does not provide decision procedure. We need to consider instead ad-hoc procedures for restricted cases. This proof of assertions is introduced by the case reasoning. It is rewriting process with context enrichment and it is only made if the result contexts are equivalent to the starting context in the initial algebra.

- Basic cases :

1. **Orienting an equation :**

$$BR1 \quad \frac{E \cup \{c \Rightarrow s = t\},\ R}{E,\ R \cup \{c \Rightarrow s \rightarrow t\}} \qquad if\ s > t$$

2. **Adding an equational consequence :**

$$BR2 \quad \frac{E,\ R}{E \cup \{p_1 \,\&\, p_2 \Rightarrow s_1 = s_2\},\ R}$$

if there exists s such that $p_1 \vdash s \rightarrow_{H,R} s_1$ and $p_2 \vdash s \rightarrow_{H,R} s_2$

3. **Deleting a trivial equation :**

$$BR3 \quad \frac{E \cup \{c \Rightarrow t = t\},\ R}{E,\ R}$$

4. **Deleting an equation with an unsatisfiable condition :**

$$BR4 \quad \frac{E \cup \{c \Rightarrow t = t'\},\ R}{E,\ R} \qquad if\ E \vdash_L c = false$$

- **Simplification rules :**

1. **Case reasoning :**
Simplifying the term part of an equation :

$$SR1 \quad \frac{E \cup \{c \Rightarrow s = t\},\ R}{E \cup \{c \,\&\, p_i \Rightarrow s_i = t,\ i = 1,..n\},\ R} \quad if \begin{cases} \forall\, i = 1,..n,\ p_i \vdash s \rightarrow_{H,R} s_i \\ E \vdash_L p_1 \vee \ --- \ \vee p_n = true \end{cases}$$

$$SR2 \quad \frac{E \cup \{c \Rightarrow s = t\},\ R}{E \cup \{c \,\&\, p_i \Rightarrow s = t_i,\ i = 1,..n\},\ R} \quad if \begin{cases} \forall\, i = 1,..n,\ p_i \vdash t \rightarrow_{H,R} t_i \\ E \vdash_L p_1 \vee \ --- \ \vee p_n = true \end{cases}$$

Simplifying the left-hand side of a rule :

$$SR3 \quad \frac{E,\ R \cup \{c \Rightarrow s \rightarrow t\}}{E \cup \{c \,\&\, p_i \Rightarrow s_i = t,\ i = 1,..n\},\ R} \quad if \begin{cases} \forall\, i = 1,..n,\ p_i \vdash s \rightarrow_{H,R} s_i \\ E \vdash_L p_1 \vee \ --- \ \vee p_n = true \end{cases}$$

Simplifying the right-hand side of a rule :

$$SR4 \quad \frac{E,\ R \cup \{c \Rightarrow s \rightarrow t\}}{E,\ R \cup \{c \,\&\, p_i \Rightarrow s \rightarrow t_i,\ i = 1,..n\}} \quad if \begin{cases} \forall\, i = 1,..n,\ p_i \vdash t \rightarrow_{H,R} t_i \\ E \vdash_L p_1 \vee \ --- \ \vee p_n = true \end{cases}$$

Simplifying the context part of an equation :

$$SR5 \quad \frac{E \cup \{c \,\&\, c' \Rightarrow s = t\},\ R}{E \cup \{c_i \,\&\, p_i \,\&\, c' \Rightarrow s = t,\ i = 1,..n\},\ R} \quad if \begin{cases} \forall\, i = 1,..n,\ p_i \vdash c \rightarrow_{H,R} c_i \\ E \vdash_L p_1 \vee \ --- \ \vee p_n = true \end{cases}$$

Simplifying the context part of a rule :

$$SR6 \quad \frac{E,\ R \cup \{c \,\&\, c' \Rightarrow s \rightarrow t\}}{E,\ R \cup \{c_i \,\&\, p_i \,\&\, c' \Rightarrow s \rightarrow t,\ i = 1,..n\}} \quad if \begin{cases} \forall\, i = 1,..n,\ p_i \vdash c \rightarrow_{H,R} c_i \\ E \vdash_L p_1 \vee \ --- \ \vee p_n = true \end{cases}$$

2. **Contextual rewriting :**
Simplifying the term part of an equation :

$$SR1' \quad \frac{E \cup \{c \,\&\, p \Rightarrow s = t\},\ R}{E \cup \{c \,\&\, p \Rightarrow s_1 = t\},\ R} \qquad if\ p \vdash s \rightarrow_{H,R} s_1$$

$$SR2' \quad \frac{E \cup \{c \,\&\, p \Rightarrow s = t\},\ R}{E \cup \{c \,\&\, p \Rightarrow s = t_1\},\ R} \qquad if\ p \vdash t \rightarrow_{H,R} t_1$$

Simplifying the left-hand side of a rule :

$$SR3' \quad \frac{E,\ R \cup \{c \,\&\, p \Rightarrow s \rightarrow t\}}{E \cup \{c \,\&\, p \Rightarrow s_1 = t\},\ R} \qquad if\ p \vdash s \rightarrow_{H,R} s_1$$

Simplifying the right-hand side of a rule :

$$SR4' \quad \frac{E,\ R \cup \{c \,\&\, p \Rightarrow s \rightarrow t\}}{E,\ R \cup \{c \,\&\, p \Rightarrow s \rightarrow t_1\}} \quad if\ p \vdash t \rightarrow_{H,R} t_1$$

Simplifying the context part of an equation :

$$SR5' \quad \frac{E \cup \{c \,\&\, p \,\&\, c" \Rightarrow s = t\},\ R}{E \cup \{c' \,\&\, p \,\&\, c" \Rightarrow s = t\},\ R} \quad if\ p \vdash c \rightarrow_{H,R} c'$$

Simplifying the context part of a rule :

$$SR6' \quad \frac{E,\ R \cup \{c \,\&\, p \,\&\, c" \Rightarrow s \rightarrow t\}}{E,\ R \cup \{c' \,\&\, p \,\&\, c" \Rightarrow s \rightarrow t\}} \quad if\ p \vdash c \rightarrow_{H,R} c'$$

Proposition 1 *[Bou88]*
The rule BR2 of basic cases which consists of adding an equational consequence to the system is a generalisation of the critical pair calculus.

Remark 5 *In the completion procedure which is presented later, we use the rule BR2 only in the particular case of critical pair calculus.*

One can find some of these inference rules in [Gan87a]. Indeed, the author presents a completion procedure for conditional reductive systems. He only deals with the second part of inference rules, namely the contextual rewriting in our terminology. Moreover, in the completion procedure, the author includes the treatement of non-reductive equations.

In the following section, we prove that the "initial algebra" approach is included in the "Log-algebra" approach if we consider conditional specification which is faithfull with respect to booleans, namely both initial algebras coincide if the specification behavior is faithfull with respect to booleans. By this way, in order to establish the correctness of the inference rules in the former approach, it is enough to prove that the inference rules are correct with respect to the deduction system L. In this paper, we only prove the correctness. The completeness proof of the inference rules is more complicated and it will be the aim of an other paper.

4.2 Correctness of the inference rules with respect to the deduction system L

In this section, we give firstly the equivalence lemma of both approaches. Secondly, we present the deduction system L and we give some lemmas which are useful for proving the correctness of the inference rules. Lastly, we state the proof theory that we consider and we prove the correctness of the rules.

Lemma 3 *[Bou88] Let $SP = (\Sigma,\ E)$ be a conditional specification and let $LogAlg_{SP}$ be the Log-algebra class which corresponds to SP. Let I_{SP} be the initial algebra of the class Alg_{SP}. We assume that SP is faithfull with respect to booleans, namely SP is complete and consistent with respect to booleans. Then I_{SP} is also the initial algebra of the class $LogAlg_{SP}$.*

This lemma proves that for a conditional specification SP, the initial algebra of the Log-algebra class which corresponds to SP coincides with the initial algebra of SP provided that this specification is faithfull with respect to booleans. Since we adopt in this paper the "initial algebra" approach, it is then enough to prove the correctness of the inference rules with respect to the deduction system L. In what follows, we present this deduction system [Nav87].

4.2.1 The deduction system L

$$L_1 \quad \frac{}{E \vdash_L \lambda\, X.\, t = t}$$

$$L_2 \quad \frac{E \vdash_L \lambda\, X.\, t_1 = t_2}{E \vdash_L \lambda\, X.\, c \Rightarrow t_1 = t_2}$$

$$L_3 \quad \frac{E \vdash_L \lambda\, X.\, c \Rightarrow t_1 = t_2,\ E \vdash_L \lambda\, X.\, c' \Rightarrow c = true}{E \vdash_L \lambda\, X.\, c' \Rightarrow t_1 = t_2}$$

$$L_4 \quad \frac{E \vdash_L \lambda\, X.\ c \Rightarrow t_1 = t_2}{E \vdash_L \lambda\, (X - \{x\}) \cup Y.\ \sigma(c) \Rightarrow \sigma(t_1) = \sigma(t_2)}$$

with $x \in X$, $\sigma\ :\ X \rightarrow T(F,\ (X - \{x\}) \cup Y),\ x \rightarrow r \in T(F,\ Y),\ u \rightarrow u\ \ (u \neq x)$

$$L_5 \quad \frac{E \vdash_L \lambda\, X.\ c \Rightarrow t_1 = t_2,\ E \vdash_L \lambda\, X.\ c \Rightarrow t_2 = t_3}{E \vdash_L \lambda\, X.\ c \Rightarrow t_3 = t_1}$$

$$L_6 \quad \frac{E \vdash_L \lambda\, X.\ c \Rightarrow t_i = t'_i,\ i = 1,\ ---,\ n}{E \vdash_L \lambda\, X.\ c \Rightarrow f(t_1,\ ---,\ t_n) = f(t'_1,\ ---,\ t'_n)}$$

for each $f \in F$ with arity = n.

$$L_7 \quad \frac{E \vdash_L \lambda\, X.\ c \Rightarrow t = t'}{E \vdash_L \lambda\, (X \cup Y).\ c \Rightarrow t = t'}$$

for each finite set of variables Y.

$$L_8 \quad \frac{E \vdash_L \lambda\, X.\ c_1 \Rightarrow t = t',\ E \vdash_L \lambda\, X.\ c_2 \Rightarrow t = t'}{E \vdash_L \lambda\, X.\ (c_1\ \vee\ c_2) \Rightarrow t = t'}$$

$$L_9 \quad \frac{}{E \vdash_L \lambda\, X.\ false \Rightarrow t = t'}$$

$$L_{10} \quad \frac{}{E \vdash_L \lambda\, \{x\}.\ x \Rightarrow x = true}$$

Lemma 4 *[Nav87] The theory which is defined by the rule system L and the axioms of E is the set of conditional equations $c \Rightarrow t = t'$ which are deducible from the axioms of E by applying several times the deduction rules of L.*

Lemma 5 *[Nav87] The deduction system L defines an equivalence relation on the sets of conditional equations equipped with same context namely the three following rules can be deduced from the system of rules L :*

$$reflexivity \quad \frac{}{E \vdash_L \lambda\, X.\ c \Rightarrow t = t}$$

$$symmetry \quad \frac{E \vdash_L \lambda\, X.\ c \Rightarrow t_1 = t_2}{E \vdash_L \lambda\, X.\ c \Rightarrow t_2 = t_1}$$

$$transitivity \quad \frac{E \vdash_L \lambda\, X.\ c \Rightarrow t_1 = t_2,\ \lambda\, X.\ c \Rightarrow t_2 = t_3}{E \vdash_L \lambda\, X.\ c \Rightarrow t_1 = t_3}$$

Lemma 6 *[Nav87] The following rules are deducible from the rule system L.*

$$L_{11} \quad \frac{E \vdash_L \lambda\, X.\ (c \Rightarrow c') = true}{E \vdash_L \lambda\, X.\ c \Rightarrow c' = true}$$

$$L_{12} \quad \frac{E \vdash_L \lambda\, X.\ c \Rightarrow c' = true}{E \vdash_L \lambda\, X.\ (c \Rightarrow c') = true}$$

$$L_{13} \quad \frac{E \vdash_L \lambda\, X.\ c" \Rightarrow c = c'}{E \vdash_L \lambda\, X.\ c" \Rightarrow (c \Leftrightarrow c') = true}$$

$$L_{14} \quad \frac{E \vdash_L \lambda\, X.\ c" \Rightarrow c = c'}{E \vdash_L \lambda\, X.\ c" \Rightarrow (c \Leftrightarrow c') = true}$$

$$L_{15} \quad \frac{E \cup \{\mu(c) = true\} \vdash_L \lambda\, X.\ \mu(t) = \mu(t')}{E \vdash_L \lambda\, X.\ c \Rightarrow t = t'}$$

Lemma 7 *[Nav87] The following rule is deducible from the system L.*

$$\frac{E \vdash_L \lambda\, X.\ c \Rightarrow t = t',\ E \vdash_L \lambda\, X.\ c = c'}{E \vdash_L \lambda\, X.\ c' \Rightarrow t = t'}$$

4.2.2 Proof theory

Let SP be a conditional specification which is faithfull with respect to booleans. Let the terms s, t in $T(F, X)$, let c be a boolean term in $T(F, X)$ and let E, R respectively be a set of equations and a set of rules. From the lemma 3, we know that in order to prove that in the context c, s and t are equal in the initial algebra I_{SP} of SP, it is enough to prove that the equation $c \Rightarrow s = t$ is deducible by the system L from the sets E and R namely,

$$E \cup E(R) \vdash_L c \Rightarrow s = t \Rightarrow I_{SP} \models s = t \quad (c)$$

where $E(R)$ is the set of equations which correspond to the non-oriented rules of R. In the following, we confuse both notions of equation and rewrite rule, ie we note R instead of $E(R)$, and we consider the congruence $(\leftrightarrow_E \cup \rightarrow_R \cup \leftarrow_R)^*$.

Let the inference rule $\mathcal{R} = \frac{E,\ R}{E',\ R'}$; In order to prove the correctness of $\mathcal{R}$ with respect to the deduction system L, we consider the formal systems $(L,\ E \cup R)$ and $(L,\ E' \cup R')$ generated by the language of the specification and respectively the sets E and R and the sets E' and R'. $\mathcal{R}$ is correct with respect to L if the formal system $(L,\ E \cup R)$ is equivalent to the formal system $(L,\ E' \cup R')$ in the class $LogAlg_{SP}$, ie if $(L,\ E' \cup R')$ is correct with respect to $(L,\ E \cup R)$ namely, $E \cup R \vdash_L E' \cup R'$ and if $(L,\ E' \cup R')$ is complete with respect to $(L,\ E \cup R)$ namely, $E' \cup R' \vdash_L E \cup R$.
In the proof of the lemma 8, we consider the relation "$\_ \vdash \_ \rightarrow_R \_$" without any hierarchy restriction. This proof is also valid for "$\_ \vdash \_ \rightarrow_{H,R} \_$" since "$\_ \vdash \_ \rightarrow_{H,R} \_$" is contained in "$\_ \vdash \_ \rightarrow_R \_$".

Lemma 8 *[Bou88] Correctness of "$\_ \vdash \_ \rightarrow_R \_$" with respect to L :*
If $c \vdash s \rightarrow_R t$ then $R \vdash_L \lambda X.\ c \Rightarrow s = t$.

Remark 6 *In the following, we work with countable set of variables and we do not write it explicitly in the inference rules. Furthermore, we consider specifications without any void sort. Indeed, as it is shown in [GM84], the equational deduction in the many-sorted case is only correct if we deal with this type of specifications.*

The following proves the correctness of the inference rule SR1 with respect to the deduction system L according to the proof theory previously described. The correctness of all inference rules is stated in an internal report [Bou88].
Simplifying the term part of an equation :

$$SR1 \quad \frac{E\ \cup\ \{c \Rightarrow s = t\},\ R}{E\ \cup\ \{c\ \&\ p_i \Rightarrow s_i = t,\ i = 1,..n\},\ R} \quad if \begin{cases} \forall\, i = 1,..n, p_i \vdash s \rightarrow_{H,R} s_i \\ E \vdash_L p_1 \vee\ ---\ \vee p_n = true \end{cases}$$

Proof : let $S_1 = E \cup \{c \Rightarrow s = t\} \cup R$ and $S_2 = E \cup \{c\ \&\ p_i \Rightarrow s_i = t, i = 1..n\} \cup R$;

1. We prove that the formal system $(L,\ S_2)$ is correct with respect to the system $(L,\ S_1)$. We have to check that $S_1 \vdash_L S_2$. Since $E \cup R \subset S_1$ then $S_1 \vdash_L E \cup R$. it is enough to check that $S_1 \vdash_L c\ \&\ p_i \Rightarrow s_i = t$ for each i = 1,..n ; on the one hand, for each i,

$$L_{11} \quad \frac{E \vdash_L (c\ \&\ p_i \Rightarrow p_i) = true}{E \vdash_L c\ \&\ p_i \Rightarrow p_i = true}$$

then $S_1 \vdash_L c\ \&\ p_i \Rightarrow p_i = true$
since $p_i \vdash s \rightarrow_{H,R} s_i$ then by lemma 8, $R \vdash_L p_i \Rightarrow s = s_i$ then $S_1 \vdash_L p_i \Rightarrow s = s_i$

$$L_3 \quad \frac{S_1 \vdash_L p_i \Rightarrow s = s_i,\ S_1 \vdash_L c\ \&\ p_i \Rightarrow p_i = true}{S_1 \vdash_L c\ \&\ p_i \Rightarrow s = s_i}$$

on the other hand, for each i = 1,..n,

$$L_{11} \quad \frac{E \vdash_L (c\ \&\ p_i \Rightarrow c) = true}{E \vdash_L c\ \&\ p_i \Rightarrow c = true}$$

then $S_1 \vdash_L c\ \&\ p_i \Rightarrow c = true$

$$L_3 \; \frac{S_1 \vdash_L c \Rightarrow s = t, \; S_1 \vdash_L c \;\&\; p_i \Rightarrow c = true}{S_1 \vdash_L c \;\&\; p_i \Rightarrow s = t}$$

by symmetry, $S_1 \vdash_L c \;\&\; p_i \Rightarrow t = s$

$$L_5 \; \frac{S_1 \vdash_L c \;\&\; p_i \Rightarrow t = s, \; S_1 \vdash_L c \;\&\; p_i \Rightarrow s = s_i}{S_1 \vdash_L c \;\&\; p_i \Rightarrow s_i = t}$$

and this is true for each i =1,..n.

2. We prove that (L, S_2) is complete with respect to (L, S_1) namely $S_2 \vdash_L S_1$;
since $E \cup R \subset S_2$ then $S_2 \vdash E \cup R$; it is enough to check that $S_2 \vdash_L c \Rightarrow s = t$

$$L_{11} \; \frac{E \vdash_L (c \;\&\; p_i \Rightarrow p_i) = true}{E \vdash_L c \;\&\; p_i \Rightarrow p_i = true}$$

then $S_2 \vdash_L c \;\&\; p_i \Rightarrow p_i = true$
since $p_i \vdash s \rightarrow_{H,R} s_i$ then by lemma 8, $R \vdash_L p_i \Rightarrow s = s_i$ then $S_2 \vdash_L p_i \Rightarrow s = s_i$

$$L_3 \; \frac{S_2 \vdash_L p_i \Rightarrow s = s_i, \; S_2 \vdash_L c \;\&\; p_i \Rightarrow p_i = true}{S_2 \vdash_L c \;\&\; p_i \Rightarrow s = s_i}$$

$$L_5 \; \frac{S_2 \vdash_L c \;\&\; p_i \Rightarrow s = s_i, \; S_2 \vdash_L c \;\&\; p_i \Rightarrow s_i = t}{S_2 \vdash_L c \;\&\; p_i \Rightarrow t = s}$$

by symmetry, $S_2 \vdash_L c \;\&\; p_i \Rightarrow s = t$

$$L_8 \; \frac{S_2 \vdash_L c \;\&\; p_1 \Rightarrow s = t, \; S_2 \vdash_L c \;\&\; p_2 \Rightarrow s = t}{S_2 \vdash_L c \;\&\; p_1 \vee c \;\&\; p_2 \Rightarrow s = t}$$

we applies the inference rule L_8 until $i = n$,

$$L_8 \; \frac{S_2 \vdash_L c \;\&\; p_1 \vee \;---\; \vee c \;\&\; p_{n-1} \Rightarrow s = t. \; S_2 \vdash_L c \;\&\; p_n \Rightarrow s = t}{S_2 \vdash_L c \;\&\; p_1 \vee \;---\; \vee c \;\&\; p_n \Rightarrow s = t}$$

then $S_2 \vdash_L c \;\&\; (p_1 \vee \;---\; \vee p_n) \Rightarrow s = t$
on the other hand, $E \vdash_L p_1 \vee \;---\; \vee p_n = true$ then $S_2 \vdash_L p_1 \vee \;---\; \vee p_n = true$

$$\frac{S_2 \vdash_L p_1 \vee \;---\; \vee p_n = true}{S_2 \vdash_L c \;\&\; (p_1 \vee \;---\; \vee p_n) = c}$$

by lemma 7,

$$\frac{S_2 \vdash_L c \;\&\; (p_1 \vee \;---\; \vee p_n) \Rightarrow s = t, \; S_2 \vdash_L c \;\&\; (p_1 \vee \;---\; \vee p_n) = c}{S_2 \vdash_L c \Rightarrow s = t}$$

4.3 Conditional completion algorithm

This section consists of the completion algorithm for hierarchical conditional rewriting systems. The algorithm uses a set of equations and an ordering $>$ on terms to orient the equations into rewrite rules. It stops with failure if any equation cannot be oriented or if either $E \vdash_L p_1 \vee \;---\; \vee p_n = true$ or $E \vdash_L c = false$ cannot be proved valid. If it stops with success, it provides rewriting system which is confluent on ground terms.
The simplification procedures are expressed by sequence of inference rule applications according to fixed priority order. In what follows, we explain how this order is choosen in order to improve the performance of the completion algorithm, then we give the principal algorithm and the simplification procedures.

Proposition 2 *[Bou88] Let the equation $c \;\&\; p \Rightarrow s = t$ such that $c \vdash s \rightarrow_{H,R} s'$ and $p \vdash c \rightarrow_{H,R} c'$. Then we have to simplify in the term part before simplifying in the context part namely, the simplification rule SR1' must have priority superior to the priority of the simplification rule SR5'.*

Proposition 3 1. *Similarly, we can prove that the simplification rule SR2' must have priority superior to the priority of the rule SR5'.*
For the same reasons, the simplification rules SR3' and SR4' must have priority superior to the priority of the rule SR6'.

2. *In the case reasoning, the same application ordering of the inference rules is necessary.*

Now, why applying the rules of the context rewriting before the rules of the case reasoning? The reason in obvious because it is useless to enrich the context of a term in order to rewrite it when this context only is enough for this rewriting.

4.3.1 Principal algorithm

This algorithm is constituted from principal part and from two auxiliary procedures, that are used respectively to simplify an equation (eq-simplification) and a rewrite rule (rl-simplification) by the current rewriting system.

Completion(E, R, >) :
initialisation : E is the initial set of equations, R is the final set of rules ; $R := \emptyset$
if $E \neq \emptyset$ then

1. take an equation e of E ; $E := E - \{e\}$
2. if e is trivial ie if we can apply the inference rule BR3, then apply BR3 ; completion(E, R, >)
3. if the condition of e is trivial, ie if we can apply the inference rule BR4 then apply BR4 ; completion(E, R, >)
4. (E_1, R) := eq-simplification($\emptyset \cup \{e\}$, R)
5. for each equation e' of E_1 do
   a) if we cannot orient e', ie if we cannot apply the inference rule BR1, then stop with failure else apply BR1 ; let r be the rule which results from the orientation of e'
   b) $R := R \cup \{r\}$
   c) (E_2, R) := rl-simplification($\emptyset$, $R \cup \{r\}$)
   d) $E := E \cup E_2$
   end-for

else

1. if all rules of R are marked, then return R, stop with success
2. choose a rule r' from R with a fair strategy and mark it
3. for each marked rule r_1 in R do
   compute all critical pairs between r_1 and r', ie apply the inference rule BR2
   end-for
4. add all computed critical pairs to E
5. completion(E, R, >)

end-if
end-completion.

4.3.2 Simplification procedures

The procedures eq-simplification and rl-simplification have both as arguments a pair constituted from a set of equations and a set of rules. They also provide as result a set of equations and a set of rules after they have performed some inference rules in order to normalize either an equation (eq-simplification), or a rule (rl-simplification) by the current rewriting system.

eq-simplification(E, R) :

1. while we can simplify the equation in the term part by the contextual rewriting (ie apply the inference rules SR1' or SR2') then simplify it else continue end-while ;
2. if we can simplify the equation in the context part by the contextual rewriting (ie apply the inference rule SR5') then simplify it and return to 1 else continue ;
3. if we can simplify the equation in the term part by the case reasoning (ie apply the inference rules SR1 or SR2) then simplify it and return to 1 else continue ;
4. if we can simplify the equation in the context part by the case reasoning (ie apply the inference rule

SR5) then simplify it and return to 1 else end.
end eq-simplification.

rl-simplification(E, R)

1. while we can simplify the rule in the term part by the contextual rewriting (ie apply the inference rules SR3' or SR4') then simplify it else continue end-while ;
2. if we can simplify the rule in the context part by the contextual rewriting (ie apply the inference rule SR6') then simplify it and return to 1 else continue ;
3. if we can simplify the rule in the term part by the case reasoning (ie apply the inference rules SR3 or SR4) then simplify it and return to 1 else continue ;
4. if we can simplify the rule in the context part by the case reasoning (ie apply the inference rule SR6) then simplify it and return to 1 else end.

end rl-simplification.

The proof of correctness of the procedure is still in progress. In principle, under the hypotheses of the confluence theorem (th 3), if the completion procedure stops with success then the result is a rewriting system which is confluent on ground terms.
Sketch of the proof : If the completion procedure stops with success, the set of equations is empty. This means that all critical pairs between the rules of R have been simplified by the procedures eq-simplification and rl-simplification. If these procedures are correct and if they end, then the completion procedure provides effectively a system which is confluent on ground terms.
The termination of these procedures is ensured by the termination of R.
Let us prove that all contextual critical pairs converge ; for this, let us prove that all contextual critical pairs form complete coherent sets of contextual normal forms.
Let any critical pair $c \Rightarrow s = s'$; the simplification of this critical pair by the procedure eq-simplification has for result a set of equations of the form $\{c_k \Rightarrow s_i = s'_j.\ k \in K.\ i \in I,\ j \in J\}$. This set constitutes the set of contextual normal forms of both terms of the critical pair. It is complete since when any context is enriched in the case reasoning, the condition $E \vdash_L p_1 \vee --- \vee p_n = true$ guarantees the preservation of the completeness of any set of contextual terms.
If this critical pair converges, then each of them is deleted either by the inference rule BR2, or by the inference rule BR3 which express exactly the coherence of a set of contextual normal forms. Since the set of normal forms of the critical pair is coherent and complete then the critical pair converges. By theorem 3, since all critical pairs define sets of contextual normal forms which are complete and coherent, then R is confluent on ground terms.
However, the confluence theorem does not include the simplification of rules. In order to establish complete proof of the completion procedure, we have to consider this case of simplification.

5 Example

The example defines the infimum function of elements which sort is equipped with an order predicate $\leq$.

Example 4 *Let $SP = (SP_0,\ SP_1)$ be the infimum specification :*
$SP_0 = (S_0,\ F_0,\ E_0)$
$S_0 = \{bool,\ elem\}$
$F_0 = \{true,\ false\ : \rightarrow bool,\ \&\ :\ bool \times bool \rightarrow bool,$
$\quad 0\ : \rightarrow elem,\ s\ :\ elem \rightarrow elem,\ \doteq,\ \leq\ :\ elem \times elem \rightarrow bool\}$
$E_0 = \{(0\ \doteq\ 0) = true,\ (0\ \doteq\ s(x)) = false,$
$\quad (s(x)\ \doteq\ 0) = false,\ (s(x)\ \doteq\ s(y)) = (x\ \doteq\ y),$
$\quad ((x \leq y) = true)\ \vee\ ((x \leq y) = false) = false,$
$\quad ((x \leq y) = true)\ \&\ ((x \leq y) = false) = true\}$
$SP_1 = (S_1,\ F_1,\ E_1)$
$S_1 = S_0,$
$F_1 = F_0 + \{inf\ :\ elem \times elem \rightarrow elem\}$
$E_1 = E_0 + \{((x \leq y) = true) \Rightarrow inf(x,\ y) = x,$
$\quad ((x \leq y) = false) \Rightarrow inf(x,\ y) = y\}$

Remark 7 *We use two equality symbols, = for join both members of an equation and $\doteq$ which is the equality in the elem sort.*

We give the example in its complete form. However, we present in what follows the completion process in its parameterized form, namely with only the two last equations of E_0 and the equations of E_1. Indeed, as it is shown in [Gan87b], the completion process can be used in the parameterized frame of conditional specifications.
The completion comes as follows : all the equations are trivial and then we can apply successively the inference rule BR1 and orient these equations in rewrite rules.
In the second step, we mark one by one the rewrite rules and we compute the critical pairs between these rules. The only critical pairs are computed between the conditional rules of R_1, namely
$((x =< y) = true)$ & $((x =< y) = false) \Rightarrow x = y$
$((x =< y) = false)$ & $((x =< y) = true) \Rightarrow y = x$
These critical pairs are eliminated because their conditions reduce to false by the last rule of R_0.

Remark 8
1. *At this point, we have to notice that the use of the equality symbol raises some problems. Indeed, using = in boolean expressions would give equations which are not syntactically correct ; for instance, $(s(x) = s(y)) = (x = y)$. That is the reason why we introduce specific symbol $\doteq$. Now we need to express the relationship between $\doteq$ and =. That is done by the rule $x \doteq y \Rightarrow x = y$. Since the rule is conditional, it can not be used in E_0. In [Nav87], in order to express the equality in every specification sort, the author uses similar axiom with the following axiom $(x \doteq x) = true$.*
2. *This example runs on REVEUR4 in its parameterized form and in its complete form. REVEUR4 is the implementation software of the contextual rewriting with two levels. It has been written at Nancy in CLU and it runs on VAX and SUN3 under UNIX system. Actually, it achieves two principal functions, the proof of the confluence on ground terms and the proof of the equality by contextual rewriting [Zha84], [RZ84], [BR87].*

6 Conclusion

The work presented in this paper is divided into two parts. In the former, we compare between two approaches that formalize the conditional theory. These approaches are different by their syntax and by the kind of algebras that they define. However, they share common features. Both approaches have emphasized the need of working in algebras whose the boolean carrier is not ordinary.
The second part of the paper presents a completion procedure for hierarchical conditional systems. This procedure is used in order to get conditional systems confluent on ground terms. We also find common features in this second part. This part has been presented within hierarchical frame. However, we can join both parts because it is possible to define both rewriting processes in each approch. Combining these rewriting dealings allows to use judiciously the advantages of each of them.
A near continuation of this work is to state the completeness proof of the inference rules and a complete proof of the completion procedure which we have presented.
Many problems need to be considered. They include use of first order formulas and study of confluence with variables. Moreover, it will be nice to develop a theory of rewriting modulo equations as in the case of rewriting systems without any condition and use of partially defined functions.

I sincerely thank Jean-Luc Rémy, Hélène Kirchner and Pierre Lescanne for their valuable comments and suggestions.

References

[BDH86] L. Bachmair, N. Dershowitz, and J. Hsiang. Orderings for equational proofs. In *Proc. Sym. Logic in Computer Science*, pages 346–357, Boston (Massachusetts USA), 1986.

[Bou88] W. Bousdira. *A completion procedure for hierarchical conditional equations.* 88-R-021, Centre de Recherche en Informatique de Nancy, 1988.

[BR87] W. Bousdira and J.L. Rémy. Complétion des systèmes de réécriture conditionnelle. In Actes des journées GROSPLAN, editor, *revue BIGRE+GLOBULE*, Aix-en-Provence, 1987.

[BR88] W. Bousdira and J.L. Rémy. Hierarchical contextual rewriting with several levels. In R. Cori and M. Wirsing, editors, *Lecture Notes in Computer Science*, pages 193–206, 5th Synposium on Theoretical Aspects of Computer Science, Springer-Verlag, Bordeaux France, 1988.

[Gan87a] H. Ganzinger. A completion procedure for conditional equations specifications. In *to appear in Journal of Symbolic Computation*, 1987.

[Gan87b] H. Ganzinger. Ground term confluence in parametric conditional equational specifications. In F.J. Brandenburg, G. Vidal, and M. Wirsing, editors, *Lecture Notes in Computer Science No 247*, pages 286–298, 4th annual Symposium on Theoretial Aspects of Computer Science, Springer-Verlag, Passau RFA, 1987.

[GM84] J.A. Goguen and J. Meseguer. *Completeness of many-sorted equational logic*. Technical Report CSLI-84-15, Center for the Study of Language and Information Stanford University, 1984.

[Hue80] G. Huet. Confluent reductions : abstract properties and applications to term rewriting systems. *Journal of the Associations for Computing Machinery*, 27(4):797–821, 1980. Preliminary version of the 18th Symposium on Foundations of Computer Science.

[JW86] J.P. Jouannaud and B. Waldmann. Reductive conditional term rewriting systems. In M. Wirsing, editor, *Elsevier Science Publishers*, 3rd IFIP Conf. on Formal description of Programming Concepts, Ebberup Denmark, 1986.

[Kap84] S. Kaplan. Conditional rewrite rules. *Theoretical Computer Science*. 3:175–193, 1984.

[Kap87] S. Kaplan. Simplifying conditional term rewriting systems : unification, termination and confluence. *to appear in Journal of Symbolic Computation*, 1987.

[KB70] D. Knuth and P. Bendix. Simple word problems in universal algebras. In J. Leech, editor, *Computational Problems in Abstract Algebra*, pages 263–297, Pergamon Press, Elmsford N.Y, 1970.

[KR87] S. Kaplan and J.L. Rémy. Completion algorithms for conditional rewriting systems. In H. Ait-Kaci and M. Nivat, editors, *Proceedings of the MCC-INRIA Colloquium on the Resolution of Equations in Algebraic Structures*, MCC and INRIA, Austin Texas, 1987.

[Nav87] M. Navarro. *Técnicas de Reescritura para especificaciones condicionales*. Thèse Doctorale, Barcelone, 1987.

[NO84a] M. Navarro and F. Orejas. On the equivalence of hierarchical and non-hierarchical rewriting on conditional term rewriting systems. In *Eurosam Conference*, Oxford, 1984.

[NO84b] M. Navarro and F. Orejas. Parameterized horn clause specifications : proof theory and correctness. In *Lecture Notes in Computer Science*, pages 202–216, TAPSOFT'87, Springer-Verlag 249, Pisa Italy, 1984.

[RZ84] J.L. Rémy and H. Zhang. Reveur4 : a system for validating conditional algebraic specifications of abstract data types. In *ECAI*, 6th ECAI Conference, Pisa Italy, 1984.

[Zha84] H. Zhang. *REVEUR4 : Etude et mise en œuvre de la réécriture conditionnelle*. Université de Nancy 1, 1984.

An effective method for handling initial algebras

Hubert Comon*

Abstract

Given an arbitrary Term Rewriting System R, we compute a "conditional grammar" of the language of ground terms which are irreducible for R. Such a presentation provides a powerful tool for reasoning in non-free algebras. We sketch here two applications:

1. The decision of inductive reducibility (which is a key concept in automating proofs by induction)
2. The automatic transformation of non-free, many sorted algebraic specifications into free order-sorted specifications.

1 Introduction

Equational methods are widely used in Computer Science, mostly for the specifications of Abstract Data Types. The so-called *initial algebra semantics* [GTW78] defines an Abstract Data Type as the quotient algebra $T_{F,E}$ of the free initial algebra T_F over $=_E$, the congruence relation generated by the set of equations E. Allowing a computer program to handle this quotient algebra requires that we have a suitable representation for it.

Sometimes, it is possible to work in T_F. For example, the word problem for $=_E$ is usually decided by compiling the equations of E into a canonical Term Rewriting System (TRS for short) $\mathcal{R}$. Unfortunately, equational deduction is not powerful enough for reasoning in this algebra. Actually, it may exist equalities that hold in $T_{F,E}$ and which are not theorems of the equational variety. This is the reason why the decision of the *protecting property* [MG85] as well as the automatization of inductive proofs [HH82,LLT86,JK86] has to use the particular fact that $T_{F,E}$ is initial.

In fact, for both problems, it appears that the key notion is the *inductive reducibility* property:

Definition 1 [JK86] *A term t is* **inductively reducible** *if, for every substitution σ from $T_F(X)$ into T_F, $t\sigma$ is reducible.*

Let us recall the background in automatization of inductive proofs. The main idea in [HH82] is to use "proofs by consistency". In this paper, the set of operators is assumed to be divided in two (disjoint) sets C, the set of *constructors* and D, the set of *defined operators*. Equations in E are assumed to be oriented into a canonical TRS $\mathcal{R}$ such that the set of irreducible ground terms, NF, exactly equals T_C. Then, for proving the inductive theorem $s = t$, the equation is added to E and a completion procedure is applied. If an equation between two terms in $T_C(X)$ is computed by the procedure, then a *disproof* is derived. Otherwise, $s = t$ is an inductive theorem (assuming that the completion does not fail).

The algorithm presented in [JK86] extends such a method by allowing "relations between constructors" or, equivalently, does not assume F to be split into constructors and defined

*Laboratoire d'Informatique Fondamentale et d'Intelligence Artificielle (LIFIA), 46 Avenue Félix Viallet, 38031 Grenoble, France. E-mail: comon@lifia.imag.fr

operators. Then, a disproof is generated when an equation $c = d$ is derived where $c > d$ and c is not inductively reducible.

Inductive reducibility is also strongly linked with sufficient completeness. Indeed, assuming that E is oriented into a canonical TRS such that every term in T_C has a normal form in T_C (see sufficient conditions for this requirement in [Com86,KNZ85]), the specification of f is sufficiently complete with respect to C iff $f(x_1, \ldots, x_n)$ is inductively reducible [Com86,KNZ85,JK86].

This shows that, in the general case (relations between constructors), inductive reducibility is the key concept.

Inductive reducibility has been widely studied, among others, let us cite [NW82,Pla85,KNZ85, JK86,Thi84,Com86]. Except in [Thi84,Com86], the method is based upon "test sets". Although such methods allowed to establish the decidability of the inductive reducibility [Pla85], an effective algorithm is still lacking. Indeed, in test sets methods, the idea is to "test" the reducibility only on a finite subset of all the ground instances of t. It has been shown that there exists such finite sets for which the reducibility test implies the inductive reducibility property. However, the method cannot be used as an effective decision procedure, since the test set is huge, even for very simple examples. Our approach is different and is effective in spite of the intrinsic complexity of the problem [KNZ86]. We give an outlook of a new decision procedure of the inductive reducibility property in the section 4.

Plaisted's algorithm is based upon some properties similar to "pumping lemmas" in language theory. This suggested that there was a formal language approach of the problem. Indeed, the first aim of this paper is to give a "description" of NF, the set of ground terms which are not reducible by $\rightarrow_{\mathcal{R}}$, the reduction relation associated with $\mathcal{R}$[1]. Such a description (a "NF-grammar") is nothing but a regular tree grammar when every left hand side of a rule in $\mathcal{R}$ is linear. Then, our inductive reducibility test is based upon a "cleaning" algorithm of the grammar. For the emptiness decision (which amounts to inductive reducibility decision) such methods are known to be more efficient than those using "pumping lemmas".

The computation of an NF-grammar was already studied in [CR87]. We give in this paper a generalized algorithm which makes use of "equational problems simplification" [KL87,CL88, Com88]. We give in section 2 an outlook of the results used in the following and show in section 3 how to compute a NF-grammar.

Our method has many other applications. Among others, we give in section 5 an application to the automatic ("canonical") transformation of non-free many-sorted algebras into free order-sorted algebras. Such a transformation is very useful. Indeed, as shown in [GM87], order-sorted algebra solves the "constructor-selector" problem. In the same way, we show that the equality of the integers cannot be correctly defined using a many-sorted specification, but, using the order-sorted specification produced by our algorithm, this becomes easy since there are no more equations between the "constructors".

2 Reduction of equational problems

In order to avoid confusions, the syntactic equality as well as the equality of equational problems will be denoted by $\equiv$. In the following, $\mathcal{R}$ will denote a TRS constructed on $T_F(X)$ and LHS will be the set of its left hand sides. An *equation* is a pair of terms denoted by $t = u$ and a *dis-*

[1] The subscript $\mathcal{R}$ will be omitted in the following, when there is no ambiguity

equation is a pair of terms denoted by $t \neq u$. In addition $\top$ and $\perp$ are two particular equations [2].

Roughly speaking, an *equational problem* is a conjunction of disjunctions of equations and disequations whose variables are split into three sets: the *principal unknowns* which are the free variables of the problem, the *auxiliary variables* which are existentially quantified and the *parameters* which are universally quantified. For example:

$$\mathcal{P}_1 \equiv \forall y_1, y_2 \,:\, x_1 = f(x_2, 0) \wedge (y_1 \neq x_1 \vee f(y_1, y_2) \neq f(x_2, x_1))$$

is an equational problem with parameters y_1 and y_2 and (principal) unknowns x_1 and x_2.

A *grounding substitution* σ is an endomorphism of $T(F, X)$ such that $Dom(\sigma) = \{x \in X | x\sigma \not\equiv x\}$ is finite and $\forall x \in Dom(\sigma)$, $x\sigma \in T(F)$.

A grounding substitution σ *validates* an equation $t = u$ if $t\sigma \equiv u\sigma$. It *validates* a disequation $t \neq u$ if $t\sigma \not\equiv u\sigma$. Moreover, every grounding substitution validates $\top$ and none validates $\perp$. This definition is extended to conjunctions of disjunctions of equations and disequations in a straightforward way [CL88].

A *solution* of an equational problem $\mathcal{P} \equiv \exists z_1, \ldots, z_k, \forall y_1, \ldots, y_n \,:\, P$ is a substitution σ whose domain is the set of principal unknowns of $\mathcal{P}$ and such that there exists a grounding substitution ϕ on $z_1, \ldots, z_k$ such that for every grounding substitution θ on $\{y_1, \ldots, y_n\}$, $\theta\phi\sigma$ validates P.

For example, the above problem $\mathcal{P}_1$ has the solution $\sigma_1 = (x_1 \leftarrow f(0,0);\ x_2 \leftarrow 0)$ since $x_1\sigma_1 \equiv f(x_2, 0)\sigma_1$ and, for every grounding substitution θ, either $y_1\theta \not\equiv f(0,0)$ or $f(y_1, y_2)\theta \not\equiv f(0, f(0,0))$.

Such problems generalize those of [Col84], extending the definitions (and results) by allowing "universally" quantified variables.

Various sets of inference rules can be given for solving equational problems [KL87,CL88, Com88,Mah88]. We recall some of them[3] in figure 1.

On the above example, the following transformations may be performed:

$$\begin{array}{lll}
\mathcal{P}_1 & \mapsto_{CD} & \forall y_1, y_2 \,:\, x_1 = f(x_2, 0) \wedge (y_1 \neq x_1 \vee y_1 \neq x_2 \vee y_2 \neq x_1) \\
 & \mapsto_{EP} & \forall y_2 \,:\, x_1 = f(x_2, 0) \wedge (x_1 \neq x_2 \vee y_2 \neq x_1) \\
 & \mapsto_{EP} & x_1 = f(x_2, 0) \wedge x_1 \neq x_2 \\
 & \mapsto_{Re} & x_1 = f(x_2, 0) \wedge x_2 \neq f(x_2, 0) \\
 & \mapsto_{Oc} & x_1 = f(x_2, 0)
\end{array}$$

This last problem is irreducible.

Theorem 1 [CL88,Com88] [4] *The non-deterministic application of the rules given in figure 1 to any equational problem terminates.*

We have moreover a "correctness" and "completeness" result for the above set of rules:

Theorem 2 [CL88,Com88] *The set of solutions of an equational problem P is equal to the union of the sets of solutions of its irreducible forms.*

---

[2] $\top$ and $\perp$ are neutral elements for $\wedge$ and $\vee$ respectively

[3] In these rules, P is a conjunctions of disjunctions of equations and disequations, d is a disjunction of equations and disequations, $t, u, \ldots$ are terms, $x, \ldots$ are unknowns and $y, \ldots$ are parameters. Moreover, the notation $\vec{y}$ stands for $y_1, \ldots, y_m$.

[4] The results given in [CL88,Com88] are actually more general since in these papers the sets of rules are larger

Explosion

$$(E)\quad \forall\vec{y} : P \;\mapsto\; \exists z_1, \ldots z_k \forall\vec{y} : P \wedge x = f(z_1, \ldots, z_k)$$

If

1. x is not a parameter
2. P contains a disequation $x \neq u$ where u contains an occurrence of a parameter
3. $z_1, \ldots, z_k$ are "fresh" variables and $f \in F$.
4. No other rule may be applied.

Elimination of parameters

$$(EP_1)\quad \forall y_1, \ldots, y_m : P \wedge (d \vee y_i \neq u) \;\mapsto\; \forall y_1, \ldots, y_m : P \wedge d(y_i \leftarrow u)$$

$$(EP_2)\quad \forall\vec{y_1}, y, \vec{y_2} : P \;\mapsto\; \forall\vec{y_1}, \vec{y_2} : P \qquad \text{If } y \text{ does not occur in } P$$

Clashes and Decompositions

$$(C_1)\quad f(t_1, \ldots, t_n) = g(u_1, \ldots u_m) \;\mapsto\; \bot \qquad \text{If } f \not\equiv g$$

$$(C_2)\quad f(t_1, \ldots, t_n) \neq g(u_1, \ldots u_m) \;\mapsto\; \top \qquad \text{If } f \not\equiv g$$

$$(D_1)\quad f(t_1, \ldots, t_n) = f(u_1, \ldots u_n) \;\mapsto\; t_1 = u_1 \wedge \ldots \wedge t_n = u_n$$

$$(D_2)\quad f(t_1, \ldots, t_n) \neq f(u_1, \ldots u_n) \;\mapsto\; t_1 \neq u_1 \vee \ldots \vee t_n \neq u_n$$

Replacement

$$(R)\quad x = t \wedge P \;\mapsto\; x = t \wedge P(x \leftarrow t)$$

If x is an unknown, x does not occur in t and x does occur in P
Moreover, if t is a variable, then t must occur in P

Elimination of trivial equations and disequations

$$(T_1)\quad t = t \;\mapsto\; \top \qquad (T_2)\quad t \neq t \;\mapsto\; \bot$$

Occur checks

$$(O_1)\quad x \neq u \;\mapsto\; \top \qquad (O_2)\quad x = u \;\mapsto\; \bot \qquad \text{If } x \in Var(u) \text{ and } x \not\equiv u$$

Non deterministic choice

$$(Nc)\quad P \wedge (P_1 \vee P_2) \;\mapsto\; P \wedge P_1 \qquad \text{If } P_1 \text{ or } P_2 \text{ does not contain any parameter}$$

Figure 1: Transformation rules for equational problems

Now, it remains to describe what are the problems we are interested in and what are their irreducible forms. Roughly speaking, we are interested in the computation of the ground instances of a term t which are not ground instances of any term in a finite set LHS. This will give us the ground instances of t which cannot be reduced at the top by a TRS whose left hand sides are LHS. In other words, assuming that terms with variables represent the (possibly infinite) set of their ground instances, we want to compute the complement of such a set in another set which is represented in the same way. However, in some cases, it is not possible to express the difference set $\{t\} - \{t'\}$ of the ground instances of t which are not ground instances of t' with the help of terms only. For example, if $F = \{0, succ, f\}$, then $\{x\} - \{succ(y)\}$ can be written $\{0, f(x_1, x_2)\}$ but $\{f(x_1, x_2)\} - \{f(x, x)\}$ cannot be written as the union of the ground instances of a finite set of terms.

Actually, we have to use finite sets of "constrained terms" rather than finite sets of terms. Indeed, finite sets of constrained terms provide a class of sets of ground terms which is closed under finite intersection, finite union and complement. Let us introduce this concept:

Definition 2 *A* **constrained term** *is a pair (t, d) where t is a term and d a conjunction of disequations in normal form whose variables are contained in those of t. When d is empty, we may write t instead of (t, d).*

Now, the complement of a set LHS of terms in a constrained term (t, d) can be easily expressed using equational problems: this is the set of terms $t\sigma$ where σ is a solution of the problem $\forall y_1, \ldots, y_m : \bigwedge_{l \in LHS} l \neq t$ if $Var(LHS) = \{y_1, \ldots, y_m\}$. However, we need that t be not reducible "at any position" and thus have to introduce more sophisticated complement problems.

Definition 3 *The complement problem (w.r.t. LHS) $E(t, d)$ of a non-variable constrained term (t, d) is defined by:*

- if $t \equiv f(t_1, \ldots t_n)$ and t is linear,

$$E(t, d) \equiv \forall y_1, \ldots, y_m : x_1 = t_1 \wedge \ldots \wedge x_n = t_n \wedge (\bigwedge_{l \in LHS} t \neq l) \wedge d$$

- otherwise, $E(t, d) \equiv \forall y_1, \ldots, y_m : \bigwedge_{p \in Pos(t)} (\bigwedge_{l \in LHS} t/p \neq l)$.

In both cases, $\{y_1, \ldots, y_m\}$ is the set of variables of LHS, $t_1, \ldots, t_n$ contain no occurrence of the y_i's and $x_1, \ldots, x_n$ are variables which occur only once in the problem. $IR(t, d)$ denotes the set of irreducible forms of the problem $E(t, d)$.

Other complement problems (related to "complete sets of positions") may be chosen. The distinction between linear and non-linear terms is not necessary since we could take the second definition in both cases. However, it would lead to complicated NF-grammars in the section 3, even in the linear case.

The properties of $IR(t, d)$ which are necessary in the proofs of our results are described in the next proposition.

Proposition 1 [5] *For every constrained term (t, d), each problem $\mathcal{P} \in IR(t, d)$ is either $\top$, $\bot$ or $\mathcal{P} \equiv (x_1 = w_1 \wedge d_1) \wedge \ldots \wedge (x_n = w_n \wedge d_n) \wedge z_1 \neq v_1 \wedge \ldots \wedge z_k \neq v_k$ with the following properties:*

(i) $x_1, \ldots, x_n$ are either the left hand sides of the equations in $E(t, d)$ (when t is linear) or the set of variables of t, and occur only once in $\mathcal{P}$

(ii) $(w_1, d_1) \ldots, (w_n, d_n)$ are linear constrained terms such that any two of them do not share any variable

[5] The proofs of the propositions and theorems given in this paper can be found in [Com88].

(iii) $z_1, \ldots, z_k$ *are variables, each one occurring in some* w_i
(iv) $v_1, \ldots, v_k$ *are terms whose variables are contained in the set of variables of the* w_i*'s*
(v) $\forall i, j,\ z_i \in Var(w_j) \Rightarrow Var(v_i) \cap Var(w_j) = \emptyset$
(vi) If t is linear, then $\forall i,\ depth(w_i) \leq Max_{l \in LHS}(depth(l))$
(vii) $k \leq \alpha \times \beta$ *where* α *is the number of non-variable positions of t and* β *is the number of non-linear terms in LHS.*

Note that, because of property (vii), $k = 0$ when every lhs of a rule is linear. i.e. : there are no disequations in $\mathcal{P} \in IR(t, d)$. Moreover, the w_i's are linear terms whatever (t, d) is (linear or not).

3 The language of ground normal forms

Now, we are able to show how to compute a grammar for the language of ground normal forms. Let us first show quickly how this is related to equational problems simplification.

3.1 The relationship between ground normal forms and equational problems

A term $t \in T_F(X)$ is *reducible* at position p if there exists a rule $l \rightarrow r$ in $\mathcal{R}$ such that the subterm t/p of t matches l, i.e. there exists $\sigma \in \Sigma$ such that $l\sigma \equiv t/p$.
Let NF be the set of terms, called ground normal forms, which cannot be reduced at any position. If (t, d) is a constrained term, $NF_{t,d}$ is the set of terms $t\sigma \in NF$ such that σ is a solution of d. Note that t is inductively reducible iff $NF_t = \emptyset$.

Proposition 2 *For every constrained term* (t, d)*,* $NF_{t,d}$ *is the set of terms* $t\sigma$ *where* σ *is a solution in* NF *of* $E(t, d)$*.*

Indeed, if σ is a solution of the complement problem $E(t, d)$, then $t\sigma$ cannot be matched (at the root) by any lhs of $\mathcal{R}$. Moreover, if the solutions are in NF, then $t\sigma$ cannot be matched at any position. Such a (recursive) definition of NF suggests treating it as a language. Indeed, in the next subsection, we show how to compute a "grammar" of this language.

3.2 Computation of a grammar of NF_t

Let us show how the algorithm works on a very simple example. Let F $= \{0, p, s\}$ and the rules be:

$$p(s(x)) \rightarrow x \qquad s(p(x)) \rightarrow x$$

This is a specification of the integers. The TRS is canonical.

Assume that we are looking for $NF_{s(x)}$. Using the proposition 2, we have to solve the problem:

$$\forall y : x = x_1 \wedge s(x_1) \neq p(s(y)) \wedge s(x_1) \neq s(p(y))$$

whose solutions are $x = 0$ and $x = s(x_2)$. (Equations whose left hand side is x_1 can be removed using a rule we did not mention in section 2).
This means that $NF_{s(x)} = \{s(0)\} \cup \{s(s(x_2))\sigma \mid s(x_2)\sigma \in NF\}$. This may be written nicely:

$$NF_{s(x)} \rightarrow s(0) \mid s(NF_{s(x_2)})$$

This last formulation is nothing but a grammar generating the language $NF_{s(x)}$. In this example, we got a regular tree grammar (as in [GB85]). The emptyness and finiteness (and, in particular the inductive reducibility) may thus be decided in the usual way.

However, this example is very simple since the only "non terminal" occurring in a left hand side of a grammar rule is $NF_{s(x_2)}$ which is, up to renaming, equal to a left hand side of a grammar rule. Obviously this is not always the case. In general, it is necessary to *complete* the presentation, adding the right hand sides which are not equal (up to renaming) to a left hand side. Moreover the "grammar rules" may be, in general, conditional ones, because of eventual disequations in the solved forms. Let us formalize these ideas.

Definition 4
A **presentation** *is a (finite) set of pairs* $((t,d), R(t,d))$ *where* (t,d) *is a constrained term and*
- *If* t *is a variable then* $R(t)$ *is the set of equations* $t = t'$ *where* $t' \in \{f(\vec{x}),\ f \in F\}$ *and* $\vec{x}$ *is a set of "new" variables.*
- *Otherwise* $R(t,d) = IR(t,d)$.

Moreover, all the terms t *in pairs* $((t,d), R(t,d))$, *except possibly one, must be linear.*

In the following, presentations will be denoted by Π,Π_0,... For each pair $(ct, R(ct)) \in \Pi$, ct is a *left hand side* of Π called a *non terminal.* The set of all these terms is denoted by $NT(\Pi)$. In the same way, a *right hand side* of a problem P in $R(t,d)$ is a constrained term (u,δ) such that u is a right hand side of an equation in P and δ is the set of disequations of P whose variables are contained in those of P. Each right hand side of a problem in some $R(t,d)$ is then a *right hand side* of Π. A *presentation of NF* is a presentation Π such that $NT(\Pi)$ contains at least one variable of each sort of S. When $(x_1 = w_1 \wedge d_1) \wedge \ldots \wedge (x_n = w_n \wedge d_n) \wedge d$ is a problem in $IR(t,d)$ where $t \equiv f(t_1, \ldots, t_n)$ is a linear term, it will be convenient to write

$$NF_{t,d} \rightarrow f(NF_{w_1,d_1}, \ldots, NF_{w_n,d_n})\ IF\ d$$

In the integer's example, $\{(x, \{x = 0, x = p(x_1), x = s(x_1)\})\}$ is a presentation of NF which may be written

$$NF \rightarrow NF_0 \mid NF_{s(x)} \mid NF_{p(x)}$$

Let us show a non-linear example (which will be developed in the next subsection). This is a definition of $+$ on the integers mod 2: $S = \{int2\}$, $F = \{0 :\rightarrow int2;\ s : int2 \rightarrow int2;\ + : int2 \times int2 \rightarrow int2\}$ and the rules are:

$$s(s(0)) \rightarrow 0 \qquad 0 + x \rightarrow x \qquad x + 0 \rightarrow x \qquad x + x \rightarrow 0$$

In order to compute $NF_{x_1+x_2}$, we solve the problem

$$\forall y :\ x_1 + x_2 \neq y + y \wedge x_1 + x_2 \neq 0 + y \wedge x_1 + x_2 \neq y + 0$$

This gives, for example, the solved form : $\exists w_1, w_2 :\ x_1 = s(w_1) \wedge x_2 = s(w_2) \wedge w_1 \neq w_2$ which leads to the conditional rule :

$$NF_{x_1+x_2} \rightarrow NF_{s(w_1)} + NF_{s(w_2)} \quad \text{IF } w_1 \neq w_2$$

Similarly the other solved forms give the derivation rules:

$$\begin{array}{lllll} NF_{x_1+x_2} & \rightarrow & NF_{w_1+w_2} + NF_{s(w_3)} & \mid NF_{w_1+w_2} + NF_{w_3+w_4} & \text{IF } w_1 \neq w_3 \\ & & \mid NF_{s(w_3)} + NF_{w_1+w_2} & \mid NF_{w_1+w_2} + NF_{w_3+w_4} & \text{IF } w_2 \neq w_4 \end{array}$$

Two non-terminals are *equivalent* if they don't differ up to a renaming of their variables. The goal is now to compute the "definitions" of the non terminals which are in the right hand sides and which are not equivalent to a left hand side.

Definition 5
A presentation Π *is a* **NF-grammar** *if every right hand side of* Π *is equivalent to a left hand side of* Π.

To "complete" a presentation is then done in a straightforward way: while there exists some right hand side r in Π which is not equivalent to a left hand side, compute $R(r)$ and add $(r, R(r))$ to Π. In the above example, we would thus have to compute $R(s(w_3))$. However, all other right hand sides are equivalent to $NF_{x_1+x_2}$. Of course, the new computations may add some new right hand sides. Then the termination is not obvious.

Theorem 3 *The above completion algorithm terminates and produces a NF-grammar.*

Both correctness and termination can be proved using proposition 1. The depth of the terms in a left hand side of Π may indeed be bound by a constant depending only on Π_0 and LHS.

If every non-terminal of Π is a linear term, then the presentation is said to be *linear*. The NF-grammar obtained by completing a presentation Π is denoted Π^*.

Proposition 3 *If Π is a linear presentation and if every term in LHS is linear, then Π^* is linear too.*

Note that a linear NF-grammar is a regular tree grammar. Thus, if every term in LHS is linear, then NF is a regular tree language. This result is also a direct consequence of the stability of regular tree languages by complement.

Moreover, we may compute a complete presentation of NF_t, starting the completion with the presentation $\Pi_0 = \{(t, IR(t))\}$.

3.3 An example

We use the above non-linear example of integers modulo 2. This is, of course, a non-linear example. We are interested in the inductive reducibility of $x_1 + x_2$. (Or, equivalently, in the full definition of $+$). The starting presentation Π_0 is given above. Then, we have to compute $R(s(x))$, solving the problem $s(x) \neq s(s(0))$. This leads to the solved forms $x = 0$, $\exists w_1, w_2 : x = w_1 + w_2$, $\exists w : x = s(s(w))$, $\exists w_1, w_2 : x = s(w_1 + w_2)$. And, turning them into grammar rules, we get:

$$NF_{s(x)} \rightarrow s(NF_0) \mid s(NF_{x_1+x_2}) \mid s(NF_{s(s(x))}) \mid s(NF_{s(x_1+x_2)})$$

Now, the right hand sides $NF_{s(s(x))}$, $NF_{s(x_1+x_2)}$, NF_0 are not equivalent to any left hand side and the completion goes on. Finally, we get the following additional rules:

$$\begin{array}{lcl} NF_0 & \rightarrow & 0 \\ NF_{s(s(x))} & \rightarrow & s(NF_{s(s(x))}) \mid s(NF_{s(x_1+x_2)}) \\ NF_{s(x_1+x_2)} & \rightarrow & s(NF_{x_1+x_2}) \end{array}$$

Now the presentation is an NF-grammar since every non-terminal in a right hand side is equivalent to a left hand side.

4 Decision of inductive reducibility

Given a term t, the previous algorithm may be used for computing a complete presentation of NF_t. Deciding inductive reducibility is thus a particular case of "cleaning" such presentations (i.e. removing useless non-terminals). Although such an algorithm is well known for regular tree grammars, it is not obvious in the general case.

Theorem 4 *The emptyness problem is decidable for a NF-grammar.*

The algorithm and it's proof of correctness would need a full paper by themselves (see [Com88]). We shall only sketch them.
We define the sequence of *covering sets* as follows:
Each C_i is a map which associates to each non-terminal a set of ground terms belonging to the language it generates. C_0 maps each non-terminal to an empty set and C_n maps each non terminal to the set of terminals they can derive in n steps. Obviously, the language generated by N is not empty iff there exists some index i such that $C_i(N)$ is not empty.
However, only some terms in the covering sets are actually useful for the emptyness decision.
Let k be an integer. Let G_n^k be defined by: G_0^k maps each non-terminal to an empty set and $G_{n+1}^k(N)$ is the set of terms which can be derived in $n+1$ steps from N *provided that* $G_n^k(N)$ *contains less than* k *elements* and is equal to $G_n^k(N)$ otherwise.
For Example, in the linear case, it is sufficient to compute G_n^1, halting the computation when $G_{n+1}^1 = G_n^1$. This is the classical "cleaning" algorithm for context-free word grammars. Of course the algorithm terminates in this case, since, for each non-terminal and for every n, $G_n^k(N)$ has less than k elements.
In the general case, $k = 1$ is not sufficient. But there exists a constant k having the same property. This comes from the fact that a conjunction of disequations in normal form is always satisfiable when its variables are allowed to be substituted by a sufficiently large number of values.

Let us show how our algorithm works on the previous example:

	G_1^k	G_2^k	G_3^k
$x_1 + x_2$	–	–	–
$s(x)$	–	$f(0)$	$f(0)$
0	0	0	0
$s(s(x))$	–	–	–
$s(x_1 + x_2)$	–	–	–

$G_3^k = G_4^k$ and the algorithm stops. Thus $x_1 + x_2$ is inductively reducible (as well as $s(s(x))$ and $s(x_1 + x_2)$).

Note that the computation does not depend on k in this example, since, for every non terminal in this NF-grammar, the generated language is finite. This is obviously not always the case.

5 Automatic transformation of many-sorted equational specifications

We only sketch this application on an example. (S, F', E) is a specification of the integers together with an equality function. $S = \{int, bool\}$, $F = \{0 : \rightarrow int;\ succ, pred : int \rightarrow int;\ true, false : \rightarrow bool\}$, $F' = F \cup \{eq : int \times int \rightarrow bool\}$, $E = \{succ(pred(x)) = x;\ pred(succ(x)) = x\}$. We have the following negative result:

Proposition 4 *There is no finite set of equational axioms E' (on $T(F', X)$) containing E and such that $\forall t, t' \in T_{F'}$, $(t =_{E'} t' \Leftrightarrow eq(t,t') =_{E'} true)$ and $(t \neq_{E'} t' \Leftrightarrow eq(t,t') =_{E'} false$.*

This means that, given an equational many-sorted specification of eq, either $true = false$ can be deduced from the equations or there are at least three distinct booleans.

This example shows that many-sorted specifications are not powerful enough. This problem, as well as the "constructor-selector" problem, can be solved using *order-sorted specifications* [GM87].

Now, using the language of ground normal forms, it is possible to give $T_{F,E}$ a "canonical" free order-sorted algebra structure and therefore solve our problem. We assume now that the TRS associated with the specification is canonical and that its left hand sides are linear. It is then possible to compute a linear presentation Π^* of NF (proposition 3). Let S' be the set $NT(\Pi^*)$ and $\leq$ the subsumption ordering on S' (recall that the element of S' are non equivalent terms). Let $t \equiv f(t_1, \ldots, t_n) \in S'$. A problem $\mathcal{P} \in IR(t)$ can be written $\mathcal{P} \equiv x_1 = u_1 \wedge \ldots \wedge x_n = u_n$. Moreover, every term u_i is equivalent to a term $v_i \in S'$. For every symbol operator f, let then $\tau(f)$ be the set of strings $(v_1, \ldots v_n, t) \in S'^* \times S'$ obtained in the same way as above. Let F' be the set of function symbols, together with the typing function τ.

Proposition 5 *NF is an $(S', \leq, F')$-algebra.*[6]

Since NF is also an (S, F)-algebra in a straightforward way (for every operator f and every terms $t_1, \ldots, t_n \in NF$, $f_{NF}(t_1, \ldots, t_n)$ is the normal form of $f(t_1, \ldots t_n)$), we have a "canonical" representation of $T_{F,E}$ as a free order-sorted algebra.

In the example of the integers, the presentation Π^* may be written:

$$\begin{array}{rcl|l|l} NF & \rightarrow & NF_0 & NF_{succ(x)} & NF_{pred(x)} \\ NF_0 & \rightarrow & 0 & & \\ NF_{succ(x)} & \rightarrow & succ(NF_0) & succ(NF_{succ(x)}) & \\ NF_{pred(x)} & \rightarrow & pred(NF_0) & pred(NF_{pred(x)}) & \end{array}$$

The set of sorts S' is more usually written $S' = \{int, pos, neg, zer\}$. The subsumption ordering gives the sorts ordering: $pos \leq int$; $neg \leq int$; $zer \leq int$. In the same way F' is written:

$$0 : \; \rightarrow zer \qquad pred : \begin{array}{ccc} zer & \rightarrow & neg \\ neg & \rightarrow & neg \end{array} \qquad succ : \begin{array}{ccc} zer & \rightarrow & pos \\ pos & \rightarrow & pos \end{array}$$

In such a specification, it is very easy to correctly specify the equality: it is now possible to give the equation $eq(succ(x_{pos}), pred(y_{pred})) = false$ (for example).[7]

6 Conclusion

We have shown that the presentations of NF provide a powerful tool for dealing with initial algebras. The two applications we described are only the *first* two applications which came in mind. For example, some programming languages use an algebraic approach for the structure of types. They use then the notion of "constructors" for functions definition ([Tho86]). When some equations between constructors are allowed, the presentation of NF provide a clean operational semantics [Sch88] . It remains also to extend the results of section 5 to the non-linear case.

[6] Let us recall that this property holds for canonical left linear term rewritig systems

[7] Of course, *pred* is no longer defined on the positive integers. However, it is now possible to define the function *pred'* with the relations $pred'(succ(x_{pos})) = x_{pos}$, $pred'(0) = pred(0)$, $pred'(pred(x_{neg})) = pred(pred(x_{neg}))$. This new presentation provides a complete definition of *pred'*. The new presentation is thus split into "constructors" and "defined operators", allowing a complete axiomatization of *eq*.

References

[CL88] H. Comon and P. Lescanne. *Equational Problems and Disunification*. Research Report Lifia 82 Imag 727, Univ. Grenoble, May 1988.

[Col84] A. Colmerauer. Equations and inequations on finite and infinite trees. In *FGCS'84 Proceedings*, pages 85–99, November 1984.

[Com86] H. Comon. Sufficient completeness, term rewriting systems and anti-unification. In *Proc. 8th Conf. on Automated Deduction, Oxford, LNCS 230*, pages 128–140, Springer-Verlag, 1986.

[Com88] H. Comon. *Unification et Disunification: Théorie et Applications*. Thèse de Doctorat, I.N.P. de Grenoble, France, 1988.

[CR87] H. Comon and J.-L. Remy. *How to Characterize the Language of Ground Normal Forms*. Research Report 676, INRIA, June 1987.

[GB85] J. H. Gallier and R. V. Book. Reductions in tree replacement systems. *Theoretical Computer Science*, 37:123–150, 1985.

[GM87] J. Goguen and J. Meseguer. Models and equality for logical programming. In *Proc. CFLP, Pisa, LNCS 250*, Springer-Verlag, March 1987.

[GTW78] J. A. Goguen, J. W. Thatcher, and E. G. Wagner. An initial algebra approach to the specification, correctness and implementation of abstract data types. In *Current Trends in Programming Methodology, vol. 4*, pages 80–149, Prentice Hall Int., 1978.

[HH82] G. Huet and J.-M. Hullot. Proofs by induction in equational theories with constructors. *Journal of Computer and System Sciences*, 25(2), 1982.

[JK86] J.-P. Jouannaud and E. Kounalis. Automatic proofs by induction in equational theories without constructors. In *Proc. 1st IEEE Symp. Logic in Computer Science, Cambridge, Mass.*, June 1986.

[KL87] C. Kirchner and P. Lescanne. Solving disequations. In *Proc. 2nd IEEE Symp. Logic in Computer Science, Ithaca, NY*, pages 347–352, 1987.

[KNZ85] D. Kapur, P. Narendran, and H. Zhang. *On Sufficient Completeness and Related Properties of Term Rewriting Systems*. Research Report, General Electric Company, October 1985. Preprint.

[KNZ86] D. Kapur, P. Narendran, and H. Zhang. Complexity of sufficient completeness. 1986. To appear in TCS.

[LLT86] A. Lazrek, P. Lescanne, and J.-J. Thiel. *Proving Inductive Equalities. Algorithms and Implementation*. Research Report, CRIN, Nancy, France, 1986. To appear in Information and Control.

[Mah88] M. J. Maher. Complete axiomatization of the algebra of finite, rational and infinite trees. January 1988. Draft Paper.

[MG85] J. Meseguer and J. Goguen. Initiality, induction and computability. In M. Nivat and J. Reynolds, editors, *Algebraic Methods in Semantics*, chapter 14, Cambridge Univ. Press, 1985.

[NW82] T. Nipkow and G. Weikum. A decidability result about sufficient completeness of axiomatically specified abstract data types. In *Proc. 6th GI. Conf.*, Springer-Verlag, 1982.

[Pla85] D. Plaisted. Semantic confluence tests and completion methods. *Information and Control*, 65:182–215, 1985.

[Sch88] Ph. Schnoebelen. *Refined Compilation of Pattern-Matching for Functional Languages*. Research Report Lifia 71 Imag 715, Univ. Grenoble, April 1988. Submitted for publication.

[Thi84] J. J. Thiel. Stop loosing sleep over incomplete specifications. In *Proc. ACM Symp. Principles of Programming Languages*, 1984.

[Tho86] S. Thompson. Laws in Miranda. In *Proc. ACM Conf. Lisp and Functional Programming*, Cambridge, Mass., August 1986.

BOOLEAN VALUED MODELS AND INCOMPLETE SPECIFICATIONS

by Bernd I. Dahn[1)]

Introduction

Logic programming has favoured universal Horn theories for several reasons. An advantage of consistent universal Horn theories, compared with arbitrary consistent theories, is the fact that they possess a "typical" model - the least Herbrand model. More precisely, in order to verify that a sentence of the form $\exists x_1 \dots \exists x_n A$, where A is a conjunction of atomic formulas (i.e. positive literals) is a consequence of a universal Horn theory $\mathcal{T}$, it suffices to prove that this sentence holds in the least Herbrand model of $\mathcal{T}$.

Moreover this distinguished model of $\mathcal{T}$ can be characterized algebraically as the initial object in the category of all models of $\mathcal{T}$ which has homomorphisms as morphisms.

It is the aim of this paper to show, that every consistent universal theory has a Boolean valued model with similar properties. This model will be described and it's principal properties will be discussed. Technical details and proofs can be found in [2]. Boolean valued models provide a natural tool for modelling incompletely specified objects, such as those arising from resolution proofs of existential formulas from non-Horn theories. Green's procedure [3] for extracting such specifications from proofs is refined to yield a complete specification of an object in a Boolean valued model.

Throughout the paper we fix some countable signature σ having at least one constant symbol and a consistent universal theory $\mathcal{T}$ of signature σ. σ_f denotes the signature obtained from σ by deleting all relational symbols. Unless stated otherwise all formulas are assumed to be first order formulas without identity.

Boolean Valued Models

A Boolean algebra $\mathfrak{B}$ is said to be *complete* if every nonempty set X of elements of $\mathfrak{B}$ has a least upper bound sup(X) in $\mathfrak{B}$ (and hence also a greatest lower bound inf(X)). We put $\sup(\emptyset) = 0$ and $\inf(\emptyset) = 1$.

We mention that in any complete Boolean algebra $\mathfrak{B} = (B, 0, 1, \cap, \cup, -)$

$$y \cap \sup(X) = \sup(\{ y \cap x : x \in X \}) \text{ and } y \cup \inf(X) = \inf(\{ y \cup x : x \in X \})$$

(see [6] for the required results on complete Boolean algebras). The natural partial ordering in

1) Sektion Mathematik der Humboldt-Universität, DDR-1086 Berlin, PSF 1297

Boolean algebras will be denoted by $\leq$. If an ambiguity could arise, we shall explicitly mention the structure from which we use the partial ordering.

A *Boolean valued structure* $\mathfrak{A} = (\mathfrak{A}_f, \mathfrak{B}, [\ ])$ *of signature* σ *with universe* U consists of an interpretation $\mathfrak{A}_f$ of σ_f with underlying set U, a complete Boolean algebra $\mathfrak{B}$ and a map $[\ ]$ assigning to each atomic sentence A of signature $\sigma(U)$ an element $[A]$ of $\mathfrak{B}$. The map $[\ ]$ is extended in a canonic way to all sentences of the signature $\sigma(U)$ putting especially

$$[\exists x A(x)] = \sup(\{[A(u)] : u \in U\}) \text{ and}$$

$$[\forall x A(x)] = \inf(\{[A(u)] : u \in U\}).$$

A is said to *hold in* $\mathfrak{A}$ ($\mathfrak{A} \models A$) if $[A] = 1$.

A natural way to construct a Boolean valued model of $\mathcal{T}$ is as follows. Let $\mathfrak{B}_0$ denote the Lindenbaum algebra of all quantifier free sentences of $\mathcal{T}$., i.e. the universe of $\mathfrak{B}_0$ consists of all equivalence classes of quantifier free sentences of signature σ modulo $\mathcal{T}$. It is well known (see [4]) that each Boolean algebra $\mathfrak{B}$ has a unique (up to isomorphism) extension to a complete Boolean algebra which is minimal among all complete Boolean algebras extending $\mathfrak{B}$. Let $\mathfrak{B}^*$ be this completion of $\mathfrak{B}_0$. By $\mathfrak{A}_f$ we denote the Herbrand interpretation of σ_f. Taking for each atomic sentence A it's equivalence class modulo $\mathcal{T}$ as $[A]$, we obtain a Boolean valued structure $(\mathfrak{A}_f, \mathfrak{B}^*, [\ ])$ which is easily seen to be a model of $\mathcal{T}$. However, this Boolean valued structure will, in general, satisfy more existential formulas than $\mathcal{T}$ can prove.

In fact, consider a signature consisting of a single unary relation symbol r and countably many constants c_n. As theory $\mathcal{T}$ we consider the empty set. The Lindenbaum algebra of the quantifier free sentences would be isomorphic to the Lindenbaum algebra of propositional logic. If the class of A in the latter algebra would be an upper bound for all the classes of propositional variables, then A would be a tautology. By isomorphy we get that in the completion of the Lindenbaum algebra mod $\mathcal{T}$

$$\sup\{[r(c_n)] : n \in \mathbb{N}\} = 1.$$

Therefore, if we interpret the existential quantifier by the supremum, $[\exists x\, r(x)] = 1$. However, $\exists x\, r(x)$ is not derivable from the empty theory.

In order to describe a Boolean valued structure satisfying exactly the existential sentences provable from $\mathcal{T}$, we need some topological concepts.

For a set X in a topological space int(X) and cl(X) denote the interior and the closure of X respectively. Such a set X is said to be *regular open* if X = int(cl(X)). From [6], Theorem 2.6, we know the following

Lemma 1. For every topological space $\mathcal{X}$ the set of all regular open sets of $\mathcal{X}$ is a complete Boolean algebra $\mathcal{R}(\mathcal{X})$ with $0 = \emptyset$, $1 = \mathcal{X}$, and with intersection $*$, union $+$ and complement $-$ defined by

$X * Y = X \cap Y$,

$X + Y = int(cl(X \cup Y))$,

$-X = int(cl(\mathcal{X} \setminus X))$ respectively.

We observe, that by definition of the intersection, the partial ordering of a Boolean algebra of regular open sets coincides with set inclusion. Nevertheless, inf and sup can be different from set theoretic intersection and union. E.g. in the algebra of regular open sets of the real line with the usual topology the infimum of the set of all open intervalls containing 0 is the empty set, while the set theoretic intersection is $\{0\}$.

Let HMod$(\mathcal{T})$ be the set of all Herbrand models of $\mathcal{T}$. For each universal sentence A let $\mathcal{U}_A$ denote the set of all Herbrand models of $\mathcal{T} \cup \{A\}$. These sets $\mathcal{U}_A$ generate a topology on HMod$(\mathcal{T})$ making up a topological space $\mathcal{HMod}(\mathcal{T})$.

It can be shown that every set $\mathcal{U}_A$ is a regular open set in this space. Moreover, every regular open set $\mathcal{V}$ is the supremum of the $\mathcal{U}_A \subseteq \mathcal{V}$.

Now let $\mathcal{H}(\mathcal{T})$ be the Boolean valued structure $(\mathcal{U}_f(\mathcal{T}), \mathcal{R}(\mathcal{HMod}(\mathcal{T})), [\ ])$ where $\mathcal{U}_f(\mathcal{T})$ is the Herbrand interpretation of o_f and for each atomic sentence A of the signature σ $[A] = \mathcal{U}_A$. Let $\mathcal{B}(\mathcal{T}) = \mathcal{R}(\mathcal{HMod}(\mathcal{T}))$. It can be shown that $[A] = \mathcal{U}_A$ even for each universal sentence.

Theorem 2. For every existential sentence A of signature σ

$$\mathcal{H}(\mathcal{T}) \models A \text{ if and only if } \mathcal{T} \models A.$$

Especially, if $\mathcal{T} \models A \to B$, where A and B are quantifier free sentences, then $[A] \leq [B]$, hence $\mathcal{R}(\mathcal{HMod}(\mathcal{T}))$ contains an isomorphic copy of the Lindenbaum algebra of quantifier free sentences.

Let us consider some examples. In the case of propositional logic, where each propositional variable is considered as a 0-ary relation symbol, every universal sentence is equivalent to a quantifier free sentence. Since every regular open set is the supremum of the sets $\mathcal{U}_A$ which it contains, $\mathcal{B}(\mathcal{T})$ is in this case the completion of the Lindenbaum algebra of quantifier free sentences of signature σ.

In order to discuss a more complicated example, it is useful to introduce the technique of forcing which is wellknown from set theory and model theory. Details can be found in [4], [6], [2].

Let U denote the set of ground terms of signature σ. The *canonic forcing structure* of signature σ is the triple $\mathcal{U} = (\mathcal{U}_f, \mathfrak{C}, \Vdash)$ such that $\mathcal{U}_f$ is the Herbrand interpretation of signature σ, $\mathfrak{C} = (C, \leq, \emptyset)$ where C is the set of all finite sets p of universal sentences such that $\mathcal{T} \cup p$ is consistent, $\leq$ is the partial ordering on C such that $p \leq q$ if and only if $q \subseteq p$ and $\Vdash$ is the relation between elements of C and atomic sentences of signature σ defined by $p \Vdash A$ if and only if $A \in p$. The elements of C are called *forcing conditions* and $p \Vdash A$ is read "condition p forces A".

The forcing relation is extended to arbitrary sentences of the signature σ by the following definitions.

$p \Vdash \neg A$ if and only if for no $q \leq p$ $q \Vdash A$.

$p \Vdash A \wedge B$ if and only if $p \Vdash A$ and $p \Vdash B$.

$p \Vdash A \vee B$ if and only if $p \Vdash A$ or $p \Vdash B$.

$p \Vdash A \rightarrow B$ if and only if for each $q \leq p$ $q \Vdash A$ implies $q \Vdash B$.

$p \Vdash A \leftrightarrow B$ if and only if $p \Vdash A \rightarrow B$ and $p \Vdash B \rightarrow A$.

$p \Vdash \exists x A(x)$ if and only if $p \Vdash A(c)$ for come $c \in U$.

$p \Vdash \forall x A(x)$ if and only if for each $c \in U$ and $q \leq p$ there is some $r \leq q$ such that $r \Vdash A(c)$.

Definition. A nonempty set $G \subseteq C$ of conditions is said to be *generic* (respectively *Δ-generic*) if and only if

1) for all $p, q \in G$ there is some $r \in G$ such that $r \leq p$ and $r \leq q$,

2) $p \in G$ and $p \leq q$ implies $q \in G$,

3) for all sentences (respectively for all atomic sentences) A of the signature $\sigma(U)$ there is some $p \in G$ such that $p \Vdash A$ or $p \Vdash \neg A$.

For each (Δ-)generic set G let $\mathfrak{U}_G$ denote the Herbrand interpretation which is such that for each atomic formula A of the signature σ $\mathfrak{U}_G \models A$ if and only if some $p \in G$ forces A. The structures $\mathfrak{U}_G$ are called the *(Δ-)generic structures* of $\mathfrak{U}$. Note that our concept of a generic structure is similar to the one used in set theory but is different from the notion of a generic model used by Makowsky in [S].

Theorem 3. For each $p \in C$ the set of Δ-generic structures $\mathfrak{U}_G$ for $p \in G$ is exactly the set of all Herbrand models of $\mathcal{T} \cup p$.

Corollary 4. For each $p \in C$ and for each quantifier free sentence A of the signature σ

$$p \Vdash \neg\neg A \text{ if and only if } \mathcal{T} \cup p \vdash A.$$

Lemma 5. If G is (Δ-)generic, then for all (quantifier free) sentences A of the signature σ $\mathfrak{U}_G \models A$ if and only if there is some $p \in G$ such that $p \Vdash A$.

Lemma 6. For each $p \in C$ and each (universal) sentence A of the signature σ

$p \Vdash A$ if and only if $\mathfrak{U}_G \models A$ for each (Δ-)generic set G containing p.

For each set $p \in C$ let $\mathcal{U}_p$ be the set of all Herbrand models of $\mathcal{T} \cup p$.

Lemma 7. For each sentence A and for all forcing conditions p

$p \Vdash \neg\neg A$ if and only if $\mathcal{U}_p \subseteq [A]$ in $\mathcal{H}(\mathcal{T})$.

Especially, $\emptyset \Vdash \neg\neg A$ if and only if $\mathcal{H}(\mathcal{T}) \models A$.

Now, let $\mathcal{T}$ be the theory of linear orderings, formulated in a signature σ, which has binary relation symbols $<$, $\equiv$ and countably many constants c_n ($n \in \mathbb{N}$). We shall show, that $\mathcal{H}(\mathcal{T})$ is a Boolean valued model of the theory of *discrete* linear orderings *with end points*.

In order to see that $\mathcal{H}(\mathcal{T}) \models \exists x \forall y\,(y < x \vee y \equiv x)$, it suffices to prove that below each forcing condition p there is some condition q forcing this sentence. Since $\mathcal{T} \cup p$ has a finite model, since it contains only finitely many constants. If c_n denotes the maximal element of this particular linear ordering, then $q = p \cup \{ \forall y\,(y < c_n \vee y \equiv c_n) \}$ has the desired property.

$\mathcal{H}(\mathcal{T}) \models \exists x\, \forall y\,(x < y \vee y \equiv x)$ is shown in a similar way.

Now, for any c_n and p, if $p \Vdash \neg \exists x\,(c_n < x)$, then

$$p \Vdash \exists x\,(c_n < x) \rightarrow \exists x(c_n < x \wedge \forall y\, \neg(c_n < y \wedge y < x)\,).$$

Otherwise there is some $p_1 \leq p$ forcing $\exists x\,(c_n < x)$, say $p_1 \Vdash c_n < c_m$, i.e. $(c_n < c_m) \in p_1$. Again we consider a finite model of $\mathcal{T} \cup p_1$, where some c_k denotes an immediate successor of c_n. Now $q = p_1 \cup \{ \forall y\, \neg(c_n < y \wedge y < x) \}$ is a condition $\leq p$ forcing that c_n has an immediate successor.

$\mathcal{H}(\mathcal{T})$ can be characterized as an initial object in some category of Boolean valued models of $\mathcal{T}$. Since, in the context of arbitrary universal theories, positive literals are not essentially different from to negative literals, as they are for Horn theories, the morphisms in the category to consider will be more similar to isomorphic embeddings than to homomorphisms.

Let $\mathfrak{A} = (\mathfrak{A}_f, \mathfrak{B}, [\ ])$ and $\mathfrak{A} = (\mathfrak{A}_f, \mathfrak{B}\ , [\ ]\ )$ be Boolean valued structures of signature σ with universes U and U^* respectively. A *Δ-morphism* (*Σ-morphism*) from $\mathfrak{A}$ into $\mathfrak{A}^*$ is a homomorphism from $\mathfrak{A}_f$ into $\mathfrak{A}_f^*$ such that for all quantifier free (respectively existential) formulas $A(x_1, \ldots, x_n)$ of signature σ and for all $u_1, \ldots, u_n \in U$

$$\mathfrak{A} \models A(u_1, \ldots, u_n) \text{ implies } \mathfrak{A}^* \models A(h(u_1), \ldots, h(u_n)).$$

If $\mathfrak{K}$ is a class of Boolean valued structures and $\mathfrak{A} \in \mathfrak{K}$, then $\mathfrak{A}$ is said to be *Σ in $\mathfrak{K}$* if every Δ-morphism of $\mathfrak{A}$ into some structure in $\mathfrak{K}$ is a Σ-morphism. Obviously, for classes of 2-valued structures every Δ-morphism is a Σ-morphism.

$\mathcal{B}Mod(\mathcal{T})$ denotes the class of all Boolean valued models of $\mathcal{T}$.

Proposition 8. For every universal theory $\mathcal{T}$ and every Boolean valued model $\mathfrak{A}$ of $\mathcal{T}$ with universe U the following conditions are equivalent.

1) $\mathfrak{A}$ is Σ in $\mathcal{B}Mod(\mathcal{T})$.

2) For every quantifier free formula $A(x_1, \ldots, x_n)$ of signature $\sigma(U)$ such that $\mathfrak{A} \models \exists x_1 \ldots \exists x_n\, A(x\ , \ldots, x_n)$ there are $u_{11}, \ldots, u_{1n}, \ldots, u_{k1}, \ldots, u_{kn} \in U$ such that $\mathfrak{A} \models A(u_{11}, \ldots, u_{1n}) \vee \ldots \vee A(u_{k1}, \ldots, u_{kn})$.

This proposition combined with Herbrand's Theorem yields immediately the following

Theorem 9. For every consistent universal theory $\mathcal{T}$ the Boolean valued model $\mathcal{H}(\mathcal{T})$ is Σ in $\mathcal{B}Mod(\mathcal{T})$.

If $\mathfrak{K}$ is a class of Boolean valued structures then $\mathfrak{A} \in \mathfrak{K}$ is said to be *initial in $\mathfrak{K}$* if for each $\mathfrak{A}^* \in \mathfrak{K}$ there is a unique Σ-morphism from $\mathfrak{A}$ into $\mathfrak{A}^*$.

Theorem 10. For every consistent universal theory $\mathcal{T}$ the Boolean valued model $\mathcal{H}(\mathcal{T})$ is initial in $\mathcal{BMod}(\mathcal{T})$.

By the previous theorem, $\mathcal{H}(\mathcal{T})$ is also initial in the category using Δ-morphisms. From one initial model we can produce many non-isomorphic initial models by extending it's complete Boolean algebra of values. However, the following theorem shows, that initial models can be obtained only in this way. The proof of this theorem makes essential use of the properties of Σ-morphisms.

Theorem 11. Let $\mathcal{T}$ be a consistent universal theory. For each Boolean valued structure $\mathfrak{A} = (\mathfrak{A}_f, \mathfrak{B}, [\ ])$ which is initial in $\mathcal{BMod}(\mathcal{T})$ the Σ-morphism h from $\mathcal{H}(\mathcal{T})$ into $\mathfrak{A}$ can be augmented by a mapping v from $C(\mathcal{T})$ into the universe B of $\mathfrak{B}$ such that

1) for each universal formula $A(x_1, \ldots, x_n)$ of signature σ and for all $u_1, \ldots, u_n \in U(\mathcal{T})$ v maps the value of $A(u_1, \ldots, u_n)$ in $\mathcal{H}(\mathcal{T})$ to the value of $A(h(u_1), \ldots, h(u_n))$ in $\mathfrak{A}$.

2) v is an isomorphic embedding of the partial ordering of $\mathcal{B}(C(\mathcal{T}))$ into the partial ordering of $\mathfrak{B}$.

3) for all $b_1, b_2 \in |\mathfrak{B}(\mathcal{T})|$

$$b_1 \cap b_2 = 0 \text{ in } \mathfrak{B}(\mathcal{T}) \text{ if and only if } v(b_1) \cap v(b_2) = 0 \text{ in } \mathfrak{B}.$$

$\mathcal{H}(\mathcal{T})$ is up to isomorphy the only initial structure in $\mathcal{BMod}(\mathcal{T})$ with this property.

We mention that the least Herbrand model of a universal Horn theory can be characterized in a similar way, using mappings preserving the validity of existential formulas whose quantifier free part is a conjunction of atomic formulas instead of the Σ-morphisms.

Incomplete Specifications

By Herbrand's Theorem we can from a universal theory $\mathcal{T}$ in general obtain incomplete specifications of objects whose existence $\mathcal{T}$ proves. Boolean valued structures can be used to represent such incompletely specified objects.

From now on we asume that σ contains a binary relation symbol $\equiv$ and that $\mathcal{T}$ has axioms saying that $\equiv$ is a congruence relation.

Let $\mathfrak{A}=(\mathfrak{A}_f, \mathfrak{B}, [\ ])$ be a Boolean valued model of $\mathcal{T}$ with universe U and let X be a partition of unity of $\mathfrak{B}$, i.e. $X \subseteq |\mathfrak{B}|$, $\sup(X)=1$ and $b_1 \cap b_2 = 0$ for all $b_1, b_2 \in X$ such that $b_1 \neq b_2$. Let for each $b \in |\mathfrak{B}|$ be given some $c_b \in U$ and let c be a new element, i.e. $c \notin U$. Let $\mathfrak{A}_f(c)$ be the extension of $\mathfrak{A}_f$ which is free generated by c over $\mathfrak{A}_f$.

We construct a Boolean valued structure $\mathfrak{A}^*=(\mathfrak{A}_f, \mathfrak{B}, [\ ])$ putting

$$[A(c)]=\sup \{ [A(c_b)] \cap b : b \in X \} \qquad (1)$$

for each atomic formula A(x) of signature σ. Especially, we obtain that

$$\sup \{ [c \equiv c_b] : b \in X \} = 1.$$

An easy induction shows, that (1) holds in fact for arbitrary formulas of signature σ.

If $\mathfrak{A} \models A(c_1) \vee \ldots \vee A(c_n)$ and if $X = \{ b_1, \ldots, b_n \}$ is a partition of 1 such that $b_i \leq [A(c_i)]$ for all $i = 1, \ldots, n$, then $\mathfrak{A}^* \models A(c)$.

In general there are many possibilities for chosing appropriate partitions of 1, e.g. the trivial way putting $b_i = \bigcap \{ -[A(c_i)] : i < j \} \cap [A(c_j)]$. We describe a procedure that can be used in connection with a backward chaining resolution based deduction system and has the advantage, that it yields evaluable candidates for the conditions b_i already before the completion of the proof of $\exists x A(x)$. This procedure is a refinement of Greens procedure (see [3]) for extracting possible answers from resolution proofs.

For each literal L -L denotes the literal dual to L. If C is a clause or a conjunction of clauses, then -C denotes the conjunction of literals, repectively the clause, which is logically equivalent with $\neg$ C.

We assume that $\mathcal{T}$ is a consistent theory of universally quantified clauses (i.e. disjunctions of literals) and that $G(\mathfrak{x})$ is a conjunction of literals such that $\mathcal{T} \models \exists \mathfrak{x} G(\mathfrak{x})$.

We consider lists $L = [[D_1,t_1], \ldots, [D_m,t_m]]$, where $D_1, \ldots, D_m$ are clauses and $t_1, \ldots, t_m$ are sequences of terms of the same length as $\mathfrak{x}$. For j=1,...,m let $E_j = \forall(\bigwedge\{D_i : i < j\} \wedge \neg D_j \rightarrow G(t_j))$.

Such a list L is said to be a *solution*, if all these sentences E_j are consequences of $\mathcal{T}$. Note the following simple lemma.

Lemma 12. a) If L is a solution and S is a substitution, then LS is a solution.

b) If L is a solution, then each initial segment of L is a solution.

c) If L and L^* are solutions, then the list obtained by concatenating L and L^* is a solution.

d) If L is a solution and C is a clause such that L contains several members of the form [C, _], and if L^* is obtained from L by cancelling the last occurence of a [C, _], then L^* is also a solution.

There is a derivation $C_0, \ldots, C_n$ of the empty clause $C_n = \Box$ by linear resolution from $C_0 = -G$ (see [1]). During the construction of this derivation we keep track of the actual instantiation of the variables in $\mathfrak{x}$ occuring in the goal G and of an additional variable L which represents the empty list at the beginnig. Let u_i denote the value of $\mathfrak{x}$ when C_i is constructed. At this time L will be instantiated to a list L_i such that we obtain a solution when we append $[[C_i, u_i]]$ to L_i.

When the deduction system performs an input resolution step to construct C_{i+1}, i.e. one of the parents of the resolvent comes from an axiom of $\mathcal{T}$, L_{i+1} is obtained from L_i by applying to it the most general unifier used for the resolution.
If this step is not a step of input resolution, then it uses for resolution C_{i-1} one of it's ancestors C_k Let S denote the most general unifier used for the resolution. Then L_{i+1} is obtained from $L_i S$ by appending $L_k S$ and $[[C_k,u_k]]S$.

In each case the Lemma 12 shows, that we obtain a solution, when we append $[[C_{i+1}, u_{i+1}]]$ to L_{i+1}.

We note, that by Lemma 12 d) the solutions can be simplified by eliminating multiple occurences of [C, _]. In fact, if the deductive system uses an ancestor C_k to construct L_{i+1} as above, then for any variable free clause C such that L_k has a member of the form [C, _], L_i must have already a member of the same form which need not be duplicated.

Continuing this procedure, the members of L_{i+1} are further instantiated. Nevertheless, a correct solution has already been obtained at this stage and any other process running in parallel can make use of it, if it finds the constructed conditions specific enough for it's purposes.

When $[[D_1,t_1], \ldots, [D_m,t_m]]$ is the list constructed in parallel of a refutation of $-G(\mathfrak{x})$ from $\mathcal{T}$ by linear resolution, and if $\mathfrak{y}$ is the value of $\mathfrak{x}$ after this refutation, then for each Boolean valued model $\mathfrak{A}$ of $\mathcal{T}$ and for each instantiation S with elements of the universe of $\mathfrak{A}$ we have that

$$[(\bigwedge\{D_i : i < j\} \wedge \neg D_j)S] \leq [G(t_j)S] \text{ for each } j \leq m+1$$

where D_{m+1} = fail and $t_{j+1} = \mathfrak{y}S$.

The objects representing the incompletely specified solutions to the call of $G(\mathfrak{x})$ can be introduced syntactically by Skolem functions as follows.

If G has k variables and $\mathfrak{z}$ is the tuple of variables occuring in the final solution, we introduce new function symbols $f_1, \ldots, f_k$ whose arity is the length of $\mathfrak{z}$. When we add to $\mathcal{T}$ the axioms

$$\forall\mathfrak{z}((\bigwedge\{D_i : i < j\} \wedge \neg D_j)S \rightarrow \mathfrak{f}(\mathfrak{z}) = t_j) \ (j = 1, \ldots, m+1),$$

where $\mathfrak{f}(\mathfrak{z})$ abreviates $[f_1(\mathfrak{z}), \ldots, f_k(\mathfrak{z})]$, we obtain a conservative extension (Skolem extension) of $\mathcal{T}$ which can be transformed to clausal form and treated by any of the variants of resolution that have been developed for handling equations. $\forall\mathfrak{z}G(\mathfrak{f}(\mathfrak{z}))$ would be a theorem of this extension of $\mathcal{T}$.

Literature

[1] Chang, Ch.- L., Lee, R. Ch.-T.: Symbolic Logic and Mechanical Theorem Proving. New York 1973.

[2] Dahn, B. I.: An analog of the least Herbrand model for non-Horn Theories, in: B. Dahn, H. Wolter (eds.), Proc. 6th Easter Conference on Model Theory, Berlin 1988, 33-52.

[3] Green, C.: Theorem proving by resolution as a basis for question-answering systems, in: Meltzer, B., Michie, D. (eds.), Machine Intelligence, Edinburgh 1969, 183-205

[4] Jech, T. J.: Lectures in Set Theory with Particular Emphasis on the Method of Forcing, Springer Lecture Notes in Mathematics, vol. 217, 1971

[5] Makowsky, J. A.: Why Horn Formulas Matter in Computer Science: Initial Structures and Generic Examples. J. Comp. and System Science, vol. 34 (1987), 266-292

[6] Rosser, J. B.: Simplified Independence Proofs, Boolean Valued Models of Set Theory, New York 1969

TYPES, MODULARISATION AND ABSTRACTION IN LOGIC PROGRAMMING

George Dayantis[1]

ABSTRACT

A simple and efficient way that the ideas of typing, modularisation and data abstraction can be realised in a logic programming framework, which also remains within the spirit of logic programming, is proposed here. A polymorphic type system with subtypes is presented and it is shown how it can form the basis for modular structure. The ideas are presented as an extension to PROLOG, which is taken as a practical representative of logic programming. Additionally, they have been implemented in a skeleton language on top of standard PROLOG.

1. INTRODUCTION

Logic programming is supposed to simplify software production by making a clear separation between logic and control and by offering both simple declarative and operational semantics. Logic programs are easier to understand, modify and reason about, due to their declarative reading based on some familiar logic. Thus far, logic programming has given rise to a practical tool, the programming language PROLOG, which is based on the Horn clause subset of first-order logic. Despite its apparent restricted expressive power and its controversial 'impure' features, PROLOG has become very popular and has already been used for large applications. In such applications a serious deficiency of PROLOG becomes evident, namely, its lack of a) any typing facilities, which is a source of errors that are difficult to detect, and b) any structure-imposing mechanisms, which renders large programs unmanageable.

Drawing experience from the lessons learned from conventional software tools, one realises that if PROLOG or any of its successors is to be successful as a tool for serious software development, it has to somehow support the concepts of typing, modularisation and data abstraction without, of course, seriously upsetting the principles of logic programming. A number of researchers have already considered the possibilities and advantages of incorporating such features into a logic programming language. Apart from MPROLOG [Domolki & Szeredi 83], it is worth mentioning the influential work of Goguen et al with EQLOG [Goguen & Meseguer 84], the Japanese response with Himiko [Furukawa, Nakajima & Yonezawa 83], Zaniolo's object-oriented extension to PROLOG [Zaniolo 84] and the most recent proposal to 'dress' PROLOG with ML's module system [Sanella & Wallen 87] .

A rather different, simple and efficient way of incorporating the features of modularisation, parameterisation and data abstraction in a logic programming language like PROLOG is proposed here. The basis of the proposal is the notion of 'type', so a polymorphic type system with subtypes for PROLOG is studied separately in the following section, while the central notion of a 'module' and its features are detailed in section 3. Implementation issues are discussed in section 4 and a discussion and comparison with relevant work follows in section 5.

2. A POLYMORPHIC TYPE SYSTEM FOR PROLOG

Based on an untyped logic, PROLOG is naturally an untyped language. We can see it as having a single type - the term. This allows for much flexibility and is useful for fast prototyping. However, this is also a deficiency when using PROLOG for building large systems, since type errors can be detected only at run-time. An indication of how undesirable this can be is given by the traditional definition of the *append* relation, which is intended to have meaning only with lists for its arguments but from which *append([],1,1)* can be deduced. Additionally, and as far as reasoning about logic programs is concerned, an axiomatisation and explicit use of data types in logic - as firstly proposed in [Clark & Tarnlund 77] - has been proved invaluable (see also [Dayantis 87]). However, the incorporation into logic programs of explicit (run-time) type information in the form of predicates generally leads to longer and more complicated proofs and deductions (see [Walther 83]). On the other hand, the obvious and theoretically elegant extension of the logical basis of PROLOG to *many-sorted* logic is also possible. In [Smolka 86] it is shown how the semantic and deductive methods developed for untyped logic generalise to *order-sorted*[2] Horn logic. In this case, the logic interpreter has to employ a more sophisticated unification

[1] Dept. of Computing, Imperial College, London, SW7 2AZ, UK.

[2] *Order-sorted* means many-sorted with a partial ordering defined over the sorts.

algorithm, thus resulting in loss of efficiency. A more attractive solution would be to impose a type system on Horn logic at a meta-level, i.e. without explicitly moving into many-sorted logic.

Various researchers have already recognised the need for a type system for Prolog and have attempted to satisfy it [Mycroft & O'Keefe 83, Mishra 84]. The question is: can we supply PROLOG with a type system without seriously upsetting the principles of the language? And the answer is yes. Conceptually we need a jump to many-sorted logic. This means that from the user's point we require a type discipline - the user needs to decide and define the types for his predicates. And from the implementation point we only need an enhancement of the compiler with a typechecker.

The type system proposed here is an extension of the one reported in [Mycroft & O'Keefe 83]. It is presented informally in the following two subsections, while section 2.3 deals with typechecking. Due to shortage of space, we refer to [Dayantis 88] for a formal syntax and semantics as well as a more detailed presentation.

2.1. Types as sets.

The notion of a type or sort should be familiar to programmers. Intuitively, a type can be understood as a set of values; in our case as a subset of the Herbrand Universe. The phrase *having a type* or *being of a type* is then interpreted as membership in the appropriate set and is denoted by *V:T* (*V is of type T*). Thus, in order to define a type we can use the same techniques as for a set, i.e. by enumeration of its elements and/or by the use of predicates. For example, we can define the familiar type of lists with the *type definition*: '***type*** *list(T)* ***iff*** *[] L [T/list(T)]*'. Without insisting on the syntax, we interpret this type definition as: 'a term is a list of T's iff it is the empty list ([]) or of the form [X|R], where X is of type T and R is again a list of T's'. This is an inductive definition of the enumeration kind. The alternative term structures ([],[_|_]) used in such a type definition will be referred to as ***subpatterns***, while the set of all subpatterns will make up a *pattern*. Furthermore, in this example, the notions of a *generic type* and *polymorphism* are introduced. The variable *T* is a *type variable*, that can stand for any type. So with a single (generic) definition we define a family of types - lists of anything. By instantiating *T* with a specific type, say ***integer***, we get ***lists of integers*** and so on. This is what we call *parametric polymorphism*.

In order to see how predicates can be used for defining types, first notice that any (untyped) predicate implicitly specifies types for its respective arguments - the sets of values that satisfy it. A simple example would then be the definition of the familiar type of integers, which is built-in in Prolog: '***type*** ***integer*** ***iff*** *(X:any* ***with*** ***integer(X))*** *L* ***(X1:integer + X2:integer)*** *L* ***(X1:integer * X2:integer)*** *L .....*'. This reads: 'anything that satisfies the unary predicate ***integer*** or is a sum of integers or a product of integers or ... is of type integer'. The type ***any*** is the union of all types, the Herbrand Universe. Consider also the generic type of lists with a specific ***length***:

type l_list(T,N:integer) **iff** L:list(T) **with** length(L,N).

Notice that, apart from *type variables*, *value variables* - variables that stand for terms, such as *N* and *L* in the above example - are also allowed in a type definition. Free variables, like *L*, should always be annotated with their types. Any defined predicate or conjunction of these can be used after a *with* keyword with the obvious restriction that no new variables are introduced. Such predicates in type definitions are called *predicate constraints*.

The reason, of course, for structuring our universe in this way is that we may restrict the arguments of all defined relations to range over specific types. In this way, we also associate predicates with types, which we call *extended types*. Thus, we may add declarations of the form :

'***rels*** ***append(list(T),list(T),list(T))***.', '***rels*** ***length(list(T),integer)***.',

before the actual definition of the relations involved. Such declarations will be referred to as *relation declarations*. What this effectively means is that we accept partially defined relations. For example, a PROLOG variable as an argument of a relation is understood to range over the whole domain (matches anything at all), while if the relation is declared to be defined only for arguments of a specific type then it can only match with terms of this type.

2.2. Subtypes.

We can now impose some more structure in our typed Universe by considering the relationships between different types. Two types can either be disjoint or have a common subset. The most interesting relation between types is the ***subtype*** relation; that is, when a type is a subset of another type. Under this set inclusion relation (denoted by $\leq$) the set of all types forms a lattice, whose top element is the type ***any*** and the bottom one is the empty set. Thus, if *V:T1* and $T1 \leq T2$ then *V:T2*. From this follows the property that makes this relation interesting: any relation defined over a type *T2* works for members of any of its subtypes *T1*.

Obviously, any type is a subtype of *any*. Additionally, it should be easy to infer from the definitions in the previous section that *l_list(T,N) < list(T)* for all types *T* and integers *N*. Thus, one way of imposing a subtype relation *T1 < T2* is by first defining *T2* and then defining its subtype *T1* by restricting the subpatterns of *T2* with a predicate constraint. Notice that, in this case, the representation of the two types is virtually the same. In this top-down manner, one can define a type hierarchy in which a type can have more than one subtypes. Additionally, there is a way in which one can define a type hierarchy in a bottom-up fashion. This is more appropriate when a type can be a subtype of more than one type (multiple inheritance). For example, consider the following definitions, which could be part of a database:

<u>**type**</u> car <u>**iff**</u> car1 | car2 | ....
<u>**type**</u> machine <u>**iff**</u> X:car | X:robot | ....
<u>**type**</u> transport <u>**iff**</u> X:car | X:ship | ....

The exact representation of a *car* is not important here. What these definitions imply is that: *car < machine* and *car < transport*. Again here, the representation of *cars* viewed as *machines* or as *transports* remains the same.

Another way of inducing a subtype relation is introduced by allowing a kind of equality between subpatterns of different types. For example, consider the following type definitions concerning familiar geometrical objects:

<u>**type**</u> parallelogram <u>**iff**</u> p(S1:integer,S2:integer,H:integer) <u>**with**</u> S1 =< S2, H =< S1.
<u>**type**</u> square <u>**iff**</u> sq(S:integer) <u>**as**</u> p(S,S,S).

In the first definition, the type *parallelogram* is defined by giving a term representation *p(S1,S2,H)*, where *S1*, *S2* and *H* are the lengths of the two sides and the height of the parallelogram respectively (three measurements, which uniquely identify a parallelogram). Furthermore, some predicate constraints are necessary to uniquely determine the position of these three parameters in the term representation. In the second definition, the type *square* is defined by a new term representation *sq(S)*, where *S* is the (common) length of its sides. Additionally, an equivalent representation is given, after the *as* keyword, relating the square with the parallelogram: a square *sq(S)* can also be regarded as a parallelogram *p(S,S,S)*. These definitions imply *square < parallelogram*, which means that, despite the different representation, we still want *squares* to be regarded as *parallelograms* and have all the operations on *parallelograms* available for them. For example, if we define a relation *perimeter*:

<u>**rels**</u> perimeter(parallelogram,integer).
perimeter(p(S1,S2,H), N) :- N is 2*(S1+S2).

we want to be able to call *perimeter* with a *square* for its first argument, e.g. '*?- perimeter(sq(1),N).*', and have the perimeter of the square returned in its second argument.

Of course, this 'equality' between subpatterns should encompass the whole pattern of the type definition in order to induce a subtype relation. Furthermore, this can be extended to accommodate 'equality' of a subpattern with more than one subpatterns (of different types), thus allowing a type to be a subtype of more than one disjoint type (*multiple inheritance*). In conclusion, a subtype relation may not be declared directly, and it is best left to the system to infer it from the relationships between the subpatterns of the different types.

The subtype relation can also be used to restrict a type variable in a type definition. For example, in order to define lists of anything whose type is a subtype of integers, we simply use :

<u>**type**</u> int_list(T≤ integer) <u>**iff**</u> [] | [T|int_list(T)].

Under this definition the expression *int_list(real)* is not a legal type.

Although subtypes introduce some extra complication in typechecking, as we shall see in the following subsection, they also provide more expressive power and support for the basic concept of *inheritance* in *object-oriented* programming.

2.3. Typechecking.

Typechecking a logic program, with respect to the relation declarations given for it, consists of ensuring that all relations are used with the correct extended type and any answer-substitution will return the correct type. [Mycroft & O'Keefe 83] have formalised the notion of a *well-typed* PROLOG program. However, their formalisation contains some errors. Below, we give a proper formalisation for this notion, which also covers subtypes, but ignores predicate constraints (introduced with *with*), since they cannot be taken into account at compile-time. The way these are handled is discussed in section 4.

Due to the fact that in a PROLOG program all variables are local to a clause we can typecheck each clause separately. Thus:

Definition: A logic program (set of clauses) is *well-typed* iff each of its clauses is *well-typed*.

Intuitively, a clause is well-typed if, and only if, each of its literals in isolation is well-typed, i.e. consistent with its relation declaration, and each variable in the clause can be assigned a unique type. For example, if we assume the declarations:

rels P(integer), Q(list(T),integer).

then the clauses: 'P(0).' and 'P(N) :- Q(L,N).' are well-typed, while the clauses: 'P([]).' and 'P(N) :- Q(N,L).' are ill-typed.

A simple example should be sufficient to illustrate the problem introduced with subtypes. Assume the declarations: **rels** P1(integer), P2(real).

where *integer < real*, and the single clause: P1(X) :- P2(X).

Do we want this to be well-typed ? Notice that if *P1* is called with *X* instantiated (to an integer) then computation can safely proceed, but if it is called with *X* uninstantiated then there is the danger that *P2(X)* may succeed and return a real (non-integer) value for *X*, which violates the type-restrictions for *P1*. Thus, we need to carefully restrict the notion of a well-typing, taking into account the instantiation states of shared variables, when a conflict between subtypes occurs. Although this may seem to result in the loss of the full power of the logical variable, this power is not really required in the cases where such restrictions have to be imposed.

In addition, polymorphism adds another dimension to typechecking. First, it is interesting to note that when a type *T1* is an instance of a polymorphic type *T2*, then it is also the case that: *T1 < T2*, e.g. *list(integer) < list(T)*. So, the same problem that concerns subtypes remains in this case. This happens to be the aspect of polymorphism that has been incorrectly dealt with in [Mycroft & O'Keefe 83]. The extra dimension, which has been handled correctly in that paper, is that shared type parameters should be substituted only with the same type instance. For example, if we assume: **rels** P(T,list(T)), Q(integer,list(list(integer))).

then the clause: P(X,L) :- Q(X,L).

is not well-typed, since the only possible typing assigns *X:integer*, *L:list(list(integer))*, which forces the type variable *T* to unify with both *integer* and *list(integer)*, which is impossible.

Below, we formalise the notion of well-typing in two steps: Firstly, we define the notion of *possible typing*, which is independent of computation and then restrict it to that of *well-typing* by taking into account particular computations.

Let *Q* be a clause or a goal, *PT* an association of extended types to all the predicates and terms in *Q*, and *VT* an association of type/set-of-literals pairs to all the variables appearing in *Q*. Each variable *X* in *Q* is associated with a) an instance of its expected type in some of the literals of *Q* and b) the set of these literals (in which *X* assumes a common type), which we call the set of *determining* literals for *X*. For each predicate *p* of arity *k* in *Q*, *PT* will contain an element of the form: $p(t_1,\ldots,t_k)$. For each functor *f* of arity *k* in *Q*, *PT* will contain an element of the form: $f(t_1,\ldots,t_k) \to t$. For each variable *X* in *Q*, *VT* will contain an element of the form *X:t/D*.

Intuitively, *VT* is a possible typing of *Q* if, and only if, each of its literals is appropriately typed, which means that its extended type agrees with its expected one given in *PT*, while its variables can be assigned to agreeable types. The general rule for such a type agreement is that the type given for a variable in *VT* has to be a subtype of the most general type that can be inferred independently (after full term decomposition) for that variable in any of its occurrences in *Q*. This is formalised as follows:

Definition: If *Q*, *PT* and *VT* are as above, we say that *VT is a possible typing of Q under PT*, and denote it by *VT/PT |- Q*, iff:

$Q = B_0 \leftarrow B_1, B_2,\ldots, B_m$ (where m=0 if no body) and
$VT/PT \vdash_j B_i$ $(i=0,\ldots,m)$

or $Q = B_1, B_2,\ldots, B_m$ and
$VT/PT \vdash_j B_i$ $(i=1,\ldots,m)$

where:

$VT/PT \vdash_j A$ (A is a literal) iff
$A = p(a_1, \ldots, a_k)$ (where k=0 if no arguments)
$p(t_1, \ldots, t_k) \in PT$
$VT/PT \vdash_j (a_i : s_i)$ $(i=1,\ldots,k)$ and
$s_i = \theta(t_i)$, $i=1,\ldots,k$, for some substitution θ.

$VT/PT \vdash_j (u : s)$ (u is a term) iff
$u = f(a_1, \ldots, a_k)$ (where k=0 if u is a constant)
$f(t_1, \ldots, t_k) \to t \in PT$ and
$VT/PT \vdash_j (a_i : s_i)$ $(i=1,\ldots,k)$ and
$s_i = \theta(t_i)$, $i=1,\ldots,k$, and $\theta(s) = t$ for some substitution θ.

VT/PT $\vdash_j$ (X : t) (X is a variable) iff

(X : t / D) ∈ VT and Bj∈ D or

(X : s / D) ∈ VT and Bj∉ D and s < t.

In the above definition, the use of the substitution θ is necessary in order to ensure that shared type parameters get assigned the same type instance.

Notice that, for each clause, there may be none or one possible typing. In the case where there is no possible typing, there can be no well-typing either. Otherwise, the possible typing may also be a well-typing. Below, we define the conditions that a possible typing has to satisfy in order to be a well-typing with respect to a particular computation.

Definition: Let Γ be a computation (sequence of resolution steps), which instantiates a variable X to a term F in literal L (of clause Q). We say that X gets ***adequately*** instantiated in L for computation Γ if, and only if, it can be determined whether or not the type of F is a subtype of the expected type for this argument position in L, regardless of the types of any variables in F.

Definition: A possible typing VT, of a clause Q, is a *well-typing* under some computation Γ if, and only if, for each variable X in Q either i) its set of determining literals is the same as the set of all the literals it appears in - in other words, there are no subtype conflicts - or ii) it gets adequately instantiated in one of its determining literals.

A desirable property of a well-typing is that in a well-typed program no predicate can ever be called with the 'wrong' type or return a term of the 'wrong' type in one of its arguments (soundness). Since the only computation (inference) rule used to execute logic programs is SLD-resolution, we effectively need to prove that SLD-resolution preserves well-typing.

Notice that, in general, resolution does not preserve possible typing. The only case where it fails is when a number of unrelated types are subtypes of the same type. For example, consider the relations P and Q, such that: <u>rels</u> P(machine), Q(robot).

the goal: ?- P(car1).

and the clause: P(X) :- Q(X).

Both have possible typings (VT1={}, VT2={X:robot/Q}), while the resolvent: '?- Q(car1).' does not have a possible typing, since a *car* cannot be a *robot*. However, under similar circumstances, the well-typing is preserved. Following the same example, the clause is well-typed only under a computation that adequately instantiates X in Q (and not in P). Thus, P may not be called with its argument instantiated, as in the example above.

THEOREM: (Soundness of well-typing)

If $R = \ <\!-\ A_1, \ldots, A_n$ is a resolvent (goal) with a possible typing $VT1$, which is a well-typing under a computation $\Gamma 1$, and $Q = B_0 <\!- B_1, B_2, \ldots, B_m$ is a clause with a possible typing $VT2$, which is a well-typing under a computation $\Gamma 2$, such that there exists a substitution $\phi = mgu(B_0, A_1)$, and $\Gamma 1 = resolve(R, Q, \phi){:}\langle\Gamma 2'\rangle{:}\langle\Gamma 1'\rangle$, $\Gamma 2 = apply(\phi, <\!- B_1, B_2, \ldots, B_m){:}\langle\Gamma 2'\rangle$, then there exists a possible typing for the resolvent $R1 = \phi(B_1, \ldots, B_m, A_2, \ldots, A_n)$. which is also a well-typing under $\Gamma = \langle\Gamma 2'\rangle{:}\langle\Gamma 1'\rangle$.

PROOF : ... Omitted; see [Dayantis 88].

Although it may at first seem that such a definition of well-typing is of little use, since it is dependent on particular executions, we will show below how it is still possible to keep most of typechecking at compile-time and so have the confidence that 'well-typed programs do not go wrong'. Obviously, a type system like the one described here can be useful even in its own right. Let us now see how this type scheme blends with our ideas for modularity and abstraction.

3. PARAMETERISED MODULES AND DATA ABSTRACTION

A PROLOG program is just a series of clauses. This 'flatness' of PROLOG renders large programs unmanageable. Of course, the disciplined programmer can impose some structure via clause clustering and commenting. But even simple grouping facilities, such as those offered by MPROLOG [Domolki & Szeredi 83], do not offer much in the way of data abstraction. Here modularity is introduced via the meta-logical notion of a *module*.

In a similar way that the notion of a type structures a universe of values the notion of a *module* structures a universe of clauses. Semantically, a module is something like an abstract data type. It generally consists of a *signature* part and a set of standard PROLOG clauses. A module definition in its complete form centres around a single type definition, which is the main component of the signature. The set of clauses defines all relevant predicates for manipulating this type. Ideally, the defined type and its associated relations constitute a conceptual unit. In a sense, some of the ideas underlying the algebraic specification approach [Berztiss & Thatte 83] have been transferred here. The main difference is that we use clauses instead of equations for the operations (predicates instead of functions) and we make a clear separation between the *basic constructors* and the rest of the operations. The term representation for values of the TOI is given as a *pattern* (there always exists a normal form).

The syntax should be easily conveyed from the examples presented in the figures below, where syntax keywords are underlined and, following the PROLOG convention, variable names begin with a capital letter.

The signature part in its complete form consists of:
(a) The module's name including its parameters, which are enclosed in brackets and separated by ','. It follows the *module* keyword.
(b) Names of previously defined modules that this module uses (parent modules). They follow the *using* keyword and are separated by the *and* keyword.
(c) At most one *type* definition - where the name of the defined type always coincides with the module-name. It follows the *pattern* keyword.
(d) The names and arities of the predicate subpattern constructors - one for each subpattern. They follow the *constr* keyword and are sparated by bars (|).
(e) For each relation defined in the module a *relation declaration* - declaring the intended types for the arguments of the relation. They follow the *rels* keyword and are separated by commas (,).

In figure 1 below, a simple module is presented defining the familiar generic data type *queue* with only two relations on it, followed by an enrichment of it with a further relation (*circular queue*). In figure 2, two modules representing related geometrical objects are presented, an example often used in object-oriented languages.

Notes:
(1) The module-name together with the number of its parameters uniquely identifies a module.
(2) The signature part is all one Prolog 'line' (or term); thus, a '.' should be placed only at the end of it, and one after 'endmodule'. This also means that all variables in the signature are shared.
(3) The part: "*module* <modname> *pattern* <pattern>" is semantically equivalent to a type definition: "*type* <modname> *iff* <pattern>".

```
module queue(Elm)
pattern empty | q(Elm,queue(Elm))
constr emptyq/1 | addq/3
rels read(queue(Elm),Elm), dequeue(queue(Elm),queue(Elm)).
clauses
  read(q(E,empty),E).
  read(q(E1,q(E2,Q)),E) :- read(q(E2,Q),E).
  dequeue(q(E,empty),empty).
  dequeue(q(E1,q(E2,Q)),q(E1,Q1)) :- dequeue(q(E2,Q),Q1).
endmodule.

module cqueue(Elm) using queue(Elm)
rels circulate(queue(Elm),queue(Elm)).
clauses
  circulate(Eq,Eq) :- emptyq(Eq).
  circulate(Q1,Q2) :- read(Q1,E), dequeue(Q1,Q), addq(E,Q,Q2).
endmodule.
```

Figure 1

```
module parallelogram
pattern p(S1:real,S2:real,H:real) with S1 =< S2, H =< S1
constr makeparal/4
rels perimeter(parallelogram,real), area(parallelogram,real).
clauses
  perimeter(p(S1,S2,H),N) :- N is 2*(S1+S2).
  area(p(S1,S2,H),N) :- N is H*S2.
endmodule.

module square using parallelogram
pattern sq(S:real) as makeparal(S,S,S).
endmodule.
```

Figure 2

(4) Each *pattern* is known only inside the module in which it is defined. Access of this pattern from another module is possible only via the constructors, the names of which are given in the 'constr' declaration. In view of this, notice in *square's* pattern how a subpattern equality (subtype relation) has to be denoted now: instead of writing "*pattern sq(S:real) as p(S,S,S)*" we use the parallelogram constructor with its last argument missing (treating it as a function). In this way, since a pattern actually provides a data representation, we ensure representation independence and facilitate data abstraction.
(5) It is meaningful to have a module without a) a pattern declaration - when it simply enriches a previous module with more relations (fig.1), b) a 'constr' declaration - for a private type, c) any clauses; therefore no 'rels'-part either (fig.2).

A program now consists of a series of interdependent modules, which can be thought of as comprising an acyclic graph. We assume the existence of a built-in *root* module at the root of this graph containing all the built-in types (boolean,integer,real,list) and Prolog predicates. We could also provide - without much trouble - some library managing facilities for modules.

As for query evaluation, since the same relation name can be present in different and independent modules - with different meaning - a query evaluation makes sense only with respect to the environment of a specific module. A module's environment consists of the module itself together with the environments of its parents. In order to facilitate this, we introduce a new operator *in* and allow queries of the form: *?- <Goal> in <Module>*. The way query evaluation is implemented is explained in section 4 below.

4. IMPLEMENTATION

4.1. Typechecking.

A typechecking algorithm for the polymorphic type system described in section 2, has been implemented in PROLOG. Here we explain the basic ideas behind this implementation.

The well-typing definition already suggests an algorithm for determining a possible typing (which finds a *VT* given *PT*). Having found a possible typing, we are then interested in determining whether it is also a well-typing. It should be obvious, from the definition of well-typing, that when a subtype conflict occurs - i.e. for some variable X, there are literals in the clause containing X, which are not determining literals for X - the necessary type reconciliation cannot be done, since the notion of instantiation states for variables is involved. However, we can still perform as much static typechecking as possible, and trap unresolved subtype conflicts during execution, but without touching unification. How is this done?

When two consecutive appearances of a variable occur in two literals with respective assigned types *t1* and *t2*, such that *t1<t2* or *t2<t1*, then an extra (system-defined) literal is inserted between them, which will test whether the concerned variable is appropriately instantiated or uninstantiated (as required for the well-typing) when computation passes that point. If the test fails it will abort with a special type-failure message. For example, consider again the case:

rels P1(integer), P2(real).
P1(X) :- P2(X).

This clause is well-typed only when *X:integer*. Since if *P1* is called with its first argument uninstantiated, it may return a non-*integer real*, an extra literal is added in the clause's body:
P1(X) :- instantiated(X), P2(X).
where: *instantiated(X) :- not(var(X)), ! ; (write('Type failure:...'), fail).*
In practice, the code for *instantiated* has to be slightly more complicated than that, in order to capture the notion of adequate instantiation.

In this way, even though one cannot be sure that a successfully compiled program is correctly typed for all different possible computations, any type failure due to incorrect usage will be captured and reported at run-time. Additionally, this trap does not require any modification to the interpreter.

As we stressed in section 2.2 above, the advantage of introducing subtypes is that relations defined over supertypes work for subtypes as well, but not the other way around. When the term representation (pattern) for the two types is the same, as in the case of the *cars* example, then there is no problem. But when different representation is used, as in the *geometry* example, then a problem arises. For example, the square *sq(1)* passed as the first argument to *perimeter* simply won't work, because the term *sq(1)* cannot unify with the parallelogram pattern *p(S1,S2,H)*. An elegant way of getting around this problem, which also incorporates a solution for the predicate constraints, is presented below.

With each subpattern of a type definition we associate a *constructor* predicate, which is automatically defined by the system. Thus, for each subpattern, the system defines a type-accessing relation, e.g. for the *list* example it asserts:

make_list1(nil).

make_list2(A,B,cons(A,B)).

where their extended types are: *make_list1(list(T)), make_list2(T,list(T),list(T))*.
For the *geometry* example:

```
make_parallelogram(S1,S2,H,p(S1,S2,H)) :- (S1 =< S2, H =< S1, !) ;
                                           (write('Warning:Type violation'), fail).
make_square(S,sq(S)).
make_square(S,X)  :-  make_parallelogram(S,S,S,X).
```

Notice how the predicate constraints are incorporated into the above clauses and that violation of these causes a warning to the user rather than abortion.

In order to make use of these predicates, a pre-processor substitutes any occurrence of a non-variable term (other than the built-in constants and constructors), in the body of a clause (or a goal), with a variable, and adds a call to the appropriate type-accessing relation, just before the literal containing this term. For example, the goal: '*?- perimeter(sq(1),N).*' is transformed into: '*?- make_square(1,S), perimeter(S,N).*' . Notice that, in this way, all the relations defined on parallelograms become available for squares via the second clause for 'make_square'.

4.2. The module system.

The module system described in section 3 has also been implemented in Prolog. The implementation takes the form of a pre-processor that transforms modular Prolog programs to semantically equivalent ordinary PROLOG programs, while typechecking them at the same time. In order to achieve modularity each relation is augmented with an extra argument, which is the module name in which it is defined. So, identical relation names defined in different modules are naturally distinguishable. Furthermore, in this way polymorphism is also achieved (free of charge) through Prolog's unification algorithm.

As for the extra-logical predicates *assert* and *retract*, it can be easily arranged so that they have a local effect by augmenting their argument with the module name in which they are called. This also provides a clean separation of logic databases.

As for query evaluation, every query passes through a pre-processor, which typechecks it and augments it with the appropriate module name before submitting it to the standard Prolog interpreter. In order to decide on the appropriate module name for a query we have to employ the concept of the *current module*. Every single query is evaluated in the environment of the *current module*. By default the *current module* is the module that was last compiled. This can be overwritten by the user by issuing a *module-navigation* command, using the syntax: "*?- -> Module.*", which makes *Module* the current module. Additionally, one can issue a *module-specific* query using the syntax: "*?- Query* ***in*** *Module*". Such a query makes *Module* the current module temporarily (so that evaluation of *Query* takes place in its environment), but after the query is evaluated, whichever module was current before the query becomes again the current module.

Notice that, since most of the expensive computation takes place at compile-time and the resulting programs are just ordinary PROLOG programs, there is no significant overhead in the execution speed of modular PROLOG programs.

5. DISCUSSION

It was taken as a starting point that, even though PROLOG does not fulfill all the ideals of logic programming, it can still be competitively used for large scale software production provided it is enhanced with mechanisms that reinforce modularity and data abstraction. A proposal was then outlined for such an enhancement, which has also been implemented in PROLOG and whose advantages outweigh the overhead in computational efficiency.

Firstly, a polymorphic type system with subtypes has been presented, which is suitable for PROLOG. Such a system offers a way of structuring data, which is so far lacking from PROLOG, and serves a more disciplined programming paradigm. Of course, we get all the merits in freedom of expression and in traping errors by static typechecking that are offered by a polymorphic type system. Multiple inheritance, one of the concepts served by the object-oriented programming paradigm, is also supported through subtyping. In addition, this system has been implemented at little extra cost in compile-time efficiency, while at no real loss of run-time efficiency. Most of the typechecking is performed at compile-time and it is the standard PROLOG interpreter that runs the well-typed programs.

The only other attempt to provide a polymorphic type system for PROLOG is reported in [Mycroft & O'Keefe 83]. However, their system is a flat one (without subtyping) and, as it was pointed out before, their formalisation of well-typing suffers from a serious flaw. The crucial point in our formalisation is the notion of adequate instantiation. In their formalisation, the typing for the head of a clause had to correspond exactly with its relation declaration, while the typing of any other literal could be an instance of its declared extended type. Thus, they seem to assume that the head of every clause is the determining literal (in our terminology) for all variables in it. Consequently, a clause can only be used (correctly) against fully instantiated calls. Under this assumption, their formalisation is correct, but seriously restricted. For example, if we assume: rels P(T,list(T)), Q(integer,list(integer)).
then, according to their definition, the clause: 'P(X,L) :- Q(X,L).' is not well-typed,
while the clause: 'Q(X,L) :- P(X,L).' is well-typed.
In contrast, according to our definition, both these clauses can be either well-typed or not, depending on their usage. For example, the first clause is well-typed if, and only if, it is called with both arguments uninstantiated, while the second clause is well-typed if, and only if, it is called with both arguments appropriately instantiated.

Secondly, we have used this type system as a basis for a module system for PROLOG.

In view of the similarity between our module system and the one used in algebraic specifications, one could argue that our proposal offers nothing more than the algebraic approach other than the use of Horn logic for defining predicates instead of equational logic for defining functions. Indeed, such a translation between functions and predicates is rather straightforward. Moreover, sometimes functions seem to be more natural than predicates, which is one of the reasons that motivated EQLOG. Clearly, the language EQLOG has already offered a similar and, in a sense, richer modular environment, based on many-sorted logic with equality, which supports Horn logic and functional programming. However, there are a number of reasons, which make our proposal more worthwhile than the pure algebraic approach and EQLOG. The main reason is that by giving specific patterns for data type representations (normal form) we avoid all the problems of consistency and completeness. Additionally, the use of equality and term-rewriting is a major source of inefficiency and, so far as EQLOG is concerned, the mixture of both functional and relational styles, although it offers more expressive power, can also be confusing for the user.

Himiko, being confined to Horn logic only, seems to be closer to this proposal. It does not have a rigid type system and so lacks all the features and advantages summed up below. It also allows more than one type to be defined in one module, a possibility excluded in this system so as to impose maximum modularity.

There have been some attempts to introduce a form of modularity with abstraction over predicates by resorting to a higher-order logic [Miller 86, Nait 86]. Although such approaches are of theoretical interest, they are far from providing a practical language.

The recent proposal to embed PROLOG in the module system developed for ML [Sanella & Wallen 87] is yet another promising approach to modularity and abstraction. In this, signatures are separate from code, which allows more freedom but requires a considerable amount of extra syntax and consequently more expensive book-keeping. Although it ignores the issue of types, which could be easily added to it, it offers abstraction over both predicates and functions, whereas TM-PROLOG seems to offer abstraction only over functions. However, this is easily compensated in practice by the facility for separate compilation. Thus, when we want to use a different program for a predicate *P* defined in module *M1* and used in module *M2*, we can simply recompile module *M1* with the new program for *P*, which will overwrite the previous one. Thus, we can simulate abstraction over predicates at a meta-level. Of course, if we also change the type of *P* we would also have to recompile *M2* to make sure that the new type is still compatible.

Finally, it should be said that, although the terminology was borrowed from the algebraic specifications school, OOP terminology could have equally well been used. Simply rename *modules* to *classes*, *instances of modules* to *objects*, the *subtype* relationship to *inheritance*, and *query evaluation* to *message passing*. Bearing this correspondence in mind, Zaniolo's proposal, a rather partial solution, is fully superseded by this system, which also supports *multiple inheritance*. Certainly, a more careful comparison is needed between this approach, and more generally the algebraic approach to abstract data types, and object-oriented programming.

It should be noted that, although the principal concepts introduced are not new, they have been realised by putting together semantic features in a novel and simple way in accordance with the spirit of logic programming. In particular, the following are original contributions:
a) the rich polymorphic type system whose main distinguishing characteristics are: i) the use of explicit patterns, which together with the predicate constraints, achieve great expressive

power, ii) the possibility of a user-definable subtype relation between different types, which extends to support multiple inheritance.

b) the 'privacy' of type patterns to their defining modules and the use of predicate type constructors - which can be defined automatically - for accessing them from other modules, which ensures data representation independence.

Even though this proposal was specifically tuned to PROLOG, it is hoped that the ideas presented here will find their way into an enhanced logic programming language that will be of practical use for serious software production.

5. REFERENCES

[1] Berztiss, A. & S. Thatte. Specification and Implementation of Abstract Data Types. *Advances in Computers, Vol.22*, pp.295-353, 1983.

[2] Clark, K. & S. Tarnlund. A first order theory of data and programs. *Information Processing (IFIP) '77*, North-Holland, pp.939-944, 1977.

[3] Clocksin, W.F. & C.S. Mellish. *Programming in Prolog*. Springer-Verlag, 1981.

[4] Cardelli L. & P. Wegner. On understanding Types, Data Abstraction and Polymorphism. *Computing Surveys, Vol.17, No.4*, December 1985, ACM, pp.471-522, 1986.

[5] Dayantis George. Logic program derivation for a class of first-order logic relations. In: *Proc. 10th IJCAI*, Vol.1, pp.9-15, 1987.

[6] Dayantis George. 'Types, modularisation and abstraction in logic programming'. D.Phil. thesis, University of Sussex, May 1988.

[7] Domolki, B. & P. Szeredi. Prolog in practice. *Information Processing (IFIP) '83*, North-Holland, pp. 627-636, 1983.

[8] Furukawa, K., Nakajima, R. & A. Yonezawa. Modularization and Abstraction in Logic Programming. ICOT TR-022, 1983.

[9] Goguen, J. & J. Meseguer. Equalities, types and generic modules for logic programming. In: de Groot, D. & E. Lindstrom (eds.). *Logic Programming: Functions, Relations and Equations*. Prentice-Hall, 1986.

[10] Miller, D.A. A theory of modules for logic programming. *Proc. 3d IEEE Symposium on Logic Programming*, pp.448-462, 1986.

[11] Mishra, P. Towards a theory of types in Prolog. *Proc. 1st IEEE Symposium on Logic Programming*, pp.289-298, 1984.

[12] Mycroft, A. & R. O'Keefe. A polymorphic type system for Prolog. D.A.I. paper, No.211, Univ. of Edinburgh, 1983. Also in: *Artificial Intelligence, Vol.23, No.3*, pp.295-307, 1984.

[13] Nait, Abdallah M. Procedures in Horn-clause programming. in: Shapiro, E. (ed.) *Proceedings of Third International Conference on Logic Programming*, London, July 1986, Lecture Notes in Computer Science, Vol. 225, Springer-Verlag, pp.433-447, 1986.

[14] O'Keefe, R. Towards an algebra for constructing logic programs. *Proc. 2nd IEEE Symposium on Logic Programming*, pp.152-160, 1985.

[15] Sannella, D.T. & L.A. Wallen. A calculus for the construction of modular Prolog programs. (To appear:) *Proc. 4th IEEE Symposium on Logic Programming*, 1987.

[16] Smolka, G. Order-sorted Horn logic, semantics and deduction. Technical paper, FB Informatik, Kaiserslautern Univ., W.Germany, 1986.

[17] Walther, C. A many-sorted calculus based on resolution and paramodulation. *Proc. 8th IJCAI*, pp.882-891, 1983.

[18] Zaniolo, C. Object-oriented programming in Prolog. *Proc. 1st IEEE Symposium on Logic Programming*, pp.265-270, 1984.

TRANSLATING ALGEBRAIC SPECIFICATIONS TO PROLOG PROGRAMS: A COMPARATIVE STUDY

K. Drosten [1)]

Abstract

Abstract programming includes two important approaches which are called functional and relational. We choose algebraic specification and the programming language Prolog as representatives of the two approaches to study the relation between them.

1. Introduction

In the field of abstract programming or abstract specification of programs and data types, there are two important approaches ultimately based on predicate logic which we call functional and relational.

The functional approach includes functional programming and its generalization to algebraic specification. This approach makes use of an equational logic with function symbols only. The intended meaning of the functions is specified by (recursive) equations or, in the most general case, by conditional equations in form of Horn clauses. Functional evaluation is done by subterm rewriting in the equational calculus.

The relational approach includes logic programming and is based on a subset of first order predicate logic. Apart from function symbols serving as data constructors predicate symbols are needed to specify relations over the data carrier. The meaning of the predicates is defined by definite Horn clauses. Predicate evaluation is done by a theorem proving mechanism based on the resolution principle [Ro65].

The idea of logic programming goes back to the early seventies and is due to R. Kowalski [Ko74] who observed that a theorem prover for a certain subset of predicate logic could be regarded as an efficient computational mechanism. Formally, a logic program is a finite set of first order sentences in the form of definite Horn clauses. Inferring logical consequences from programs is done by SLD resolution. Unlike theorem provers whose only interest is to demonstrate logical consequence, a logic programming system has to keep track of the bindings of variables which are made during resolution because they give us the output from a running program. Correctness and completeness of SLD resolution are responsible for that exactly all correct answers are derived, thus ensuring compatibility of declarative and procedural semantics [cf. Ll84].

Logic programming has become quite popular in connection with the programming language Prolog which was first implemented by A. Colmerauer [CKRP73]. Today, there are quite a number of Prolog systems available. They all agree on a common subset, called pure Prolog, but more or less differ in the built-in predicates they offer. Some of these built-in facilities cannot be described with logic very easily. The 'cut', for instance, allows to control backtracking by removing backtrack points. Affecting backtracking, however, is hard to define in terms of logic and resolution, and this leads to the problem of giving a precise logic-based semantics to full Prolog.

The method of algebraic specification starts at the thesis that data types are algebras and offers equations as main instrument to describe algebras independently of a concrete represen-

1) current address: GMD-IPSI, Dolivostr. 15, 6100 Darmstadt

tation of data ("abstract data type"). Semantically, three major approaches are distinguished which are referred to as initial algebra semantics, terminal algebra semantics, and observational semantics. The initial approach [GTW78] plays a dominant role because it allows for executing algebraic specifications effectively. To this end, the equations are oriented from left to right and are treated as rewrite rules. Justification to this method is given by the fact that the evaluation of (ground) terms in initial algebra reduces to the computation of normal forms provided subterm replacement is confluent and terminating. Thus, term rewriting systems (TRS, for short) are an adequate tool for implementing algebraic specifications.

The programming language Eqlog [GM84] combines concepts of functional and logic programing, based on Horn clauses in a logic with equality. Thus, Prolog clauses, equations, conditional equations and mixed forms of these are expressible making pure Prolog and algebraic specification sublanguages of Eqlog. Combining two concepts in one language, however, is not quite the same as implementing one concept by means of the other. It is this latter issue that our paper addresses.

Algebraic specifications are usually executed by an interpreter for term rewriting systems such as OBJ or AFFIRM. Another way is to translate them to Prolog programs which do the execution when running on a Prolog machine. The translation from TRSs to Prolog benefits from the fact that unification in connection with a decomposition of terms is very successful to describe the mechanism of subterm rewriting. Depending on whether the decomposition occurs at run time or compile time two approaches are distinguished which are referred to as interpretational [DE84] and compilational [EY86]. In the following, both approaches are revisited and compared in detail. Besides, we also address the fine point of separate translation in the framework of parameterized data types.

2. Prerequisites

Given a binary relation R over a set M we denote by R^* the reflexive, transitive closure of R. $x \in M$ is an R-*normal form* iff there is no $y \in M$ such that xRy. The set of R-normal forms is denoted by NF_R or simply NF if no ambiguity arises. $x \in M$ is an R-*normal form of an element* $u \in M$ iff uR^*x and $x \in NF$. R is *weakly terminating* iff $\forall x \in M\ \exists y \in NF$: xR^*y. R is *strongly terminating* or *Noetherian* iff there is no infinite sequence $x_1Rx_2R \ldots Rx_nR \ldots$ R is *confluent* iff $\forall u,x,y \in M$: $[(uR^*x \wedge uR^*y) \Longrightarrow (\exists z \in M: xR^*z \wedge yR^*z)]$. Termination is responsible for the existence, confluence for the uniqueness of normal forms. For any $x \in M$, the unique R-normal form, if it exists, is denoted by $nf_R(x)$ or simply by nf(x) if no ambiguity arises.

A *signature* is a pair $\Sigma=(S,F)$ where S is a set of *sorts* and $F=(F_{w,s})_{w \in S^*, s \in S}$ is an $S^* \times S$-indexed family of disjoint sets of *function symbols*. For any function symbol $f \in F_{s1 \ldots sn,s}$, the arity of f is n and the sort of f is s; nullary function symbols are called *constants*. Instead of $f \in F_{s1 \ldots sn,s}$ we simply write f: s1...sn $\longrightarrow$ s if F is clear from the context.

Let $\Sigma=(S,F)$ a signature and $V=(V_s)_{s \in S}$ an S-indexed family of disjoint sets of *variables* such that $F \cap V=\emptyset$. The set of terms over Σ and V is denoted by $T_{\Sigma,V}$; the elements in $T_{\Sigma,V}$ are referred to as (Σ,V)-terms. The set of *ground terms* over Σ, i.e. terms without variables, is denoted by T_Σ. The sort of a term is determined by the sort of its outermost symbol. var(A) denotes the set of variables occurring in a term A. A (Σ,V)-*substitution* is a finite set $\theta \subseteq (V \times T_{\Sigma,V})$ such that, for any $X \leftarrow B$, $Y \leftarrow C \in \theta$, we have $X \neq B$, sort(X)=sort(B) and $X=Y \Longrightarrow B=C$. We denote by $A\theta$ the result of applying the (Σ,V)-substitution θ to the (Σ,V)-term A; $A\theta$ is called an *instance* of A.

Terms can be viewed as labelled trees in the following way: A (Σ,V)-term A is a partial function from N^* into $F \cup V$ such that its domain dom(A) satisfies that the empty word ε is in dom(A), and $vi \in$dom(A) iff $v \in$dom(A) and $i \in [1,$arity$(A(v))]$; moreover, $A(v) \in F_{s1 \ldots sn,s}$ implies sort(A(vi)) $\in$ si for i=1,...,n. dom(A) is referred to as the set of *occurrences* of A. A/v denotes the *subterm* of A at occurrence v, A[v←B] is the term obtained by replacing A/v by B in A where A/v and B are assumed to be terms of the same sort.

To compare occurences we make use of the following notations: anc is the ancestor relation on $\mathbb{N}^*$: x anc y iff $\exists w \in \mathbb{N}^*$: xw=y. anc≠ is the proper ancestor relation on $\mathbb{N}^*$: x anc≠ y iff x anc y and x≠y. ⊥ is the independence relation on $\mathbb{N}^*$: x⊥y iff ¬(x anc y) and ¬(y anc x).

A *term rewriting system* (TRS) is a triple T=(Σ,V,R) where Σ=(S,F) is a signature, $V=(V_s)_{s \in S}$ is an S-indexed family of disjoint sets of variables such that F∩V=∅, and $R \subseteq (T_{\Sigma,V} \times T_{\Sigma,V})$ is a binary relation on (Σ,V)-terms such that, for any A→B $\in$ R, sort(A) = sort(B), A∉V and var(B) ⊆ var(A). The elements of R are called *rewrite rules*. The *reduction relation* $\xrightarrow{R}$ is defined as a binary relation on $T_{\Sigma,V}$ by: A $\xrightarrow{R}$ B iff there is a (Σ,V)-substitution θ, a rewrite rule G→H $\in$ R and an occurence x $\in$ dom(A) such that A/x=Gθ and B = A[x←Hθ]. A TRS is said to be terminating, Noetherian, and confluent resp. iff its reduction relation is. A TRS is called *leftdominant* iff, for any rule A ⟶ B, there is no variable occurring more times in B than in A; it is said to be *rightextensive* iff, for any rule A ⟶ B, B is not a single variable. We sometimes write $A \xrightarrow[x]{R} B$ instead of $A \xrightarrow{R} B$ to indicate that A is reduced at occurrence x; in this case, we call x a *redex* of A. When applied to a (Σ,V)-term A, the mappings red, imr: $T_{\Sigma,V} \longrightarrow 2^{\mathbb{N}^*}$ and lir: $T_{\Sigma,V} \longrightarrow \mathbb{N}^*$ yield the set of *redexes*, the set of *innermost redexes* and the *leftmost innermost redex* of A resp. All reduction sequences starting from a given term can be depicted by a "reduction tree". A reduction strategy is then characterized by the way it searches the reduction trees.

We assume the reader to be familiar with the foundations of logic programming [cf. Ll84] and the programming language Prolog [CM81].

3. Interpretational Translation Schemes

TRSs are based on the idea to orient equations from left to right in order to rewrite terms to somewhat simpler expressions. The reduction sequences generated thereby correspond to derivations in the equational calculus where the symmetry rule is not applied at all and the transitivity rule is only applied at the end. Nevertheless, reduction is a sound and complete inference rule for the equational logic whenever the set of one-step replacements forms a confluent and terminating relation. The same holds for a many-sorted logic if the carrier sets of all models are non-empty.

This close relationship between equations and rewrite rules makes normal form computation an adequate method for equational inference. The following scheme is aimed at translating TRSs to Prolog programs which, on input of a term, return its normal form. The main problem coming up results from different mechanisms which are in use for TRSs ("matching") and Prolog programs ("unification") to apply rewrite rules and program clauses resp.:

(1) Unification forces variables in both subgoals and program clauses to be instantiated. Matching, on the other hand, can only instantiate variables in rewrite rules but does not affect variables in the term to be reduced.
(2) Unification can only unify full expressions, whereas matching applies to arbitrary subexpressions.

By point (1) variables have to be treated twofold: as Prolog constants, if they occur in the term to reduced, and as Prolog variables if they occur in rewrite rules. By point (2), a term will not be reduced at its root occurence before all its subterms are in normal form. The latter guarantees that terms are in normal form iff they cannot be unified with the lefthand side of any rewrite rule.

Definition 3.1
Let $T=(\Sigma,V,R)$ a finite TRS with $\Sigma=(S,F)$. A Prolog program P is called a *translation* of T iff it satisfies the following translation rule. [1]

(1) For any function symbol $f \in F_{s1\ldots sn,s}$ $(n \in \mathbb{N}_0)$, there is a clause
analyse(f(I1,...,In),N) ← analyse(I1,K1),..., analyse(In,Kn),
normalize(f(K1,...,Kn),N)

(2) For any variable $U \in V$, there is a fact
analyse(u,u) ←

(3) For any rewrite rule $A \longrightarrow B \in R$, there is a fact
rule(A,B) ←

(4) Apart from the clauses above, there are two other clauses in P, namely
normalize(X,Y) ← rule(X,Z), analyse(Z,Y)
normalize(X,X) ←
such that the first one precedes the second one.

(5) P does not have any other clauses than required by (1) - (4).

Definition 3.2
Let P a translation of $T=(\Sigma,V,R)$. S(P) denotes the sequence of elements in R corresponding to P: $A \longrightarrow B \in R$ precedes $C \longrightarrow D \in R$ in S(P) iff rule(A,B) ← precedes rule(C,D) ← in P.

Definition 3.3
Let $R \subseteq T_{\Sigma,V} \times T_{\Sigma,V}$ a finite set of rewrite rules and $S(R)=(C_1 \longrightarrow D_1,\ldots,C_n \longrightarrow D_n)$, $n=|R|$, a sequence of elements in R such that each element in R occurs exactly once in S(R). $\overset{S(R)}{\longmapsto}$ is a binary relation on $T_{\Sigma,V}$ defined as follows:

$A \overset{S(R)}{\longmapsto} B$ iff $A/x=C_i\theta$ and $B=A[x\leftarrow D_i\theta]$
where $x=\mathrm{lir}(A)$ and $A/x \neq C_j\psi$ for all (Σ,V)-substitutions ψ and $j<i\leq n$

Theorem 3.4 [DE84]
Let $T=(\Sigma,V,R)$ a finite TRS and P a translation of T. If $\overset{S(P)}{\longmapsto}$ is Noetherian then, for any (Σ,V)-term A, P terminates for ← analyse(A,N) having N instantiated to a $\overset{R}{\longrightarrow}$-normal form of A. [2]

Corollary 3.5 [DE84]
Let $T=(\Sigma,V,R)$ a finite TRS. If $\overset{R}{\longrightarrow}$ is Noetherian and confluent any translation of T terminates for the goal ← analyse(A,N) instantiating N to nf(A).

For some applications of TRSs such as the validity problem it is convenient to assume a countably infinite supply V of variables. As a consequence, it is impossible to generate one program clause per variable as required by point (2) of the translation scheme. To overcome this problem clauses for variables are to be delayed until run time.

There are other applications such as prototyping algebraic specifications which only need ground terms to be reduced. In these cases, point (2) of the translation scheme becomes obsolete; moreover, the preconditions of Theorem 3.4 and Corollary 3.5 can be weakened in a natural way. Let $\overset{S(P)}{\Vdash\!\!\longrightarrow}$ and $\overset{R}{\Vdash\!\!\longrightarrow}$ resp. denote the restrictions of $\overset{S(P)}{\longmapsto}$ and $\overset{R}{\longrightarrow}$ to $T_\Sigma x T_\Sigma$. Then, we have:

---

1) We implicitly assume that the function symbols in Σ start with a small letter and the variables in V with a capital letter. To treat the variables in V as Prolog constants in (2), we replace the starting capital letter by the corresponding small letter assuming that there is no confusion with function symbols.

2) N is assumed to be a Prolog variable and A a (Σ,V)-term whose variables are treated as Prolog constants. Treating the variables in A as Prolog variables would result in "narrowing" rather than reduction.

Theorem 3.6 [DE84]
Let T=(Σ,V,R) a finite TRS and P a translation of T. If $\overset{S(P)}{\Vdash\!\!\longrightarrow}$ is Noetherian then, given a ground term $A \in T_{\Sigma}$, P terminates for ← analyse(A,N) having N instantiated to a $\overset{R}{\mapsto}$-normal form of A.

Corollary 3.7 [DE84]
Let T=(Σ,V,R) a finite TRS. If $\overset{R}{\mapsto}$ is Noetherian and confluent any translation of T terminates for ← analyse(A,N), $A \in T_{\Sigma}$, instantiating N to nf(A).

By Corollary 3.4, our translation scheme will certainly generate correct programs whenever the reduction relation is strongly terminating. In case of weak termination, however, the target programs cannot be expected to halt on every input. To see this consider the example below:

```
nat1
  SORTS:        nat
  FUNS:         zero: ⟶ nat
                succ: nat ⟶ nat
                add: nat × nat ⟶ nat
  VARS:         M,N: nat
  RULES: (R1)   add(zero,N) ⟶ N
         (R2)   add(succ(N),M) ⟶ succ(add(N,M))
         (R3)   add(N,M) ⟶ add(M,N)
```

<u>nat1</u> is confluent and weakly terminating on ground terms. Although (R3) does not contribute to the semantics of <u>nat1</u> - the addition is already defined by (R1) and (R3) - it plays a dominant role in the translation process: It destroys the Noetherian property of <u>nat1</u> and prevents some target programs from termination. Consider the following translation of <u>nat1</u>:

```
      analyse(zero,N) ← normalize(zero,N)
      analyse(succ(I1),N) ← analyse(I1,K1), normalize(succ(K1),N)
      analyse(add(I1,I2),N) ← analyse(I1,K1), analyse(I2,K2), normalize(add(K1,K2),N)
(*)   rule(add(N,M),add(M,N)) ←
      rule(add(zero,N),N) ←
(**)  rule(add(succ(N),M),succ(add(N,M))) ←
      normalize(X,Y) ← rule(X,Z), analyse(Z,Y)
      normalize(X,X) ←
```

The program above prefers rule (R3) for reduction which is why normal forms are not found, in general. On input ← analyse(add(zero,zero),N), for instance, non-termination is indicated by the fact that satisfying the goal reduces to satisfying the identical subgoal. The situation may be cured by interchanging the clauses marked with (*) and (**). But this method does not always work. The reason why is that all programs generated by our translation scheme implement the leftmost innermost reduction strategy. Therefore, a permutation of clauses will not help in all cases where the leftmost innermost strategy certainly fails to terminate such as for {f(X) ⟶ f(f(X)), f(f(X)) ⟶ X}.

The only strategy which always finds existing normal forms is based on searching the reduction tree breadth-first. For practical reasons, however, a brute force strategy is not acceptable because it is space consuming and inefficient. The significance of the brute force strategy is only of theoretical nature. We refer to Dr88 for an interpretational translation scheme based on a brute force strategy.

4. Parameterized Data Types

The concept of parametrization is an important structuring mean supporting both modularity and reusability of abstract data types. The power of parametric TRSs is based on a universal mechanism which allows for composing arbitrarily complex programs from smaller modules. Along a specification morphism indicating how to replace the formal parts of a TRS the result of parameter passing is determined by pushout construction [Eh 82]. The problem comes up under which conditions single executable modules, i.e. confluent and terminating TRSs can be put together to complex executable systems automatically. The composition theorem (Theorem 4.1) gives a satisfactory answer for the case that parameter passing morphisms are injective and the formal parameters of all modules involved in the process are sort parameters. Whenever the composition theorem applies the idea of separately translated modules becomes feasible.

Definition 4.1
Let $T=(\Sigma,V,R)$ and $T1=(\Sigma1,V1,R1)$ TRSs with $\Sigma=(S,F)$, $\Sigma1=(S1,F1)$. T is a *sub-TRS* of T1 (written as $T\leq T1$) iff $S\subseteq S1$, $F\subseteq F1$, $V\subseteq V1$ and $R\subseteq R1$.

Definition 4.2
A *parametric TRS* is a pair $PT=(P,B)$ of TRSs where $P\leq B$. P is called *formal parameter* and B *body* of PT.

Definition 4.3
Given signatures $\Sigma=(S,F)$ and $\Sigma1=(S1,F1)$ a *signature morphism* $h{:}\Sigma \longrightarrow \Sigma1$ is a pair of mappings $h=(h_S{:}\,S \longrightarrow S1,\ h_F{:}\,F \longrightarrow F1)$ such that $f{:}\ s1 \ldots sn \longrightarrow s$ in F implies $h_F(f){:}\ h_S(s1) \ldots h_S(sn) \longrightarrow h_S(s)$ in F1.

Signature morphisms indicate the renamings and identifications of formal sorts and function symbols when replacing the formal parameter by an actual one. Signature morphisms are extended to specification morphisms to guarantee consistency of the actual parameter rules w.r.t. those of the formal parameter.

Definition 4.4
Let signatures $\Sigma=(S,F)$, $\Sigma1=(S1,F1)$, a signature morphism $h{:}\ \Sigma \longrightarrow \Sigma1$ and sets of variables $V=(V_s)_{s\in S}$, $V1=(V1_s)_{s\in S1}$ be given. A mapping $h^{\#}{:}\ T_{\Sigma,V} \longrightarrow T_{\Sigma1,V1}$ is an *extension* of h iff

(i) $h^{\#}(f) = h(f)$ for each constant f

(ii) $h^{\#}(X) \in V1_{h(sort(X))}$ for each variable X

(iii) $h^{\#}(f(A_1,\ldots,A_n)) = h(f)(h^{\#}(A_1),\ldots,h^{\#}(A_n))$ for each n-ary function symbol f $(n\geq1)$

Definition 4.5
Let $T=(\Sigma,V,R)$ and $T1=(\Sigma1,V1,R1)$ TRSs. A signature morphism $h{:}\ \Sigma \rightarrow \Sigma1$ is a *specification morphism* from T to T1 (written as $h{:}\ T \longrightarrow T1$) iff, for any rule $G\rightarrow H\in R$, there is an extension $h^{\#}{:}\ T_{\Sigma,V} \longrightarrow T_{\Sigma1,V1}$ with $h^{\#}(X) = h^{\#}(Y) \Leftrightarrow X=Y$ for all $X,Y\in(var(A) \cup var(B))$ such that $h^{\#}(G)\rightarrow h^{\#}(H)$ in R.

Definition 4.6
Given a parametric TRS $PT=(P,B)$, a TRS A (*actual parameter*), and a specification morphism $h{:}\ P \longrightarrow A$ (*parameter passing morphism*) where $P=(\Sigma_P,V_P,R_P)$, $B=(\Sigma_B,V_B,R_B)$, $A=(\Sigma_A,V_A,R_A)$ with $\Sigma_P=(S_P,F_P)$, $\Sigma_B=(S_B,F_B)$, $\Sigma_A=(S_A,F_A)$.

1. Let the mapping $h^*{:}\ S_B^* \longrightarrow [h(S_P)+(S_B-S_P)]^*$ be recursively defined by $h^*(\varepsilon)=\varepsilon$ and $h^*(s\cdot w)=$ <u>if</u> $s\in S_P$ <u>then</u> $h(s)\cdot h^*(w)$ <u>else</u> $s\cdot h^*(w)$. The *body signature renamed* by h is given by $\Sigma=(S,F)$ where
$S = h(S_P) + (S_B-S_P)$
$F = h(F_P) + \{f{:}\ h^*(w) \longrightarrow h^*(s) \mid (f{:}\ w \rightarrow s) \in F_B-F_P\}$

2. The ***parameter passing morphism*** induced by h is defined as signature morphism h_B: $\Sigma_B \longrightarrow \Sigma$ by
   $h_B(s)$ = if $s \in S_P$ then $h(s)$ else s
   $h_B(f: w \longrightarrow s)$ = if $(f: w \longrightarrow s) \in F_P$
   then $h(f): h^*(w) \longrightarrow h^*(s)$
   else $f: h^*(w) \longrightarrow h^*(s)$
3. The ***renamed body*** RENAME(PT,A,h) is defined as a TRS consisting of
   (1) the renamed body signature $\Sigma=(h_B(S_B),h_B(F_B))$
   (2) the set of variables $V=(V_s)_{s \in h_B(S_B)}$ where $V_s = \bigcup_{s=h_B(s')} V_{B,s}$.
   (3) the set of rules $R = \{h_B^*(G) \to h_B^*(H) \mid G \to H \in R_B\}$ where $h_B^*: T_{\Sigma_B,V_B} \longrightarrow T_{\Sigma,V}$ is an extension of h_B with $h_B^*(X)=X$ for any $X \in V_B$
4. The *result* APPLY(PT,A,h) of the application of PT to A by h is the sum A+RENAME(PT,A,h) with $(h(S_P),h(F_P))$ as common signature.

h_B is a specification morphism from B to RENAME(PT,A,h). Because of RENAME(PT,A,h) $\subseteq$ APPLY(PT,A,h), h_B can be extended to a specification morphism B $\longrightarrow$ APPLY(PT,A,h). When choosing p: P $\longrightarrow$ B and p_A: A $\longrightarrow$ APPLY(PT,A,h) as inclusions of sub-TRSs, the following *parameter passing diagram* commutes

$$\begin{array}{llcl} PT: & P & \overset{p}{\hookrightarrow} & B \\ & \downarrow h & & \downarrow h_B \\ & A & \overset{p_A}{\hookrightarrow} & APPLY(PT,A,h) \end{array}$$

On the analogy to Eh82, the parameter passing diagram is characterized as pushout diagram in the category of TRSs and specification morphisms.

Definition 4.7
A TRS $T=(\Sigma,V,R)$ with $\Sigma=(S,F)$ is called a *base* iff $F=\emptyset$ (and, therefore, $R=\emptyset$).

Theorem 4.1 (Composition Theorem)
Let PT=(P,B) a parametric TRS, A an actual parameter, and h: P $\longrightarrow$ A a parameter passing morphism. If P is a base and h is injective we have that
1. APPLY(PT,A,h) is weakly terminating iff A and B are weakly terminating.
2. APPLY(PT,A,h) is confluent iff A and B are confluent.
3. APPLY(PT,A,h) is Noetherian and leftdominant iff A and B are Noetherian and leftdominant.
4. APPLY(PT,A,h) is Noetherian and rightextensive iff A and B are Noetherian and rightextensive.

Proof:
Since P is a base, APPLY(PT,A,h) is the direct sum of A and RENAME(PT,A,h). Since h is injective, B and RENAME(PT,A,h) are identical up to renaming of sorts. Therefore, RENAME(PT,A,h) is weakly terminating (Noetherian, confluent, leftdominant, rightextensive) iff B is. The proof then follows from results on the direct sum of TRSs investigated recently by Toyama, Rusinowitch and Drosten (see Dr88 for reference) □

It is well-known that the direct sum need not be Noetherian even if its "summands" are. As a consequence, strong termination is not a modular property.

The composition theorem can be exploited to address the fine point of separate translation. Referring to the translation scheme in the previous chapter it would be a good idea to drop the normalize clauses until run time. In this way, separately translated TRSs can simply be combined by disjoint union to obtain a reliable prototype. The procedure, however, implicitly assumes that all TRSs involved in the process are either leftdominant or rightextensive (see Corollary 3.5 and Theorem 4.1). To get rid of the latter restriction a brute force based translation scheme has to be adapted.

5. Related Work

There are several approaches in the literature dealing with the translation of TRSs to Prolog programs but illuminating the topic under different aspects: Ba82, DE84, EM81 and EM86 are theoretically oriented and emphasize the relationship between functional and logic programming. Starting point is the common assumption that TRSs and algebraic specifications resp. are an adequate model for functional programming. The precise mathematical semantics of algebraic specifications can then serve as a basis for a formal verification of the translation schemes.

van Emden [EY86] distinguishes between two standards of correctness, a modeltheoretic one and an operational one. Given a set of equations a logic program is correct from a modeltheoretic point of view if any computed answer (success branch in an SLD tree) is a logical consequence from the equations augmented with standard equality theory and "definitional sentences" giving the intended meaning of the predicates. On the contrary, the only interest of operational correctness is that part of an SLD-tree, which is searched by the Prolog interpreter. In this sense, chapter 3 is only concerned with operational correctness.

Other approaches are of more practical interest and take the generated Prolog programs as a tool for particular applications of TRSs such as inductive proofs for data types [Pe85] or prototyping algebraically specified systems [BD81, GML84]. In these cases, the applications are well to the fore rather than the foundations of programming and this is why the verification of translation schemes is omitted.

The translation of TRSs to Prolog programs is essentially based on the fact that unification combined with a decomposition of terms gives an adequate mechanism to describe the effect of subterm rewriting. Depending on whether the decomposition takes place at run time or compile time van Emden [EY86] distinguishes between an interpretational and a compilational approach. The interpretational approach is characterized by the fact that the generated programs logically consist of two separate components. The "knowledge component" contains the rewrite rules and represents the knowledge about legal reduction steps; an explicit control ("interpreter") accesses the rewrite rules during run time, interprets them and reduces terms according to a fixed strategy. In this sense, the procedure in chapter 3 is interpretative. Unlike the interpretational approach, where the target programs have the rewrite rules directly embedded the compilational approach is based on the idea to obtain program clauses by compiling the rewrite rules. To this end, the hierarchical structure of terms ("composite function calls") is transformed to an equivalent sequence of elementary function calls. The translation process is finished when lefthand and righthand sides of all rewrite rules are completely decomposed.

The fundamentals of logic programming are substantially characterized by the fact that data constructors and predicates are strictly seperated from each other. TRSs, on the other hand, do not provide such a clean distinction between constructors and defined functions. When compiling rewrite rules into program clauses the problem comes up which function symbols to treat as constructors and which as predicate symbols. The applicability of the compilational approach is, therefore, restricted to TRSs which have no function symbols playing the role of both constructors and defined symbols. Moreover, all rewrite rules are to be of the form $f(A_1,\dots,A_n) \longrightarrow \dots$ with each A_i built up from variables and constructor symbols only. The latter condition is called "constructor discipline" and assures that Prolog clauses are obtained. Without a constructor discipline the compilational approach would produce clauses with more than one positive literal.

Example
The nat specification (see above) provides a clean distinction between constructor symbols (zero, succ) and defined symbols (add, mult). In addition, the rewrite rules satisfy the constructor discipline. The compilational approach is, therefore, applicable to nat and would generate the following clauses

plus(zero,N,N) ←
plus(succ(M),N,succ(Z)) ← plus(M,N,Z)
times(zero,N,zero) ←
times(succ(M),N,Z) ← times(M,N,Y), plus(Y,N,Z)

When running the above program terms are to be transformed to equivalent sequences of elementary function calls to be reduced correctly. In this way, a bottom-up evaluation becomes mandatory. For example, mult(add(zero,succ(zero)), add(zero,zero)) has to be transformed to the goal ← plus(zero,succ(zero),M), plus(zero,zero,N), times(M,N,Z) ***

Example
The specification of one and more element queues gives a realistic example where the compilational approach does not apply to. The point here is that cat (concatenation of lists) is a defined symbol also serving as data constructor to build up lists of arbitrary length.

natqueue

SORTS:	nat, queue
FUNS:	zero: → nat
	succ: nat → nat
	new: nat → queue
	cat: queue × queue → queue
VARS:	X,Y,Z: queue
RULES:	cat(X,cat(Y,Z)) → cat(cat(X,Y),Z)

***

The grade of a translation scheme depends on how far the requirements for universality and efficiency are matched. The same criteria also fit to compare the whole subclasses of interpretational and compilational translation schemes one with another. Although both universality and efficiency can differ from scheme to scheme within the same subclass one can make out overall bounds being inherent to the interpretational and compilational method: The idea of interpreting rewrite rules during program execution is responsible for that the decomposition of terms cannot be done before run time. Obviously, much more efficient programs are obtained when performing the decomposition at compile time. Reversely, the aspect of universality is in favour of the interpretational approach because the latter can employ any reduction strategy in principle and is therefore applicable to arbitrary TRSs. The compilational approach, on the other hand, is bound to the constructor discipline and a bottom-up evaluation of terms. Advantages and disadvantages are collected in the following table.

	interpretational	compilational
code generation	inefficient (decomposition at run time)	efficient (decomposition at compile time)
applicability	universal (flexible reduction strategy, brute force in the worst case)	restricted (bottom-up evaluation, compliance with the constructor discipline)

An interpretational translation scheme considers the reduction process as a succession of decomposition and reduction steps, the reduction strategy being determined by the sequence in which the single steps follow each other. The method of Bandes [Ba82] is aimed at incorporating this program logic into code generation. The main idea is to have an integrated component performing the reduction and decomposition steps and to have the control seperated which is responsible for the reduction strategy. Based on such a modular design, a translation scheme is obtained in Ba82 which integrates innermost reductions, outermost reductions and a mixed form of both. The following diagram gives an outline of the logical structure of the generated programs.

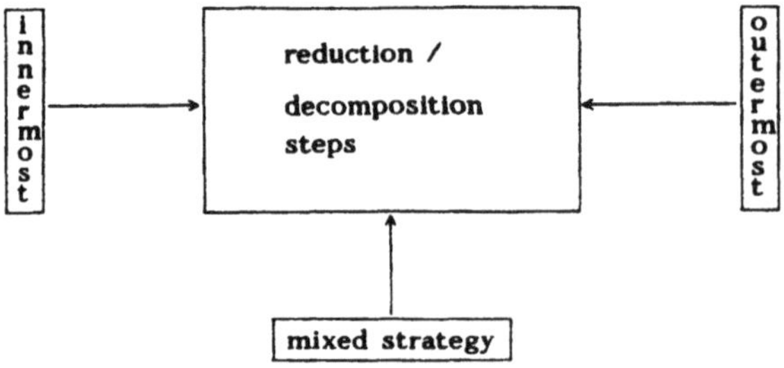

When running the program the user may decide for that strategy which is best suited to find normal forms reliably and efficiently.

References

Ba82 Bandes, R.G.: *Algebraic Specification and Prolog*, Technical Report 82-12-02, Dept. of Comp. Science, Univ. of Washington, Seattle, 1982

BD81 Bergman, M. / Deransart, P.: *Abstract Data Types and Rewriting Systems: Application to the Programming of Abstract Data Types in Prolog*, Proc. 6th CAAP (E. Astesiano / C. Boehm, eds.), LNCS 112, Springer-Verlag, Berlin, 1981

CKRP73 Colmerauer, A. / Kanoui, H. / Roussel, P. / Pasero, R.: *Un Systeme de Communication Homme - Machine en Francais*, Rapport de Recherche, Groupe d'Intelligence Artificielle, Univ. d'Aix-Marseille, 1973

CM81 Clocksin, W.F. / Mellish, C.S.: *Programming in Prolog*, Springer-Verlag, Berlin, 1981

DE84 Drosten, K. / Ehrich, H.-D.: *Translating Algebraic Specifications to Prolog Programs*, Informatik-Bericht Nr. 84-08, TU Braunschweig, 1984

Dr88 Drosten, K.: *On Extensions of Term Rewriting Systems and their Application to Rapid Prototyping of Algebraic Specifications* (in German), Dissertation, TU Braunschweig, 1988

Eh82 Ehrich, H.-D.: *On the Theory of Specification, Implementation, and Parametrization of Abstract Data Types*, Journal ACM 29(1), 1982

EY86 van Emden, M.H. / Yukawa, K.: *Equational Logic Programming*, Technical Report, Dept. of Comp. Science, Univ. of Waterloo, 1986

GLM84 Gaube, W. / Lockemann, P.C. / Mayr, H.C.: *ORS-Spezifikationslabor: Generierung von Prolog Programmen aus Definitionen abstrakter Datentypen*, Technical Report 15/84, Univ. Karlsruhe, 1984

GM84 Goguen, J.A. / Meseguer, J.: *Equality, Types, Modules, and (Why not?) Generics for Logic Programming*, Journal of Logic Programming 1(2), 1984

GTW78 Goguen, J.A. / Thatcher, J.B. / Wagner, E.G.: *An Initial Algebra Approach to the Specification, Correctness and Implementation of Abstract Data Types*, Current Trends in Programming Methodology IV (R.T. Yeh, ed.), Prentice Hall, Englewood Cliffs, 1978

Ko74 Kowalski, R.: *Predicate Logic as a Programming Language*, Information Processing 74 (J.L. Rosenfeld, ed.), North-Holland, Amsterdam, 1974

Ll84 Lloyd, J.W.: *Foundations of Logic Programming*, Springer-Verlag, Berlin, 1984

Pe85 Petzsch, H.: *Automatic Prototyping of Algebraic Specifications Using Prolog*, Technical Report, Lehrstuhl für Informatik II, RWTH Aachen, 1985

Ro65 Robinson, J.A.: *A Machine-Oriented Logic Based on the Resolution Principle*, Journal ACM 12(1), 1965

Rewrite Rule Systems
for Modal Propositional Logic

Annie Foret *

Abstract

The aim of this paper is to explain our new results relating modal propositional logic and rewrite rule systems. More precisely, we give *complete term rewriting systems* for the modal propositional systems known as K,Q,T and S5.

These systems are presented as extensions of Hsiang's system for classical propositional calculus [7]. We have checked local confluence with the rewrite rule system K.B. developped by the Formel project at INRIA[2,4]. We prove that these systems are noetherian, and then infer their confluence from Newman's lemma.

Therefore each term rewriting system provides a new automated *decision procedure* and defines a *canonical form* for the corresponding logic. We also show how to characterize the canonical forms thus obtained.

1 Introduction

Our study of modal logic is motivated by its use in the logic of programs, such as temporal logic or dynamic logic and its application to Artificial Intelligence e.g. knowledge representation or natural language. We also emphasize topology and intuitionistic logic [3,12].

Several decision procedures are already known for modal logic: we particularly mention semantic methods like the tableau method, natural deduction (cf [5,13]), and modal resolution which has recently been developped by P. Enjalbert and L. Farinas [1].

We shall only consider *normal modal systems*, the system K is in a precise sense the weakest normal modal system. These systems are based on classical propositional calculus, CPC. We briefly recall their definitions as Hilbert systems.

The language of modal logic

The *CPC-formulae* are constructed from the following symbols and operators:

- the constants (0) and (1), which represent truth and falsity
- a denumerable set of propositional variables, which we write alternatively as: $Vx, Vy, Vz, Vx_1, Vx_2, \ldots$ or $p, q, r, p_1, \ldots$
- the connectives $\cdot$, $\vee$, $\Rightarrow$, $\neg$, $\oplus$ and $\equiv$ which denote respectively the classical conjunction,disjunction, implication, negation, exclusive or and equivalence (we may restrict to $\vee$ and $\neg$).

*University Paris 6 and INRIA Domaine de Voluceau Rocquencourt, 78153 Le Chesnay FRANCE.

The *modal formulae* are obtained by adding two modal unary operators denoted as L and M, which are read as "necessarily" and "possibly" respectively. We take L as primitive, and define M by:

$$M(x) = \neg L(\neg x)$$

Normal modal systems

A modal system S is a class of formulae, whose members are called theorems. We write ($\vdash_S x$) to express that x is a theorem of S. As is usual for Hilbert systems, S is presented by axioms and inference rules.

A modal system S is *normal* iff

- S contains all CPC-theorems,
- S contains the formula (axiom K)

$$L(x \Rightarrow y) \Rightarrow (L(x) \Rightarrow L(y))$$

- S satisfies the following *inference rules*:
  - If x is a theorem, so is every substitution-instance of x.
  - If x and $x \Rightarrow y$ are theorems, so is y (Modus Ponens).
  - If x is a theorem, so is $L(x)$ (Necessitation).

These rules define the modal system K. The other systems are extensions of K, obtained by adding axioms as follows.

• the system Q is K plus the axiom :	$L(x) \Rightarrow M(x)$
or equivalently:	$L(0) \equiv 0$
• the system T is K plus the axiom (T):	$L(x) \Rightarrow x$
• the system S4 is T plus the axiom :	$L(x) \Rightarrow L(L(x))$
• the system S5 is T plus the axiom :	$\neg L(x) \Rightarrow L(-L(x))$

Term Rewriting systems

Our approach is based on Hsiang's system for CPC. We recall this system where $\cdot$ and $\oplus$ are taken as primitive (the dual system is obtained with $\vee$ and $\equiv$). It is confluent modulo the associativity-commutativity of $\cdot$ and $\oplus$.

```
*cp
x ⊕ x ↪ 0
x · (y ⊕ z) ↪ (x · y) ⊕ (x · z)
x · 0 ↪ 0
x · x ↪ x
x ⊕ 0 ↪ x
x · 1 ↪ x
```

Notations. We call *CPC-normal forms* the canonical forms obtained by this system $*cp$. We say that two formulae are *AC-distinct* when they are distinct modulo the associativity-commutativity of $\cdot$ and $\oplus$, and we write *modulo AC* for modulo the associativity-commutativity of $\cdot$ and $\oplus$.

The *CPC-normal forms* may be characterized inductively as follows:

- the constants (0), (1) and every conjunction of distinct variables are in CPC-normal form
- if s is the exclusive disjunction of terms b_i,
  where each b_i is the constant (1) or the conjunction of distinct variables
  and for all $i \neq j$: b_i and b_j are AC-distinct
  then s is in CPC-normal form
- there are no other CPC-normal forms

We give the example of equivalent clausal normal forms reduced to the same CPC-normal form (clausal normal forms are not canonical forms):

$$(-x \vee y) \cdot (-y \vee z) \cdot (-z \vee x) = (-x \vee z) \cdot (-z \vee y) \cdot (-y \vee x)$$

these two terms are both rewritten as:

$$x \oplus y \oplus z \oplus (x \cdot y) \oplus (y \cdot z) \oplus (z \cdot x) \oplus 1$$

2 The Systems K and Q

The results for K and Q are very similar. An equational presentation for K is obtained by adding to CPC:

$$L(x) \cdot L(y) = L(x \cdot y)$$

$$L(1) = 1$$

and we obtain Q by adding to K the equation:

$$L(0) = 0$$

Proposition 1 *The systems *cp-K and *cp-K-Q obtained by orienting the above equations from left to right are locally confluent and terminating:*

*cp-K

$x \oplus x \hookrightarrow 0$
$x \cdot (y \oplus z) \hookrightarrow (x \cdot y) \oplus (x \cdot z)$
$x \cdot 0 \hookrightarrow 0$
$x \cdot x \hookrightarrow x$
$x \oplus 0 \hookrightarrow x$
$x \cdot 1 \hookrightarrow x$
$L(x) \cdot L(y) \hookrightarrow L(x \cdot y)$
$L(1) \hookrightarrow 1$

*cp-K-Q

$x \oplus x \hookrightarrow 0$
$x \cdot (y \oplus z) \hookrightarrow (x \cdot y) \oplus (x \cdot z)$
$x \cdot 0 \hookrightarrow 0$
$x \cdot x \hookrightarrow x$
$x \oplus 0 \hookrightarrow x$
$x \cdot 1 \hookrightarrow x$
$L(x) \cdot L(y) \hookrightarrow L(x \cdot y)$
$L(1) \hookrightarrow 1$
$L(0) \hookrightarrow 0$

To prove the termination, we may use the following interpretation on integers greater than 2:

$$\begin{aligned} I(x \cdot y) &= 2 \times I(x) \times I(y) \\ I(x \oplus y) &= I(x) + I(y) + 1 \\ I(Lx) &= 2 \times I(x) \\ I(Vx) = I(0) = I(1) &= 2 \end{aligned}$$

Canonical forms in K and Q

Definition 1 *We define the notion of* K-normal form *and the auxiliary notion of* L-K-component *inductively as follows:*

- if s is a CPC-formula in CPC-normal form then it is in K-normal form.
- if a is in K-normal form
  and each b_i is a conjunction of distinct variables
  and for every $(i \neq j)$ b_i and b_j are not composed of the same variables
  then l defined by:
  $$l = (L(a) \cdot b_1) \oplus (L(a) \cdot b_2) \oplus \ldots \oplus (L(a) \cdot b_n)$$
  is a L-K-component . The term a is called the *principal term* of l.
- if a is a CPC-formula in CPC-normal form
  and if all l_i are L-K-components with AC-distinct principal terms
  then the following terms s_1 and s_2 are in K-normal forms:
  $$s_1 = a \oplus l_1 \oplus l_2 \oplus \ldots \oplus l_n$$
  $$s_2 = l_1 \oplus l_2 \oplus \ldots \oplus l_n$$
- there are no other K-normal forms and no other L-K-components than those generated by the above rules .

We similarly define the notions of *Q-normal form* and *L-Q-component* just by rejecting 0 as an acceptable principal term for L-Q-components.

Proposition 2 *The K-normal forms and the Q-normal forms are exactly the normal forms obtained by the corresponding rewriting systems *cp-K and *cp-K-Q.*

This proposition is easy to verify. These forms are thus canonical, in the sense that each class of equivalent terms has a unique representative of this form.

3 The System T

We obtain an equational presentation of T by adding to K the necessity-rule:

$$L(x) \cdot x = L(x)$$

In this case, however, we cannot get a confluent rewriting system so easily as for K and Q, by merely adding the above equation properly oriented. Consider for example the problems involving distributivity and the necessity-rule:

$$((L(x)\cdot x_1)\oplus(L(x)\cdot x_2)\oplus\ldots\oplus(L(x)\cdot x_n))=L(x)$$

where $x=x_1\oplus x_2\oplus\ldots\oplus x_n$

or the following permutative equivalence:

$$L(x1\oplus x2\oplus 1)\cdot x1=L(x1\oplus x2\oplus 1)\cdot x2$$

with these two terms still equivalent to

$$L(x1\oplus x2\oplus 1)\cdot x1\cdot x2$$

This last example suggests that we should somehow orientate the necessity-rule from right to left. To keep the termination property, we are led to introduce a new operator C, so as to distinguish the occurrences of L where the rule has already been applied. This amounts to translating the system T into the system K, where the modal operators C and L are respectively those of K and T, by the rule

$$L(x)\hookrightarrow C(x)\cdot x$$

We show how the above permutative problem gets solved:

$$\begin{aligned}L(x_1\oplus x_2\oplus 1)\cdot x_1 &\hookrightarrow C(x_1\oplus x_2\oplus 1)\cdot(x_1\oplus x_2\oplus 1)\cdot x_1\\ &\hookrightarrow C(x_1\oplus x_2\oplus 1)\cdot x_1\cdot x_2\end{aligned}$$

The following property states the translation correctness.

Definition 2 *We define a translation Tr from T to K, and a "converse" translation rt from K to T by the following equations, where the necessity operators are L for T and C for K:*

$$\begin{aligned}Tr(x\cdot y)&=Tr(x)\cdot Tr(y) & rt(x\cdot y)&=rt(x)\cdot rt(y)\\ Tr(x\oplus y)&=Tr(x)\oplus Tr(y) & rt(x\oplus y)&=rt(x)\oplus rt(y)\\ Tr(Lx)&=C(Tr(x))\cdot Tr(x) & rt(Cx)&=L(rt(x))\\ Tr(Vn)&=Vn & rt(Vn)&=Vn\\ Tr(1)&=1 & rt(1)&=1\\ Tr(0)&=0 & rt(0)&=0\end{aligned}$$

Lemma 1 $T\vdash rt(Tr(x))=x$

Proposition 3 $T\vdash x=y$ iff $K\vdash Tr(x)=Tr(y)$

To prove the "if" part, we use the translation rt, and the previous lemma. The "only if" part has been proved by induction on the length of an equational derivation.

Proposition 4 *The following rewriting system which codes the above translation is locally confluent and terminating:*

*cp-T

$x \oplus x \hookrightarrow 0$
$x \cdot (y \oplus z) \hookrightarrow (x \cdot y) \oplus (x \cdot z)$
$x \cdot 0 \hookrightarrow 0$
$x \cdot x \hookrightarrow x$
$x \oplus 0 \hookrightarrow x$
$x \cdot 1 \hookrightarrow x$
$C(x) \cdot C(y) \hookrightarrow C(x \cdot y)$
$C(1) \hookrightarrow 1$
$L(x) \hookrightarrow C(x) \cdot x$

To prove the termination, we apply the interpretation given for L in *cp-K to the operator C and add the following interpretation of L:

$$I(L(x)) = 8 \times I(x) \times I(x)$$

We now give a formal characterization of the normal forms obtained by the above rewriting system *cp-T. We first define a *saturation condition* on terms.

Definition 3 *A formula x of K is said to be* saturated *iff*

$$K \vdash Tr(rt(x)) = x$$

Definition 4 *We define the* TK-normal forms *as the saturated K-normal forms, and the* T-normal forms *as the translation by rt of the TK-normal forms.*

In other words the T-normal forms are the fix-points (modulo AC) of the following transformation:
-replace every occurrence of Ly by $Ly \cdot y$ (just once)
-normalize by the system given for K : *cp-K

Proposition 5 *The TK-normal forms are the normal forms obtained by the rewriting system *cp-T from the T-formulae and the T-normal forms are canonical in T.*

Note that we might have defined a similar translation from T to Q and that saturated Q-normal forms and saturated K-normal forms coincide.

Generalization

We briefly describe a generalization of the method we have developped for the system T.

Given a set of equations E, and another equation $G = D$, let $\mathcal{A}$ denote the equational algebra defined by $E \cup \{G = D\}$, with an alphabet Σ and a set of variables $\mathcal{V}$.

We suppose that the equation $G = D$ is such that G has the form $L(x_1, \ldots, x_n)$ where L is an operator of arity n and where all x_i are variables and D is constructed from the variables x_i of G, the term $L(x_1, \ldots, x_n)$, and the alphabet $\Sigma - \{L\}$.

We may then introduce a new operator C of arity n and consider the equational algebra $\mathcal{B}$, with alphabet $\Sigma \cup \{C\} - \{L\}$, defined by the set of equations E' obtained from E by replacing L by C.

We then define - as we did for T- a translation rt from $\mathcal{B}$ to $\mathcal{A}$ rewriting C in L, and a translation Tr from $\mathcal{A}$ to $\mathcal{B}$ rewriting $L(x_1, \ldots, x_n)$ in the term D where every L has been replaced by C.

We obtain the following property:

if	$\forall g = d \in E \cup \{G = D\}$	$\mathcal{B} \vdash Tr(g) = Tr(d)$
then	$\mathcal{A} \vdash x = y$ ssi $\mathcal{B} \vdash Tr(x) = Tr(y)$	

Which implies that if a complete rewriting system is known for $\mathcal{B}$, it gives a decision procedure for $\mathcal{A}$ as well.

4 The System S5

We obtain an equational presentation of S5 by adding to T the following equations:

$$\begin{aligned} L(L(x)) &= L(x) \\ L(L(x) \oplus 1) &= L(x) \oplus 1 \end{aligned}$$

Transformation of S5-formulae to formulae of modal degree one

We recall the inductive definition of the modal degree $d(x)$ of a formula x:

$$\begin{aligned} d(Lx) &= 1 + d(x) \\ d(x \cdot y) = d(x \oplus y) &= sup\{d(x), d(y)\} \\ d(Vx) = d(0) = d(1) &= 0 \end{aligned}$$

When classical conjunction, disjunction and negation are taken as primitive, it is a well-known result that the equations above allow to transform any S5-formula to an equivalent S5-formula of modal degree one or zero. This transformation then leads to the notion of *modal conjunctive normal form*.

We may also try to perform this transformation with conjunction and exclusive disjunction taken as primitive, so as to obtain a rewriting system based on the previous ones. We thus obtain the following rules:

$$\begin{aligned} L(L(x)) &\hookrightarrow L(x) \\ L(L(x) \cdot y) &\hookrightarrow L(x \cdot y) \\ L(L(x) \oplus y) &\hookrightarrow L(x) \vee L(-y)) \cdot (L(x) \vee L(y)) \\ L((L(x) \cdot y) \oplus z) &\hookrightarrow (-L(x) \vee L(-z)) \cdot L(-y \vee -z) \cdot (L(x) \vee L(y \vee z)) \end{aligned}$$

However, the system obtained by adding these rules to *cp-T, is not terminating.

We give a solution which amounts to reducing any S5-formula to a modal conjunctive normal form and then to normalize it in the system T. The correctness of this method comes from the identitity of S5-theorems and T-theorems of modal degree zero or one .

A complete rewrite system for S5

In the following system, $T1$ is introduced to transform any S5-formula constructed with the classical connectives of conjunction, disjunction and negation, (respectively denoted as $\times$, $+$ and $-$, in the rewrite system) to an equivalent formula of degree one and constructed with $\cdot$ and $\oplus$.

The auxiliary operators $T2$ and $T3$ are introduced to make the completion easier.

The ternary operator K sorts the clauses in a disjunction, its first argument is a term to be sorted, the second one is the list of terms $L(x)$ or $-L(x)$ and the third one is the list of clauses of modal degree 0. The operator : is a list operator, but has the same meaning as disjunction.

Proposition 6 *The following system *cp-abv-s5 is locally confluent and terminating*

*cp-abv-s5

$x \oplus x \hookrightarrow 0$	(classical rules)
$x \cdot (y \oplus z) \hookrightarrow (x \cdot y) \oplus (x \cdot z)$	
$x \cdot 0 \hookrightarrow 0$	
$x \cdot x \hookrightarrow x$	
$x \oplus 0 \hookrightarrow x$	
$x \cdot 1 \hookrightarrow x$	
$xvy \hookrightarrow (x \cdot y) \oplus x \oplus y$	(abbreviation)
$C(x) \cdot C(y) \hookrightarrow C(x \cdot y)$	(rules for K)
$C(1) \hookrightarrow 1$	
$T1(L(x)) \hookrightarrow T2(K(x,0,0))$	(main translation: T1)
$T1(x + y) \hookrightarrow (T1(x) \cdot T1(y)) \oplus T1(x) \oplus T1(y)$	
$T1(-(x)) \hookrightarrow T1(x) \oplus 1$	
$T1(V(x)) \hookrightarrow V(x)$	
$-(-(x)) \hookrightarrow x$	
$K(x + V(y), z, u) \hookrightarrow K(x, z, V(y) : u)$	(sorting operator: K)
$K(x + -(V(y)), z, u) \hookrightarrow K(x, z, -(V(y)) : u)$	(K(x,y,z) means L(x + y + z))
$K(V(x), y, z) \hookrightarrow K(0, y, V(x) : z)$	((x:y) means (x + y))
$K(-(V(x)), y, z) \hookrightarrow K(0, y, -(V(x)) : z)$	
$K(x + L(y), z, u) \hookrightarrow K(x, L(y) : z, u)$	
$K(x + -(L(y)), z, u) \hookrightarrow K(x, -(L(y)) : z, u)$	
$K(L(x), y, z) \hookrightarrow K(0, L(x) : y, z)$	
$K(-(L(x)), y, z) \hookrightarrow K(0, -(L(x)) : y, z)$	
$K(x + -(y + z), u, v) \hookrightarrow K(x + -(y), u, v) \times K(x + -(z), u, v)$	
$K(x + (y + z), u, v) \hookrightarrow K((x + y) + z, u, v)$	
$K(-(x + y), z, u) \hookrightarrow K(-(x), z, u) \times K(-(y), z, u)$	
$K(x + 0, y, z) \hookrightarrow K(x, y, z)$	
$T2(x \times y) \hookrightarrow T2(x) \cdot T2(y)$	(translation when K(x,y,z)
$T2(K(0,0,x)) \hookrightarrow T3(x) \cdot C(T3(x))$	is sorted)
$T2(K(0, x : y, z)) \hookrightarrow (T1(x) \cdot T2(K(0,y,z))) \oplus T1(x) \oplus T2(K(0,y,z))$	
$T2(K(x + -(0), y, z)) \hookrightarrow 1$	
$T3(0) \hookrightarrow 0$	(translation of ":")
$T3(x : y) \hookrightarrow (T1(x) \cdot T3(y)) \oplus T1(x) \oplus T3(y)$	

The following interpretation on integers proves the termination:

$$
\begin{aligned}
I(x \cdot y) &= 2^2 \times I(x) \times I(y) & I(x : y) &= 2^4 \times I(x) \times I(y) \\
I(x \oplus y) &= I(x) + I(y) + 1 & I(Kxyz) &= (2^2 \times I(y) \times I(z))^{I(x)} \\
I(Cx) &= 2 \times I(x) & I(L(x)) &= 2^{5 \times I(x)} \\
I(T1x) &= 2 \times I(x) & I(-x) &= 2^2 \times I(x) \\
I(T2x) &= 2 \times I(x) & I(x \times y) &= 2^2 \times I(x) \times I(y) \\
I(T3x) &= 2 \times I(x) & I(0) = I(1) &= 2 \\
I(x \vee y) &= 2^2 \times I(x) \times I(y) & I(Vx) &= 2^4 \\
I(x + y) &= 2^4 \times I(x) \times I(y)^2 & &
\end{aligned}
$$

The following proposition states the correctness and completeness of the above rewriting system:

Proposition 7 *$S5 \vdash x = y$ if and only if $T1(x) =_{*cp-s5} T1(y)$*

The proof is made by induction on the length of a derivation. The "only if" part is a long examination by cases. The "if" part is obtained by defining a converse translation rt such that $S5 \vdash rt(T1(x)) = x$.

Proposition 8 *The canonical forms obtained are the saturated K-normal forms of modal degree one. They may be characterized as the K-normal forms of modal degree one such that each LK-component l written as:*

$$l = (L(a) \cdot b_1) \oplus (L(a) \cdot b_2) \ldots L(a) \cdot b_n$$

satisfies the condition that $b_1 \oplus b_2 \ldots b_n$ is the CPC-normal form of:

$$a \cdot (b_1 \oplus b_2 \ldots b_n)$$

The proof can be made directly, based on semantic Kripke structures or it can be an application of the results for T.

Examples

KB: normalize

---

term: (T1 (L (V x)))
T1(L(V(x))) ↪ C(V(x)) · V(x)
Time: 1 s. [GC: 0 s.]

---

term: (T1 (L (L (V x))))
T1(L(L(V(x)))) ↪ C(V(x)) · V(x)
Time: 9 s. [GC: 3 s.]

CONCLUSION

These results provide a new decision procedure but they also provide a uniform framework for investigating modal structures: they give at one and the same time canonical forms and a procedure to reduce to them, which allows a precise way of reasoning on terms .

As concerns term rewriting techniques, whereas the technique developped for S5 seems to be specific, our method for the system T may be applied to a class of structures.

References

[1] P. Enjalbert et L. Farinas Del Cerro. *1986 Modal Resolution In Clausal Form.* Greco de programation du C.N.R.S.

[2] F.Fages. *1984 Le système KB Présentation et Bibliographie, mise en oeuvre* Rapport INRIA, France.

[3] D.M. Gabbay. *1981 Semantical Investigations in Heyting's Intuitionnistic Logic.* Dordrecht, Reidel.

[4] G.Huet and D. Oppen. *January 1980 Equations and Rewrite Rules:a survey.* Technical Report 15, Stanford Verification Group.

[5] G.E. Hughes et M.J. Cresswell. *1968 An Introduction to Modal Logic.* Methuen & Co.,London.

[6] G.E. Hughes et M.J. Cresswell. *1984 A Companion to Modal Logic.* Methuen & Co.,London.

[7] J.Hsiang *Nov. 1982 Topics in Automated Theorem Proving and Program Generation.* Ph.D. Thesis, Univ.of Illinois at Urbana-Champaign.

[8] J.C.C. McKinsey and A. Tarski *1948 Some Theorems about the Sentential Calculi of Lewis and heyting.* JSL Vol. 13 pp. 1-15.

[9] S.A. Kripke.*1974 Semantical analysis of intuitionistic logic I.* Formal systems and recusive functions, ed. J.N. Crossley and M.A.E. Dummett p.92-130, North Holland, Amsterdam.

[10] R.E. Ladner. 1977 The computational complexity of provability in systems of modal propositional logic SIAM J. C. 6 (3)

[11] H. Rasiowa et R. Sikorsi. *1968 The Mathematics of Metamathematics.* Warszowa.

[12] H. Rasiowa. *1974 An Algebraic Approach To Non-Classical Logics.* North-Holland, Amsterdam.

[13] J.J. Zeman. *Modal Logic : The Lewis-Modal Systems.* Clarendon Press, Oxford.

A CHARACTERIZATION OF IRREDUCIBLE SETS MODULO LEFT-LINEAR TERM REWRITING SYSTEMS BY TREE AUTOMATA

Z. Fülöp and S. Vágvölgyi
Research Group on Theory of Automata
Hungarian Academy of Sciences
H-6720 Szeged, Somogyi u. 7., Hungary

J. H. Gallier and R. V. Book showed that the set of irreducible terms modulo a left-linear term rewriting system can be recognized by a deterministic bottom-up tree automaton. However, this device is too general to characterize irreducible sets: there are recognizable term sets which cannot be the set of irreducible terms modulo any term rewriting system. We introduce the class of one-state deterministic top-down tree automata with prefix look-ahead, which exactly recognizes the class of irreducible terms modulo left-linear term rewriting systems.

A one-state deterministic top-down tree automaton with prefix look-ahead is a system $A=(\Sigma,q,P)$, where

(a) Σ is a ranked alphabet,

(b) q is the state of A,

(c) P is a finite set of rules of the form

$\langle q\sigma\to(q,\dots,q);L\rangle$ with $m\geq 0$, $\sigma\in\Sigma_m$, where $L=tT_\Sigma=\{t(t_1,\dots,t_n)\mid t_i\in T_\Sigma$ for $1\leq i\leq n\}$ for some linear tree $t\in T_\Sigma(X_n)$. Moreover, for any different rules $\langle q\sigma\to(q,\dots q);L_1\rangle$ and $\langle q\sigma\to(q,\dots,q),L_2\rangle$, $L_1\cap L_2=\emptyset$ holds.

The tree language recognized by A is $L(A)=\{t\in T_\Sigma\mid q(t)\overset{*}{\Rightarrow}t\}$, where $\overset{*}{\Rightarrow}$ is the reflexive-transitive closure of $\Rightarrow$ defined as follows: for $u,v\in T_\Sigma(qT_\Sigma)$, $u\Rightarrow v$ if and only if

(i) $u=c(q(\sigma(t_1,\dots,t_m)))$ for some $c\in T_\Sigma(\{x_1\})$, x_1 occurs once in c, $m\geq 0$, $\sigma\in\Sigma_m$ and $t_1,\dots,t_m\in T_\Sigma$,

(ii) $v=c(\sigma(q(t_1),\dots,q(t_m)))$,

(iii) the rule $\langle q\sigma\to(q,\dots,q);L\rangle$ is in P,

(iv) $\sigma(t_1,\dots,t_m)\in L$.

Our main result is the following theorem.

Theorem: For any set T of terms, T is the set of irreducible terms with respect to a left-linear term rewriting system if and only if T can be recognized by a one-state deterministic top-down tree automaton with prefix look-ahead.

FUNCTIONAL DATA TERM MODELS AND SEMANTIC UNIFICATION

Hans-Joachim Goltz [1)]

Abstract

A unified language which combines logic and functional programming is a good candidate for a basic programming language for the knowledge representation. A new semantic of such a language is presented. The interpretation is based on functional data term models. The universe of such a model is a set of data terms defined by constructor symbols. The symbols of defined functions are interpreted as partial functions on data terms. The chosen axioms for the equality do not contain the reflexivity axiom. Terms can only be equal if the terms are defined terms. A new unification algorithm for the presented semantic is given and the completeness of it is proved.

1. Introduction

The abstract form of the representation of declarative knowledge is the representation based on some set of objects, relations between objects and functions on objects. The basic representation of objects are data terms. Therefore, an important approach to a basic programming language for knowledge representation is the development of a unified language which combines logic and functional programming. Horn clauses extended by rewrite rules have been proposed by several authors (see, e.g. [2],[3],[5]) as a natural way of combining logic and functional programming, because of the similarities between rewrite rules and Horn clauses. In most of existing languages of this kind computation is performed by using extended unification. We present a new model theoretic interpretation of such a language. The language is an integration of a functional language into a logic programming language based on definite clauses. The function symbols are divided into constructor symbols and "defined function" symbols. Our functional language is similar to the constructor-based functional language in [3] (see also [10]). Functions can be defined by rewrite rules, where the terms of the left-hand side have to satisfy the following condition: if $f(t_1,...,t_n)$ is such a term then $t_1,...,t_n$ are data terms and f is a symbol for a defined function. This restriction is called "constructur discipline" by O'Donnell ([10]). The new semantic interpretation is based on functional data term models. The universe of such a model is the set of ground data terms (data terms are inductively defined by constructor symbols) and is not the set of ground terms. The "defined function" symbols are interpreted as partial functions on data terms. In contrast to other approaches our approach is based on equational theory without the reflexivity axiom. Terms can only be equal if the terms are "defined terms". Furthermore, a new unification algorithm is given, and the correctness and the completeness of it (with respect to the presented semantic) is proved.

We assume the reader to be familiar with the usual notations and basic results of logic programming and term rewriting (see, e.g. [8],[7]).

---

1) Central Institute of Cybernetics and Information Processes, PF 1298, Kurstr. 33, Berlin, 1086, GDR

2. Basic Definitions and Notations

A <u>signature</u> is determined by a set of function symbols FC and a set of predicate symbols PRED (with $FC \cap PRED = \emptyset$), and is denoted by $\Sigma = (FC, PRED)$. Let FC be such a finite set and $V = \{x, y, z, x_1, \ldots\}$ be a set of variables, where FC is the union of a set F of symbols for defined functions and a set C of constructor symbols ($F \cap C = \emptyset$, $FC \cap V = \emptyset$). We assume that C contains at least one function symbol of arity 0. Functions of arity 0 are also called constants. The set of <u>data terms</u> T(C,V) and the set of <u>terms</u> T(FC,V) are defined inductively by

- each constant symbol of C and each variable symbol is a data term;
- if $c \in C$ is an n-ary constructor symbol and $t_1, \ldots, t_n$ are data terms then $c(t_1, \ldots, t_n)$ is a data term;
- each data term is a term;
- if $f \in FC$ is an n-ary function symbol and $t_1, \ldots, t_n$ are terms then $f(t_1, \ldots, t_n)$ is a term.

A term $t = f(t_1, \ldots, t_n)$ (or $t = f$) is called <u>C-term</u> if $f \in C$ and t is called <u>F-term</u> if $f \in F$. The set of variables in a term t (in a set M of terms) is denoted by Var(t) (Var(M)). We say a term (data term) t is <u>ground</u> if Var(t) is empty. The set of <u>ground data terms</u> is denoted by T(C) and the set of <u>ground terms</u> is denoted by T(FC).

Let PRED be a finite set of predicate symbols with fixed arity. Atomic formulas, formulas, literals, definite clauses and Herbrand models are usually defined. Definite clauses are represented in the form of rules $p \leftarrow q_1, \ldots, q_m$, where p is the positive literal of the clause. The existential closure of a formula φ is denoted by $\exists\bar{x}\varphi$.

A <u>substitution</u> σ is defined by a mapping from V to T(FC,V) such that the set $dom(\sigma) = \{x \in V: \sigma(x) \neq x\}$ is finite. The composition of two substitutions σ and τ is defined by $\sigma \circ \tau(x) = \tau(\sigma(x))$. A substitution σ is a <u>data term substitution</u> if $\sigma(x)$ is a data term for each variable x. A term t is an <u>instance</u> (<u>data term instance</u>) of a term $u \in T(FC,V)$ if t is a ground term and there is a substitution σ (data term substitution σ) such that $t = \sigma(u)$. Correspondingly, an instance (data term instance) of an atomic formula is defined.

A <u>term rewriting system</u> is a set R of ordered pairs of terms (rewrite rules) such that $Var(r) \subseteq Var(u)$ for all $(u,r) \in R$. As usual the reduction relation $\longrightarrow$ and the normal form of a term (with respect to R) are defined. We use $\xrightarrow{*}$ to denote the reflexive and transitive closure of $\longrightarrow$. A term rewriting system R is said to be <u>canonical</u> if there is a unique normal form (denoted by $t\downarrow$) for each term t.

3. Functional Data Term Models

Let FC, F, C, PRED be given sets of function symbols, symbols for defined functions, constructor symbols and predicate symbols, respectively. We assume that the set PRED includes the binary predicate symbols $=$, $\equiv$, $\Rightarrow$. Let us assume that BA is the following set of basic axioms.

(A1) for each constant symbol $a \in FC$: $a \equiv a$

(A2) for each n-ary function symbol $f \in FC$ $(n > 0)$:

$$f(x_1, \ldots, x_n) \equiv f(y_1, \ldots, y_n) \longleftarrow x_1 \equiv y_1, \ldots, x_n \equiv y_n$$

(B1) for each constant symbol $a \in C$: $a = a$

(B2) for each n-ary constructor symbol $c \in C$ $(n > 0)$:

$c(x_1,\dots,x_n) = c(y_1,\dots,y_n) \longleftarrow x_1 = x_1,\dots,x_n = y_n$

(B3) for each n-ary function symbol $f \in F$:

$f(y_1,\dots,y_n) = z \longleftarrow f(x_1,\dots,x_n) \Rightarrow x,\ x_1 = y_1,\dots,x_n = y_n,\ x = z$

$z = f(y_1,\dots,y_n) \longleftarrow f(x_1,\dots,x_n) \Rightarrow x,\ x_1 = y_1,\dots,x_n = y_n,\ x = z$

A progam P consists of a program for function definitions and a program for predicate definitions. A program PF for function definitions is a set of rewrite rules $f(t_1,\dots,t_n) \Rightarrow u$, where $f \in F$, $u \in T(FC,V)$, $t_1,\dots,t_n \in T(C,V)$ (the t_i's are data terms!) and $Var(u) \subseteq Var(f(t_1,\dots,t_n))$. It is assumed that PF is a canonical term rewriting system. For instance, the function 'append' (concatenation of two lists) can be defined by

append([],z) $\Rightarrow$ z
append([x|y],z) $\Rightarrow$ [x|append(y,z)]

A program PP for predicate definitions is a set of definite clauses, where the predicate symbols $\Rightarrow$ and $\equiv$ are not included and the predicate symbols of the positive literals are not identical to $=$. Each definite clause $p(t_1,\dots t_n) \longleftarrow q_1,\dots,q_m$ can be transformed into a homogeneous definite clause $p(x_1,\dots,x_n) \longleftarrow x_1 = t_1,\dots,x_n = t_n,\ q_1,\dots,q_m$, where $x_1,\dots,x_n$ are new variables and are not included in the original clause (see also [4]). Let PH be the set of homogeneous definite clauses obtained from PP by this transformation.

Obviously, the set BA $\cup$ PF $\cup$ PH is a set of definite clauses. It is well known that for any set of definite clauses there is a least Herbrand model satisfying these clauses (see, e.g. [8]). Let H_0 be the set of atomic formulas (without variables) determining the least Herbrand model of BA $\cup$ PF $\cup$ PH and let $H_=$, $H_\equiv$, and $H_\Rightarrow$ be the subsets of H_0 consisting of the atomic formulas with the predicate symbols $=$, $\equiv$, and $\Rightarrow$, respectively. Since PH is a set of homogeneous definite clauses, $u_1 = t_1,\dots,u_n = t_n, p(u_1,\dots,u_n) \in H_0$ implies $p(t_1,\dots,t_n) \in H_0$ for each n-ary predicate symbol $p \in PRED\setminus\{=,\equiv,\Rightarrow\}$ and ground terms $t_1,\dots,t_n,u_1,\dots,u_n$.

Since PF is assumed to be canonical, there is a unique normal form for each term of T(FC,V). We are interested in functions $T(C)^n \longrightarrow T(C)$. For instance, the normal form of append([1,2],[3]) is the data term [1,2,3], but the normal form of append(1,2) is the term append(1,2) which is not a data term. Therefore, we introduce the notation 'defined term'. A ground term t is called <u>defined</u>, if $t\downarrow$ is a data term and (if $t = f(t_1,\dots,t_n)$) $t_1,\dots,t_n$ are also defined terms.

A <u>functional data term model</u> $\mathbf{A}$ for a program P = PF $\cup$ PP is an algebraic structure of the signatur $\Sigma_1 = (FC, PRED_1)$, where $PRED_1 = PRED\setminus\{\equiv,\Rightarrow\}$, satisfying the following conditions:

(C1) the universe of $\mathbf{A}$ is T(C) (or is isomorphic to T(C));

(C2) for each n-ary symbol for a defined function $f \in F$ there is a partial function $f_{\mathbf{A}}: T(C)^n \longrightarrow T(C)$ defined by

(i) the domain of $f_{\mathbf{A}}$ is the set $Dom(f) = \{(t_1,\dots,t_n) : t_i \in T(C),\ f(t_1,\dots,t_n)\downarrow \in T(C)\}$ and

(ii) if $(t_1,\dots,t_n) \in Dom(f)$ then the value of $f_{\mathbf{A}}(t_1,\dots,t_n)$ is the term $f(t_1,\dots,t_n)\downarrow$;

(C3) for each defined term $t \in T(FC)$ the interpretation value(t) is defined inductively as follows:

(i) for each data term $t \in T(C)$, value(t) is identical to t

(ii) suppose that $t = f(t_1,..t_n)$; if $f \in C$ then value(t) is defined by $f(\text{value}(t_1),\ldots,\text{value}(t_n))$, and if $f \in F$ then value(t) is defined by $f_{\mathbf{A}}(\text{value}(t_1),\ldots,\text{value}(t_n))$;

(C4) for each n-ary predicate symbol $p \in PRED_1$ the interpretation $p_{\mathbf{A}}$ is defined by $p_{\mathbf{A}} = \{(t_1,\ldots,t_n) \in T(C)^n : p(t_1,\ldots,t_n) \in H_0\}$;

(C5) for each n-ary predicate symbol $p \in PRED_1$ and for all ground terms $t_1,\ldots,t_n \in T(FC)$, the validity of $p(t_1,\ldots,t_n)$ in $\mathbf{A}$ is defined as follows: $\mathbf{A} \models p(t_1,\ldots,t_n)$ iff $t_1,\ldots,t_n$ are defined terms and $(\text{value}(t_1),\ldots,\text{value}(t_n)) \in p_{\mathbf{A}}$

(C6) if $t,u \in T(FC)$ then $\mathbf{A} \models t = u$ iff t,u are defined terms and value(t) is identical to value(u);

(C7) for each n-ary predicate sympol p of $PRED_1$ and for all terms $t_1,\ldots,t_n \in T(FC)$: $\mathbf{A} \models p(t_1,\ldots,t_n)$ iff $p(t_1,\ldots,t_n) \in H_0$

We have to show that there is such functional data term model for a program P. Obviously, it can be defined an algebraic structure $\mathbf{A}_0$ of the signature Σ_1 satisfying the conditions (C1)-(C5). Note that the value of each defined term $t \in T(FC)$ is a data term and that value(t) is identical to $t\downarrow$. It is easy to see that such algebraic structure $\mathbf{A}_0$ satisfies the conditions (C6) and (C7) if the following theorems are proved.

THEOREM 1. If $t,u \in T(FC)$, then $t = u \in H_=$ iff t,u are defined terms and $t\downarrow$, $u\downarrow$ are identical.

A proof of Theorem 1 follows from Lemma 1, Lemma 4 and Lemma 5, which are proved in section 4.

THEOREM 2. If $t_1,\ldots,t_n \in T(FC)$ and $p \in PRED_1$, then $p(t_1,\ldots,t_n) \in H_0$ iff $t_1,\ldots,t_n$ are defined terms and $p(t_1\downarrow,\ldots,t_n\downarrow) \in H_0$.

Proof. Let p be an n-ary predicate symbol of $Pred_1$ and let $t_1,\ldots,t_n \in T(FC)$. Obviously, $H_1 = H_0 \setminus (H_= \cup H_{\equiv} \cup H_{\Longrightarrow})$ is the least Herbrand model of PH, and $p(t_1,\ldots,t_n) \in H_0$ iff $p(t_1,\ldots,t_n) \in H_1$. Firstly, assumed that $p(t_1,\ldots,t_n) \in H_1$. Since H_1 is a least Herbrand model of a set of homogen clauses there are $u_1,\ldots,u_n \in T(FC)$ such that $\{t_1 = u_1,\ldots,t_n = u_n\} \subseteq H_=$. Because of Theorem 1 the terms $t_1,\ldots,t_n$ are defined terms and $t_i\downarrow$ is identical to $u_i\downarrow$ (for each i). Obviously, $\{t_1\downarrow = u_1,\ldots,t_n\downarrow = u_n\} \subseteq H_=$. Consequently, $p(t_1\downarrow,\ldots,t_n\downarrow)$ belongs to H_1. In the other case, if $p(t_1\downarrow,\ldots,t_n\downarrow) \in H_1$ and $t_1,\ldots,t_n$ are defined terms, then from $t_i\downarrow = u_i \in H_=$ follows $t_i = u_i \in H_=$. Thus, analogously to the first case, it can be shown that $p(t_1,\ldots,t_n) \in H_1$. □

COROLLARY 1. If $\mathbf{A}_0$ and PH are defined as above then $\mathbf{A}_0$ is a model of PH.

EXAMPLE. Let C = {s,0} be a set of constructor symbols and F = {+,-} be a set of symbols for defined functions, where 0 is a constant symbol, s is a unary function symbol and +,- are binary function symbols. Furthermore, let $<$ be a binary predicate symbol. We regard the following program P_N:

$$x + 0 \Longrightarrow x \qquad\qquad 0 < s(x)$$
$$x + s(y) \Longrightarrow s(x + y) \qquad\qquad s(x) < s(y) \longleftarrow x < y$$
$$x - 0 \Longrightarrow x$$
$$s(x) - s(y) \Longrightarrow x - y$$

The standard model of natural numbers $\mathbb{N}$ with +, -, $<$ and their standard interpretation is a functional data term model for P_N. Note that for all ground terms $t,u \in T(C)$ the term $t - u$ is not a defined term if $t < u$, and that there are not computation rules for the combination of + and - in some term (e.g. $s(s(0)) + (0 - s(0)$ is not a defined term).

4. Properties of the Equational Relation

Let H_0, $H_=$, $H_\equiv$, $H_\Rightarrow$ be the sets defined in section 3 (with respect to a given program $P = PF \cup PP$). Note that PF is assumed to be a canonical rewrite system.

LEMMA 1. If t and u are ground terms of T(FC) then $t \equiv u \in H_\equiv$ iff the terms t and u are (syntactically) identical.

The propositon of this lemma can be proved inductively on the term construction and by using that $H_\equiv$ is a least set satisfying the axioms (A1) and (A2). Obviously, the relation $\equiv$ is an equivalence relation with respect to T(FC).

LEMMA 2
(i) If $t,u \in T(C)$, then $t = u \in H_0$ iff $t \equiv u \in H_0$.
(ii) The relation = is an equivalence relation with respesct to the set of ground data terms T(C).
(iii) If $t,u \in T(FC)$ and $t = u \in H_0$ then $u = t \in H_0$.

Proof. The propositions (i) and (iii) can be proved inductively on the term construction (and by using the axioms). Then, propositon (ii) follows from Lemma 1. □

LEMMA 3. If t and u two terms of T(FC,V), then $H_0 \models \exists\bar{x}(t = u)$ iff one of the following conditions is valid
(1) t and u are identical constant symbols of C (i.e. $u \equiv t \in H_0$);
(2) there is an n-ary constructor symbol $c \in C$ $(n>0)$ such that
$$H_0 \models \exists\bar{x}(t \equiv c(x_1,\dots,x_n) \wedge u \equiv c(y_1,\dots,y_n) \wedge x_1 = y_1 \wedge\dots\wedge x_n = y_n);$$
(3) there is an n-ary symbol of a defined function $f \in F$ such that
$$H_0 \models \exists\bar{x}(((t \equiv f(x_1,\dots,x_n) \wedge u \equiv y) \vee (t \equiv y \wedge u \equiv f(x_1,\dots,x_n)) \wedge \wedge f(z_1,\dots,z_n) \Rightarrow x \wedge x_1 = z_1 \wedge\dots\wedge x_n = z_n \wedge x = y) .$$

Proof. The proposition is a consequence of the definition of the relation = and the fact that H_0 is a least Herbrand model. □

In the following, we define the degree of an equation $t = u$ (denoted by degree($t = u$), where t and u are ground terms. Degrees of equations are only used in the proof of Lemma 4. Firstly, we have to define sets of equations. Let $\mathfrak{M}_0$ be the set $\{ a = a : a \text{ is a constant symbol of } C \}$, and let $\mathfrak{N}_0$ be the set $\mathfrak{M}_0 \cup \{ t \Rightarrow u : t \Rightarrow u \text{ is an instance of a rewrite rule of PF}\}$. If $k > 0$ then the set $\mathfrak{N}_k$ is defined by $\mathfrak{N}_k := \bigcup\{ \mathfrak{M}_m : m < k\}$ and $\mathfrak{M}_k$ is defined by
$$\mathfrak{M}_k := \{c(t_1,\dots,t_n) = c(u_1,\dots,u_n) : c \in C,\ n > 0,\ t_i = u_i \in \mathfrak{N}_k \text{ for each } i \leq n\} \cup$$
$$\cup \{u = f(t_1,\dots,t_n), f(t_1,\dots,t_n) = u : f \in F, \text{ there is a rule } f(v_1,\dots,v_n) \Rightarrow v \in \mathfrak{N}_0 \text{ with } v = u \in \mathfrak{N}_k \text{ and } v_i = t_i \in \mathfrak{N}_k \text{ for each } i \leq n\}.$$
It can easily be shown that $\bigcup\{ \mathfrak{N}_m : m < \omega\}$ is the set $H_= \cup H_\Rightarrow$. If t and u are ground terms then the degree of the equation $t = u$ is defined by
$$\text{degree}(t = u) := k \text{ iff } k = \min\{ k : t = u \in \mathfrak{M}_k\}.$$

LEMMA 4. Let t,u be ground terms of T(FC). If $t = u \in H_0$ then t,u are defined terms and $t\downarrow \equiv u\downarrow \in H_0$.

Proof. The proposition is proved inductively on degree(t = u). Let degree(t = u) be equal to k. If k = 0 then the propositon is trivial. Let k > 0 and let t be a term of the form $f(t_1,\dots,t_n)$, where $f \in F$. Since $f(t_1,\dots,t_n) = u \in \mathfrak{R}_k$, there are $v_1,\dots,v_n,v \in T(FK)$ such that $\mathfrak{R}_k$ includes $f(v_1,\dots,v_n) \Rightarrow v$, $v_1 = t_1, \dots, v_n = t_n$ and $v = u$ (correctly, it can be necessary to regard this with respect to the term u). By using the assumption of the induction, the terms v_i, t_i, v, u are defined terms and $v_i\downarrow \equiv t_i\downarrow$, $v\downarrow \equiv u\downarrow \in H_0$ (for each $i \leq n$). Since $f(v_1,\dots,v_n)$-*→ -*→$f(v_1\downarrow,\dots,v_n\downarrow)$, $f(t_1,\dots,t_n)$-*→$f(t_1\downarrow,\dots,t_n\downarrow)$, $f(v_1\downarrow,\dots,v_n\downarrow)$ and $f(t_1\downarrow,\dots,t_n\downarrow)$ are identical and PF is assumed to be canonical, $t\downarrow$ and $v\downarrow$ are identical. Thus, $t\downarrow \equiv v\downarrow \in H_0$ and t is also a defined term. The proofs of the other cases are easy. □

Now we define a binary relation ≺ in the set M = {{t,u} : t,u are defined terms of T(FC)} We need this relation for the proof of Lemma 5 only. The relation ≺ is defined by:

$\{t_1,u_1\} \prec \{t_2,u_2\}$ iff there are defined terms t,u,v,w such that {t,u} is identical to $\{t_1,u_1\}$, {v,w} is identical to $\{t_2,u_2\}$ and one of the following conditions holds

(1) there are an n-ary constructor symbol c such that $v \equiv c(v_1,\dots,v_n)$, $w \equiv c(w_1,\dots,w_n)$, $t \equiv v_i$, $u \equiv w_j$ for some i and some j ;

(2) there are an n-ary functional symbol $f \in FC$ such that $v \equiv f(v_1,\dots,v_n)$, u is a data term and $t \equiv v_i$ for some i ;

(3) $u \equiv w$ and v is reducible to t .

Obviously, there does not exist an infinite chain of the form $\dots \prec \{t_3,u_3\} \prec \{t_2,u_2\} \prec \{t_1,u_1\}$.

LEMMA 5. Let $t,u \in T(FC)$ be defined terms. If $t\downarrow \equiv u\downarrow \in H_0$, then $t = u \in H_0$.

Proof. Assumed that there are defined terms $t,u \in T(FC)$ such that $t\downarrow \equiv u\downarrow \in H_0$ and $t = u \notin H_0$. We can choose terms t,u such that there do not exist defined terms v,w with this property and {v,w} ≺ {t,u}.

Case 1: t and u are C-terms ($t = c(t_1,\dots,t_n)$, $u = d(u_1,\dots,u_m)$). Then $t\downarrow$ is the term $c(t_1\downarrow,\dots,t_n\downarrow)$ and $u\downarrow$ is $d(u_1\downarrow,\dots,u_m\downarrow)$. Since $t\downarrow \equiv u\downarrow \in H_0$ we have that $t_i\downarrow \equiv u_i\downarrow \in H_0$ for each $i \leq n$ (and c = d, n = m). Because of the choice of t and u , for each i, $t_i = u_i \in H_0$. Then, Lemma 3 implies $t = u \in H_0$.

Case 2: t is a F-term (the case that u is a F-term is analogous). Let t be the term $f(t_1,\dots,t_n)$ and let r be the term $f(t_1\downarrow,\dots,t_n\downarrow)$. Since PF is assumed to be canonical the term r is a defined term and $r\downarrow \equiv t\downarrow \in H_0$. Consequently, $r\downarrow \equiv u\downarrow \in H_0$ and the term r is reducible. There is a rewrite rule $f(v_1,\dots,v_n) \Rightarrow v$ of PF and a substitution б, such that $б(v_i) \equiv t_i\downarrow$ (for each i). Let w be the term б(v). Obviously, $f(t_1\downarrow,\dots,t_n\downarrow) \Rightarrow w$ belongs to H_0. Since $t_i\downarrow \equiv (t_i\downarrow)\downarrow$ and $\{t_i,t_i\downarrow\} \prec \{t,u\}$ we have $t_i = t_i\downarrow \in H_0$ for each $i \leq n$. Furthermore, $u\downarrow \equiv t\downarrow \equiv r\downarrow \equiv w\downarrow$ and {w,u} ≺ {t,u}. Thus w = u belongs to H_0 and, by using Lemma 3, $f(t_1,\dots,t_n) = u$ belongs to H_0. □

COROLLARY 2. The relation = is a equivalence relation in the set of defined ground terms of T(FC).

5. Semantic Unification

Two terms t and u are <u>semantically unifiable</u> with respect to a program P iff there exists a data term substitution 6 (called <u>semantic unifier</u> for t and u) such that $t_1 = u_1 \in H_=$ for each data term instance $t_1 = u_1$ of 6(t) = 6(u). The following lemma is a consequence of this definition.

LEMMA 6. If 6 is a unifier for a variable x and a data term t then $6(x) \equiv 6(t)$.

A set of equations $\{t_1 = u_1,\dots,t_n = u_n\}$ can be also regard as a set of term pairs $\{(t_1,u_1),\dots,(t_n,u_n)\}$. Since the relation = is symmetric we can assume that each term pair (t_i,u_i) satisfies the following condition

if u_i is a variable then t_i is also a variable

(in the other case we can exchange these terms).

A substitution 6 is a semantic unifier for a set of term pairs if and only if, for each term (t_i,u_i) of this set, 6 is a semantic unifier for t_i and u_i. A term pair $(t_i,u_i) \in \{(t_1,u_1),\dots,(t_n,u_n)\}$ is said to be in <u>solved form</u> if t_i is a variable, t_i occurs only there in this set of term pairs, and u_i is a data term. A set of term pairs is said to be in solved form if and only if all pairs are in solved form. Obviously, if $\{(x_1,t_1),\dots,(x_n,t_n)\}$ is a set of term pairs in solved form then the substitution $\{x_1/t_1,\dots,x_n/t_n\}$ is a semantic unifier of this set. This substitution is also called the substitution represented by this set of term pairs in solved form. Since the semantic unification of a set of term pairs in solved form is a syntactic unification the following lemma is easily proved.

LEMMA 7. If 6 is a substitution represented by a set S of term pairs in solved form and if τ is a unifier for S, then there is a substitution λ such that $\tau = 6 \circ \lambda$ (i.e. 6 is more general as τ).

Let M be a set of term pairs. Formally, the set M is partitioned into two parts, the solved part S and the unsolved part U, and we denote this by M = S + U . We define now some transformation rules on sets of term pairs which will be used by the algorithm for the semantic unification. Our aim is the transformation of a set of term pairs into such set in solved form, if it is possible. In the following x denotes a variable.

(R1) Trivial reduction: $(\{(x,x)\} \cup U_1) + S \Longrightarrow U_1 + S$

(R2) C-term decomposition:

$(\{(c(t_1,\dots,t_n),c(u_1,\dots,u_n))\} \cup U_1) + S$

$\Longrightarrow (\{(t_1,u_1),\dots,(t_n,u_n)\} \cup U_1) + S,$

where c is an n-ary constructor symbol

(If n = 0 this rule is reduced to $(\{(c,c)\} \cup U_1) + S \Longrightarrow U_1 + S$)

(R3) Variable substitution:

$(\{(x,t)\} \cup U_1) + S \Longrightarrow 6(U_1) + (6(S) \cup \{(x,t)\})$, where x does not occur in t, t is a data term, and 6 is the substitution {x/t}

(R4) Outermost term rewriting:

$(\{(f(t_1,\ldots,t_n),s)\} \cup U_1) + S$

$\Longrightarrow (\{(t_1,u_1),\ldots,(t_n,u_n),(r,s)\} \cup U_1) + S$

or $(\{(s,f(t_1,\ldots,t_n))\} \cup U_1) + S$

$\Longrightarrow (\{(t_1,u_1),\ldots,(t_n,u_n),(r,s)\} \cup U_1) + S$,

where $f(u_1,\ldots,u_n) \Rightarrow r$ is a rewrite rule of PF.

(R5) C-term generation:

$(\{(x,c(t_1,\ldots,t_n))\} \cup U_1) + S$

$\Longrightarrow (\{(x_1,\sigma(t_1)),\ldots,(x_n,\sigma(t_n))\} \cup U_1) + (\sigma(S) \cup \{(x,c(x_1,\ldots,x_n))\})$,

where $c(t_1,\ldots,t_n)$ is a C-term, but is not a data term, σ is the substitution $\{x/c(x_1,\ldots x_n)\}$ and $x_1,\ldots,x_n$ are new variables.

Since we assume that a set of term pairs satisfy the condition described above the terms of each pair have to be exchanged if it is necessary.

LEMMA 8. Let M_1 be a set of term pairs obtained by applying any of the above transformation rules to a set of term pairs M. If a substitution σ is a semantic unifier for M_1 then σ is also a semantic unifier for M.

Proof. Let (R_k) be the rule such that M_1 is obtained by applying this rule (R_k) to the set M. Since $\sigma(x)$ is a data term for each variable x, the proposition holds if k = 1. If k = 2 or k = 4 the propositon can easily be shown by using the Lemma 3. Let k = 3, (x,u) be the chosen term pair of M and let σ_x be the substitution {x/u}. Since $\sigma(x) \equiv \sigma(u)$ (Lemma 6), we have $\sigma(t) \equiv \sigma(\sigma_x(t))$ for any term t. If $(r,t) \in M\backslash\{(x,u)\}$ then $(\sigma_x(r),\sigma_x(t)) \in M_1$. Thus σ is also a semantic unifier for M. Supposed that k = 5. Since $\sigma(x) \equiv \sigma(c(x_1,\ldots,x_n))$ the proposition can be shown analogously to the case k = 3. □

LEMMA 9. Let M be a set of term pairs and let σ be a semantic unifier for M, where the unsolved part of M is not empty and $dom(\sigma) \subseteq Var(M)$. Then there is a transformation rule and a substitution σ_1 such that $\sigma_1 \restriction dom(\sigma) = \sigma$ and σ_1 is a semantic unifier for M_1, where M_1 is obtained by applying this transformation rule to M.

Proof. Let (r,t) be a term pair of the unsolved part U of M = U + S. Since the data term instances of $\sigma(r) = \sigma(t)$ belong to $H_=$, there is an SLD-refutation of r = t (correctly, an SLD-refutation of $BA \cup PF \{\neg(r = t)\}$) which generates an answer substitution τ such that $\sigma = \tau \circ \lambda$ for a substitution λ (see [1]). In the first step this refutation have to use one of the axioms (B1), (B2) or (B3).
Firstly, we assume that the terms r and t are not variables. If r and t are C-terms then the propositon is easily to be shown. If r or t is an F-term, then axiom (B3) can only be used in the first step of the refutation. Assumed that r is the F-term $f(r_1,\ldots,r_m)$. Then r = t is reduced to $f(x_1,\ldots,x_m) \Rightarrow x \wedge x_1 = r_1 \wedge\ldots\wedge x_m = r_m \wedge x = t$. We can assume that a rewrite rule $f(u_1,\ldots,u_m) \Rightarrow u$ of PF is used in the next step. Thus, r = t is reduced to $u_1 = r_1 \wedge\ldots\wedge u_m = r_m \wedge u = t$ and the transformation rule (R4) can be applied to M. Because of Lemma 3 the substitution σ is a semantic unifier for M_1. The proof is analogous if t is a variable.
Now, we regard (x,t), where x is a variable. Obviously, σ is a semantic unifier for M_1 if t is identical to x. If t is a data term such that x is not occur in t, then $\sigma(x) \equiv \sigma(t)$ and the transformation rule (R3) can be

applied to M. Let σ_x be the substitution $\{x/t\}$. Since $\sigma(u) \equiv \sigma(\sigma_x(u))$ for any term u, σ is a semantic unifier for M_1. In the last case, let t be the C-term $c(t_1,\ldots,t_n)$, where t is not a data term. Obviously, $\sigma(x)$ must be a term of the form $c(r_1,\ldots,r_n)$. Therefore, the data term instances of $r_1 = \sigma(t_1),\ldots,r_n = \sigma(t_n)$ belong to $H_=$. In this case, the transformation rule (R5) can be applied to M = U + S. The set M_1 is $(\{(x_1,\sigma_x(t_1)),\ldots,(x_n,\sigma_x(t_n))\} \cup U_1) + (\sigma_x(S) \cup \{(x,c(x_1,\ldots,x_n))\})$, where $U_1 = U\backslash\{(x,t)\}$, σ_x is the substitution $\{x/c(x_1,\ldots x_n)\}$ and $x_1,\ldots,x_n$ are new variables. We define a substitution σ_1 by

$$\sigma_1(y) := \begin{cases} \sigma(y), & \text{if } y \in dom(\sigma) \\ r_i\ , & \text{if } y \equiv x_i \in \{x_1,\ldots,x_n\} \\ y\ , & \text{otherwise} \end{cases}$$

Thus we have $\sigma(x) \equiv \sigma_1(\sigma_1(x))$ and it can easily be shown that σ_1 is a semantic unifier for M_1. □

From this proof it follows that at least one transformation rule can be applied to a chosen term pair of the unsolved part.

COROLLARY 3. If no transformation rule can be applied to a set of term pairs M = U + S with U ≠ ∅, then M is not semantically unifiable.

We can now give the algorithm for the semantic unification. Assume that two terms t,u are given. The aim of the algorithm is the generation of a semantic unifier for t and u, if t and u are semantically unifiable. Initially U = {(t,u)} and S is empty, and at the end U must be empty if t and u are semantically unifiable. The kernel of this algorithm is:

```
repeat
  if some transformation rule can be applied to U + S
  then apply this rule to U + S and let U1 + S1 be the result of it;
          U := U1 ; S := S1
  else stop with failure
until U = ∅ or failure
```

The presented unification algorithm is nondeterministic both with respect to the selection of the term pair and to the choice of the rewrite rule to be used if the transformation rule (R4) is applied. The choice of a transformation rule is deterministic if a term pair is already chosen. If the algorithm stops with empty U, it generates a substitution which is represented by a set S of term pairs in solved form. Otherwise the algorithm can either stop with failure or not terminate. Using Lemma 8, Lemma 9 and Lemma 7 the correctness and the completeness of this unification algorithm can easily be proved.

THEORM 3 (correctness). Let t,u be two terms of T(FC,V). If σ is a substitution generated be the defined unification algorithm, then σ is a semantic unifier for t and u .

THEOREM 4 (completeness). Let t,u be two terms of T(FC,U). If σ is a semantic unifier for t and u, then the unification algorithm can generate a semantic unifier τ for t and u such that $\sigma = \tau \circ \lambda$ for some substitution λ (i.e. τ is more general as σ).

6. Concluding Remarks

In this paper we present a new model theoretical interpretation of a unified language which combines logic and functional programming. A semantic unifcation algorithm with respect to such an interpretation is given and the completeness of it is proved. A disadvantage of the presented approach is the restricted use of rewrite rules, such that it is impossible to represent properties of functions and relationships between functions (e.g. associativity, distributivity). Our further research includes the representation of such properties in the form of metarules. In my opinion it is very useful to distinguish between rewrite rules for the definition of functions and rewrite rules for the representation of properties of functions and relationships between functions.

One advantage of the presented algorithm for the semantic unification is the restriction of the syntactic unification to data terms (e.g. the terms append(1,2) and append(1,2) are not semantically unifiable if 'append' is the defined function for the concatenation of two lists). Hence this unification algorithm can be defined such that the choice of a transformation rule is deterministic if a term pair is already chosen. In contrast to the presented unification it is possible that, by using narrowing or the unification procedure presented in [9], there are more than one path to generate a certain solution of an equation. Thus the presented algorithm has better termination properties and can be implemented more efficiently than those other algorithms.

The presented approach is a part of our development of the unified programming language FPROLOG (Functional PROLOG) integrating logic and functional programming. The first approach to FPROLOG is presented in [6]. Currently, FPROLOG is being implemented in Standard PROLOG.

References

[1] Clark,K.L.: Predicate Logic as a Computational Formalism. Res. Report 79/59, Dep. of Computing, Imperial College, 1979.

[2] DeGroot,D., G.Lindstrom (eds.): Logic Programming: Functions, Relations and Equations. Prentice Hall 1986.

[3] Dincbas,H., P.van Hentenryck: Extended Unification Algorithms for the Integration of Functional Programming into Logic Programming. J. Logic Programming 4 (1987), 199-227.

[4] van Emden,M,H., J.W.Lloyd: A Logical Reconstruction of Prolog II. J. Logic Programming 1 (1984), 143-149.

[5] van Emden,M.H., K.Yukawa: Logic Programming with Equations. J. Logic Programming 4 (1987), 265-288.

[6] Goltz,H.-J., U.Geske, F.Wysotzki: The Rewrite Approach to Combined Logic and Functional Programming. J. New Generation Computer Science 1 (1988), 49-62.

[7] Huet,G., D.Oppen: Equations and Rewrite Rules: A Survey. In: Formal Language Theory: Perspectives and Open Problems, R.Book (Hrsg.); Academic Press 1980, 349-405.

[8] Lloyd,J.W.: Foundations of Logic Programming. Springer 1984.

[9] Martelli,A., C.Moiso, G.F.Rossi: An Algorithm for Unification in Equational Theories. In: Proc. Symp. on Logic Programming 1986.

[10] O'Donnell,M.J.: Equational Logic as a Programming Language, MIT Press, Cambridge, Mass., 1985.

MODULAR ALGEBRAIC SPECIFICATIONS

Horst Hansen, Michael Löwe[1]

Abstract

Module, import, export and detail hiding are well-known notions in software engineering. If algebraic specifications together with their operational semantics of term rewriting should be useful as a programming language, similar concepts must be developed to cope with very large specifications. Especially the concept of detail hiding is important, since many problems and their efficient solutions require hidden sorts or hidden functions.

In this paper we discuss a module concept for algebraic specifications. A module consists of an algebraic specification, the body specification, together with import and export specifications. The semantics is given by a free functor from the class of import algebras to the class of body algebras followed by a subalgebra restriction according to the export signature. A module is correct, if its semantics does not change the reexport part of any import algebra. The reexport specification is the common part of import and export specifications. This requirement is much weaker than the requirements for module correctness we find in other approaches, e.g. in [EW 86], [BEP 87]. Since the restriction construction in our approach differs from the classical construction we are able to allow incompleteness and inconsistency of the body specification with respect to the import and reexport specifications.

This makes module specifications more suitable for software engineering purposes. For instance there is no need for explicit error handling in the body specification if the export restricts data to non-error data only. The example of a specification of a process scheduler demonstrates the usefulness of this concept. On the other hand, theoretic results concerning correctness of the semantics can be carried over to the new module concept. In this paper, we show that some basic constructions on modules preserve correctness and that the composition of correct modules yields a correct module.

Introduction

Term rewriting systems like OBJ [FGJM 85], ASSPEGIQUE [BCV 85] or RAP [Hus 85] provide operational semantics for algebraic specifications (for basic notions see [EM 85]). Thus, algebraic specifications written to be evaluated by these systems can be called algebraic programs. Since most of the problem-oriented specifications cannot be performed efficiently by term rewriting, algebraic programming means to construct an algebraic specification that is correct w.r.t. the problem-oriented one and can be performed by a special system efficiently enough to satisfy the users requirements.

Usually these algebraic programs resp. parts of them contain non-problem-oriented data types, functions, interpretation strategy information or system directives that should not be made reference to by the system user or by a programmer who uses the specification as a part of a bigger program; these details should be hidden from the user. Imperative and functional programming languages, like MODULA 2 [Wir 85] resp. ML [HMM 86], offer module concepts with explicit import and export for detail hiding.

Although the need of structuring facilities for large specifications has been recognized in the algebraic specification community (compare [SW 83], [EW 86], [Gogu 86]) the concept of detail hiding is not very well investigated up to now. Many approaches like [Gogu 86] provide only syntactical features which can be used to mark certain parts of the specification as hidden parts. Viewed parameterized canons as defined in [Rei 85] can be seen as modules with import (called 'parameter') and export (called 'view'). These modules do not have functorial semantics constructing export algebras from import algebras as the notions of [Rei 85] are based on behavioural semantics. In contrast [BEP 87] or [EW 86] define modules which reflect detail hiding on the semantical, i.e. functorial level.

These modules consist of body, import and export specifications. The semantics is given by a free functor from the class of import algebras to the class of body algebras followed by a restriction according to the export specification. The restriction models detail hiding. There are many module operations defined in this approach

[1] Institut for Software and Theoretical Computer Science, Department of Computer Science, Technical University of Berlin, Fanklinstraße 28/29, D-1000 Berlin 10

like composition, union, actualization [BEP 87], partial composition and recursion [Par 87], product and iteration [Par 88] that make this concept easy to use in the process of software design. There is a concept of refinement and realization [EFHLP 87] that constitutes a framework to express development steps like implementation or optimization. And the theory of this approach to module specifications is well-elaborated, too. There are a lot of compatibility results and results about compositionality of the semantics of combined modules (a comprehensive survey will be given in [EM 88]).

But the whole theory is based on very strong and restrictive requirements every single module must satisfy: The free functor from the import to the body must be conservative, which implies consistency and completeness of the body with respect to the import. This requirement seems too restrictive for software engineering purposes. The need for detail hiding very often arises from the fact, that certain hidden operations are not complete or consistent on all data. Hiding via export is then used as a tool to get rid of these (error-)cases. [BEP 87] requires explicit "error handling" (which is very hard to achieve in many cases, compare [Gogo 87] or [GM 87]) even if the export restricts to non-error data only.

Therefore we feel encouraged to propose a different module concept with restriction semantics, the theory of which is able to handle modules that hide inconsistency and incompleteness. Modules are algebraic specifications together with import and export specifications. The common part of the import and export specifications is called reexport. The semantics is given by a free functor from the import to the body followed by a restriction construction yielding an export subalgebra of the body. The restriction used here is different from the one in [BEP 87]. The theory will be designed for modules, the semantics of which preserves the reexport part. These modules are called correct. This requirement seems to be a natural extension of the concept of persistency for parametrized algebraic specifications to the level of modules. Note that this assumption does not imply completeness or consistency of the module's free functor. Thus, theoretical results are much harder to achieve. Especially the extension lemma which plays a central role in the theory of parametric specifications [EM 85] and module specifications [BEP 87] cannot be applied in our approach. We therefore prove an analogon for non-persistent functors. Based on that, we are able to show that some basic constructions on modules, i.e. export restriction, import extension, and module extension, preserve correctness. With the help of these results, we show that the composition of two correct modules (import export match) yields a correct module.

Contents

1. Algebraic Modules : Syntax, Semantics and Correctness

In this section we introduce algebraic modules, their syntax and semantics as well as a notion of correctness of algebraic modules. These modules are intended as an abstract description of some aspects of software or program modules as known from modern high-level programming languages as e.g. MODULA 2 or ML. Algebraic modules consist of an import specification, an export specification and a body specification which are connected by an injective specification morphisms from the import to the body and another specification morphism from the export to the body. The import specification defines the class of allowable import algebras for the module while the export specification tells a user of the module what properties of the module he can rely on. The body specification describes a semantical construction relating the class of import algebras to the class of export algebras.

While in most modern programming languages the requirements on possible import algebras are just syntactical requirements given by some signature, in our approach, as well as in the one by [EW 86], these requirements can be enhanced by semantical requirements on the operations. The same is true for the export part of a module. This differs from many other approaches where properties of the export can only be stated informally by natural language comments.

Definition 1.1: Syntax of Algebraic Modules

An algebraic module $M = (i: I \to B, e: E \to B)$ consists of three algebraic specifications I, E, and B, called import specification, export specification, and body specification, together with an injective specification morphism $i:I \to B$ and a specification morphism $e:E \to B$. □

We depict modules this way:

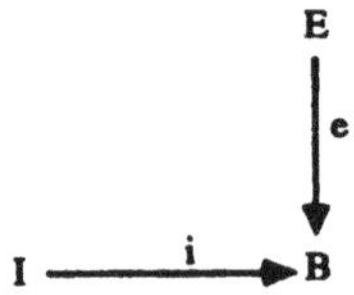

Figure 1.1

The technical requirement for $i: I \to B$ to be injective does not seem unduly strong: If we consider that the body specification defines the operations and data present in the export in terms of imported operations and data, it does not make too much sense to have some identification of operation symbols or sorts along i.

Modules as given by Definition 1.1 can easily be completed in a certain sense: A user of some algebraic module of course is interested in those parts of the import specification of the module which are also present in the export specification. Syntactically these common parts of the import and export can easily be constructed from the syntax of the algebraic module: this common part is the so called reexport part of the module. If both i and e are inclusions of specifications the reexport part is just the intersection of the import and export specifications. Technically speaking, the reexport specification in general is the pullback object of i and e in the category <u>SPEC</u> of specifications and specification morphisms.

Definition 1.2: Syntactic Completion of Algebraic Modules

Let $M = (i: I \to B, e: E \to B)$ be an algebraic module. The syntactic completion M^* of M is given by $M^* = (M, p: R \to I, r: R \to E)$, where R is the pullback object of M in the category <u>SPEC</u> and p and r are the resulting specification morphisms. □

The following figure shows the syntactic completion M^* of M:

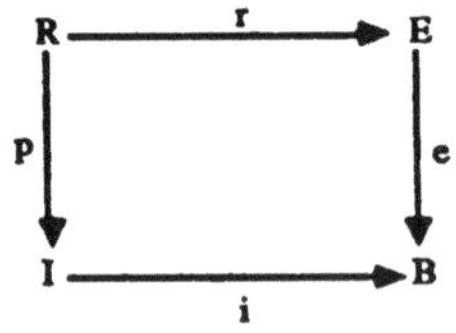

Figure 1.2

In the remainder of this paper we will not distinguish between an algebraic module M and its syntactic completion M^*, as the latter can be uniquely (up to isomorphism) constructed from the former.

The semantics of an algebraic module will be defined for syntactically completed modules. Intuitively, the semantics of the module M should be some transformation sending any import algebra to some export algebra. The semantics must reflect that the user of the module has access to the module via reexported data and exported operations only. Thus, we define the semantics of a module for any import algebra A_I to be that part of the free body algebra over A_I that is reachable from the free export algebra over reexported data of A_I. The following construction makes this more precise. Note, that the construction is different from the one of [EW 86]!

Construction 1.3: Semantics of an Algebraic Module

Let $M = (i: I \to B, e: E \to B, p: R \to I, r: R \to E)$ be an algebraic module. The next figure shows the syntactical level in the category <u>SPEC</u> as well as the semantical level in the category <u>CAT</u> of categories and functors:

[1] Throughout this paper we mainly use the terminology and framework of algebraic specifications with initial resp. free functor semantics, as found in [EM 85].

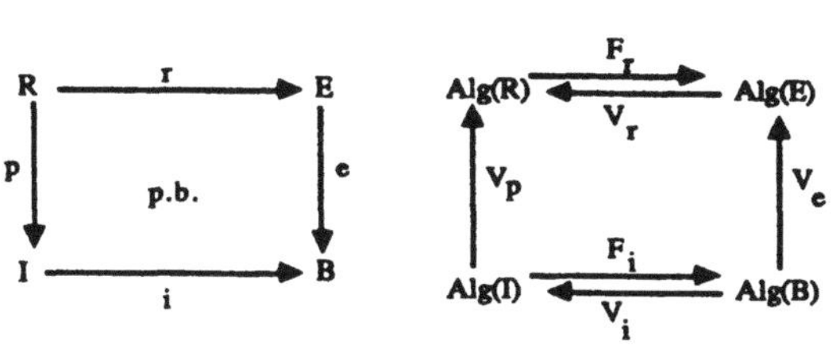

Figure 1.3

Alg(X) (X ∈ {I, E, B, R}) stands for the category of X-algebras and X-homomorphisms;
V_x, F_x (x ∈ {i, e, p, r}) for the corresponding forgetful resp. free functors between the appropriate categories.

Let $A_I \in$ Alg(I) be any import algebra. Consider the universal morphism $u_{i,A_I}: A_I \rightarrow V_iF_i(A_I)$ and apply the forgetful functor V_p to it. This yields a reexport morphism $f := V_p(u_{i,A_I}): V_p(A_I) \rightarrow V_pV_iF_i(A_I)$. Considering that the diagram of forgetful functors in figure 1.3 commutes, we have $V_pV_iF_i(A_I) = V_rV_eF_i(A_I)$. Applying the forgetful functor V_p to the import algebra A_I yields a reexport algebra $V_p(A_I)$. As F_r is a free functor, there is a universal morphism $u := u_{r,V_p(A_I)}: V_p(A_I) \rightarrow V_rF_rV_p(A_I)$.

The algebra $F_rV_p(A_I)$ constitutes the free export algebra over all data, which are reexported from the import algebra A_I. This algebra exactly describes all data and operations which are accessible to a user of the module M for any fixed import algebra A_I. On the other hand the algebra $F_i(A_I)$ consists of all data and operations freely generated over the import algebra A_I using the body specification. Applying V_e to this algebra turns it into an export algebra.

What we are interested in is that part of the export algebra $V_eF_i(A_I)$, which can be reached using the syntactical (and semantical) means, which a user of the module is allowed to know and use, i.e. the algebra $F_rV_p(A_I)$. This part in general is a subalgebra of $V_eF_i(A_I)$, which can be won this way:

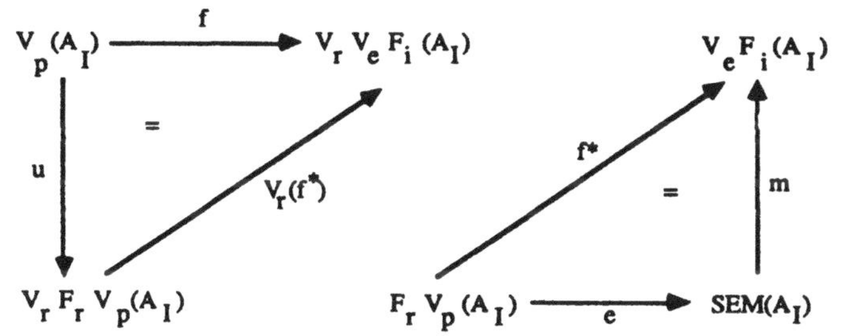

Figure 1.4

In this commuting diagram $f^*: F_rV_p(A_I) \rightarrow V_eF_i(A_I)$ is the uniquely determined extension of the "assignment morphism" f. An Epi-Mono-factorization of f^* yields the lower triangle in figure 1.4. We call the middle object of this factorization $SEM(A_I)$, because we consider this subalgebra of $V_eF_i(A_I)$ the semantics of the algebraic module M at A_I. This pointwise construction of the semantics of M extends to a functor SEM: Alg(I) → Alg(E).

Definition 1.4: Semantics of an Algebraic Module
The semantics of an algebraic module M is the functor SEM: Alg(I) → Alg(E) as defined in the previous construction. □

Not every algebraic module is a correct one in an intuitive sense. Consider an import algebra A_I and look at its reexport part $V_p(A_I)$. This is some fixed algebra for any choice of A_I. On the other hand $SEM(A_I)$ is the semantics of M at A_I, which is some export algebra, a subalgebra of $V_eF_i(A_I)$. Now look at the algebra

$V_rSEM(A_I)$, which again is some reexport algebra. $V_p(A_I)$ and $V_rSEM(A_I)$ are connected by the morphism $V_r(e)\circ u : V_p(A_I) \rightarrow V_rSEM(A_I)$, compare figure 1.4. If this morphism is not bijective, the semantics of the module changes the semantics of its reexport part. This contradicts the intuition about reexported parts and motivates the following notion of correctness:

Definition 1.5: Correctness of Algebraic Modules
An algebraic module M is correct, if for all import algebras A_I we have $V_P(A_I) \cong V_rSEM(A_I)$. □

This means that the module M leaves the reexport part of every import algebra untouched (up to isomorphism). This notion of correctness is the natural analogon of the concept of persistency for parametric algebraic specification on the level of modules. In section 3 we will provide a characterization of correct modules, which shows that we do not need persistency of the free functor F_i: Alg(I) → Alg(B) for correct modules, as required e.g. in [EW 86].

2. Example Module: A Process Scheduler

To demonstrate the usefulness of the module concept introduced above, we give a typical example that cannot be handled by the concepts of [BEP 87]. We chose a small part of an operating system: The kernel of a process scheduler. Although we are aware that applicability of any module concept should be tested by modelling large and real-world software systems, this example demonstrates many of the advantages of the new approach.

We use the following conventions for writing down examples: The body of the module is enclosed by the keywords <u>module</u> and <u>end module</u>. Parts belonging to both import and export are listed behind the keyword <u>reexport</u>. Additional import and export components are given in the <u>import</u> and <u>export</u> section respectively. Internal parts, i.e. parts that neither belong to the import nor to the export, are written down behind the keyword <u>body</u>. Specifications are written in an ACT ONE-like style (see [EM 85]).

```
module scheduler
reexport  sorts process
import    sorts process queue
          opns  empty: → process queue
                add: process, process queue → process queue
                remove: process, process queue → process queue
                tail: process queue → process queue
          eqns  (e1) remove (p, remove(p, q)) = remove(p, q)
export    sorts system
          opns  boot system: → system
                schedule: system → system
                new process: process, system → system
                kill process: process, system → system
body      opns  make: process queue → system
                first: process queue → process
                supervisor: → process
          eqns  (e2) first(add(p, empty)) = p
                (e3) first(add(p, add(p', q))) = first(add(p', q))
                (e4) remove(p,add(supervisor,q)) = add(supervisor,remove(p,q))
                (e5) schedule(make(q)) = make(add(first(q), tail(q)))
                (e6) new process(p, make(q)) = make(add(p, q))
                (e7) kill process(p, make(q)) = make(remove(p, q))
                (e8) boot system = make(add(supervisor, empty))
end module  scheduler
```

The module given above imports some basic list structure, entries of which are of sort "process". This sort is reexported because export operations of the module make reference to this sort. The additional export consists of a new sort "system" and four operations on this sort modelling some basic scheduling operations. The structure of the processes that have to be scheduled is transparent to this module. The interesting part of the module is the body part. Here we find the 'implementation' of the exported sort (by means of the operation "make") and operations (by means of e5 - e8). This implementation is heavily based on the imported list structure that is turned into a queue by an additional incompletely specified operation "first" (compare e2 and e3: there is no equation for 'first(empty)'). Another incompleteness is introduced by the process constant "supervisor". There is no equation that identifies this process with an imported one. This process, an internal feature of the module, models the system manager initially running when the system is started (see e8). It cannot be removed from the "process queue" (compare e4). This means, that this process cannot be stopped from outside the module. This process is unknown to the user of the module "scheduler".

These two cases of incompleteness are typical examples of situations that often arise in the software development process and our module concept was designed to cope with. Exploiting the fact that internals of the module are hidden from the outside world, internal operations are often designed to work properly only on certain subsets of data. The design of the module, especially of the export interface, prevents internal operations from being applied to data not in this subset. The second case deals with the problem that sometimes internal constants should not be identified with data available outside the module. These constants are intended to serve internal needs exclusively.

The module "scheduler" is correct according to 1.5 on the subclass of import algebras that contain an injective image of the free algebra that is generated by "empty" and "add" on the imported data of sort "process". The module's free functor from this subclass to the class of body-algebras is consistent and the export interface forgets about all internally added extra-processes, since there isn't any export operation that returns elements of sort "process". These two properties will turn out to be sufficient to guarantee correctness of the module (compare section 3).

3. An Extension Lemma for Non-persistent Functors

The notion of correctness of module specifications as defined in e.g. [EW 86], [BEP 87] requires the free functor between the class of import algebras and class of body algebras to be persistent, i.e. any import algebra must be isomorphic to the algebra which is constructed from the import algebra by first applying the free functor and then the forgetful functor.

In contrast to this, we only require persistency of the module's semantics. Therefore, it is allowed to specify operations in the body incompletely (compare section 2). Thus, the theory of [EM 85] is not applicable in our context. We need another basis for proving properties of the module operations we are going to introduce in section 4. This basis is provided in this section: We show some sufficient conditions for correctness of modules, characterize correctness of modules, and prove an analogon to the "Extension Lemma" as found e.g. in [EM 85]. In our version of the extension lemma the requirement of persistency is substituted by the requirement of consistency together with an additional property called relative robustness.

For the following definitions let $\mathbf{S} \underset{W}{\overset{U}{\rightleftarrows}} \mathbf{I} \underset{V}{\overset{F}{\rightleftarrows}} \mathbf{R}$ be a diagram in CAT, F a left adjoint functor of V.

Definition 3.1: Persistent and Consistent Functors

The functor F is called persistent, consistent, complete relative to the prior application of U, if the universal morphism $u_{U(A)} : U(A) \to VFU(A)$ is bijective, injective, surjective for any object $A \in \mathbf{S}$. The functor F is called persistent, consistent, complete relative to the later application of W, if after applying W to the universal morphism $u_A : A \to VF(A)$ the morphism $W(u_A):W(A) \to WVF(A)$ is bijective, injective, surjective for any object $A \in \mathbf{I}$. □

For U = ID these definitions almost specialize to the usual definition of persistency, as found e.g. in [EM 85].

Definition 3.2 : Robust and Relative Robust Functors

A functor is called robust, if it protects injective morphisms, i.e. it sends injective morphisms to injective morphisms.

A left adjoint functor F is called relative robust, if for any injective morphism $f : A \to A'$ the morphism $VF(f) : VF(A) \to VF(A')$ restricted to the image of the universal morphism $u_A : A \to VF(A)$ is injective.

A left adjoint functor F is called robust relative to later application of a functor W, if for any injective morphism $f : A \to A'$ the morphism $WVF(f) : WVF(A) \to WVF(A')$ restricted to the image of the morphism $W(u_A) : W(A) \to WVF(A)$ is injective. □

Using these definitions we can give two sufficient conditions for the correctness of algebraic modules.

Lemma 3.3: Sufficient Conditions for the Correctness of Algebraic Modules

Let M be an algebraic module (see figure 3.1). M is correct,

(1) if F_i is consistent relative to the later application of V_p and F_r is complete relative to the prior application of V_p, or

(2) if F_i is persistent relative to the later application of V_p.

Proof: We have the following semantical diagram:

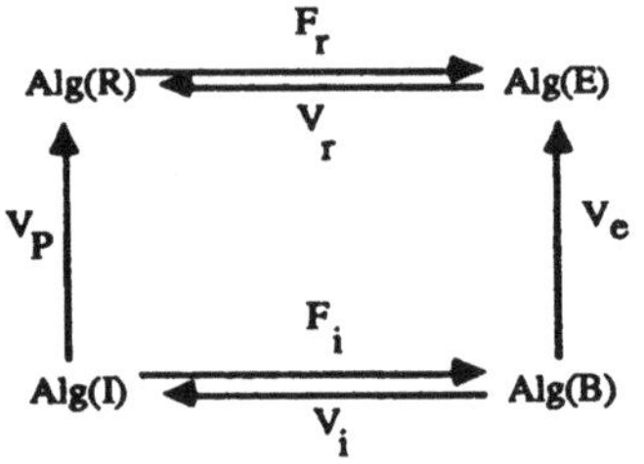

Figure 3.1

(1) As F_i is consistent relative to the later application of Vp we have an injective morphism $f := V_p(u_{i,A_I}) : V_p(A_I) \to VpV_iF_i(A_I)$ for all $A_I \in Alg(I)$. As F_r is complete relative to the prior application of Vp we have a surjective universal morphism $u := u_{r,V_p(A_I)} : Vp(A_I) \to V_rF_rV_p(A_I)$. Recalling the definition of the semantics of the module M we have the diagram

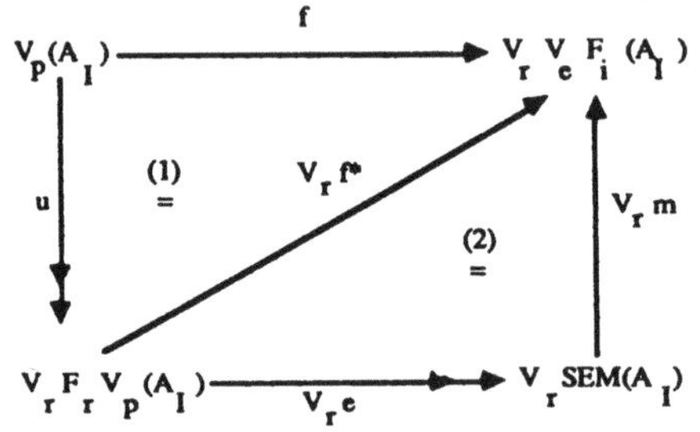

Figure 3.2

As f is injective and the triangle (1) commutes, u becomes injective. Thus it is bijective. Therefore $V_r(f^*)$ is injective and so is $V_r(e)$, as triangle (2) commutes. Therefore we have $V_p(A_I) \cong V_rSEM(A_I)$, i.e. M is correct in the first case.

(2) If F_i is persistent relative to the later application of V_p, then f becomes bijective. As the triangles (1) and (2) commute, we have $f = V_r(m) \circ V_r(e) \circ u$. As f is bijective $V_r(e) \circ u$ becomes injective and $V_r(m)$ surjective. $V_r(m)$ is injective by definition. So $V_r(m)$ is bijective and thus $V_r(e) \circ u$, as well. □

The following definition leads us to a characterization of correct modules.

Definition 3.4: Export-complete Algebraic Module

Let M be an algebraic module. M is termed export-complete, if the subobject defined by the Epi-Mono-factorization of the "assignment morphism" $f := V_p(u_{i,A_I}) : Vp(A_I) \to V_rV_eF_i(A_I)$ agrees with the subobject defined by the Epi-Mono-factorization of $V_r(f^*) : V_rF_rV_p(A_I) \to V_rV_eF_i(A_I)$. □

Proposition 3.5:

An algebraic module M is correct iff M is export-complete and F_i is consistent relative to later application of V_p.

The proof of this proposition can be found in [HL 88]. □

Proposition 3.6: Extension Lemma for Non-Persistent Functors

Let the following syntactical diagram in SPEC be a pushout. Consider the corresponding semantical diagram in CAT. Then the following statement holds: If F_i is consistent and F_h is relative robust then $F_{i'}$ is consistent relative to later application of V_h.

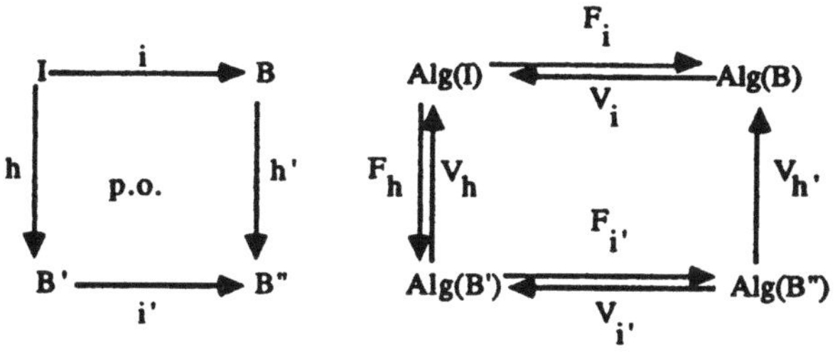

Figure 3.3

This proposition is an analogon to the extension lemma since some given property of the functor F_i is transferred to a corresponding property of $F_{i'}$. Here consistency of F_i is transferred to consistency of $F_{i'}$ relative to V_h. This result clearly cannot be achieved without additional assumptions. To prove our result we need that F_h is relatively robust (see Definition 3.2), while persistency of F_i carries over to persistency of $F_{i'}$ without additional assumptions.

Proof: Here we give only the idea of the proof of this proposition[1]. We explicitly define a functor $G : Alg(B') \rightarrow Alg(B")$, show that this functor is left adjoint to the forgetful functor $V_{i'}$ and prove that it is consistent relative to later application of V_h. In order to define G we introduce 3 endofunctors on $Alg(I)$: V_hF_h, $V_i F_i$ and $V_iF_iV_hF_h$. Using these endofunctors we construct a chain of morphisms in $Alg(I)$. We show that the colimit C_I of this chain is a fixpoint of the 3 endofunctors. The algebras $F_h(C_I)$ and $F_i(C_I)$ have the property $V_hF_h(C_I) = C_I = V_iF_i(C_I)$, so that we can build the amalgamated sum $F_h(C_I) +_{C_I} F_i(C_I)$ of these algebras[2]. This construction on objects extends to a functor G, which (on objects) is defined by $G(A_{B'}) = F_h(C_I) +_{C_I} F_i(C_I)$, where C_I is the colimit of the above chain constructed from $V_h(A_{B'})$. □

Corollary 3.7:

Let the situation be as in proposition 3.6, but consider an additional specification morphism $p : R \rightarrow I$. Then we can modify the assumptions of proposition 3.6: If F_i is consistent relative to later application of V_p and F_h is robust relative to later application of V_p, then $F_{i'}$ is consistent relative to later application of V_pV_h.

Proof: The proof of proposition 3.6 can easily be modified so that the above statement holds. □

4. Constructions on Algebraic Modules

In this section we introduce some fundamental constructions on algebraic modules. We investigate the conditions under which application of these constructions on correct modules yields correct modules. (Detailed proofs can be found in [HL 88].)

The first construction is export restriction. This construction allows to reduce the part of the module that is visible to the user. Using export restriction, it is possible to create different views of the same software component (as known from data base theory).

The second construction is called import extension. This construction allows to add extra import parts to a given module. This situation occurs in the software development process, when an improvement of the implementation of the export part, i.e. a change to the body part, cannot be performed unless some additional import is supplied.

Module extension, the third construction, corresponds to the enrichment or extension concept on the level of

[1] The details of the proof can be found in [HL 88].

[2] For details on amalgamation of algebras see [EM 85].

basic or parametric algebraic specifications. It allows to add extra export parts as well as corresponding 'implementations', i.e. extra body parts, to a given module.

The last construction we deal with in this paper is the composition of two modules. This construction is similar to composition of modules as introduced in [EW 86], while the former constructions have not been considered in that framework. Composition means to substitute the import of module M by another module M' by matching the import specification I of M with the export specification E' of M' via a specification morphism $h: I \rightarrow E'$. This composition of modules corresponds to module usage and linkage in programming languages.

Proposition 4.1: Export Restriction

Let $M = (i: I \rightarrow B, e: E \rightarrow B)$ be a correct module, $\underline{e}: E' \rightarrow E$ a specification morphism. Then the module $M' = (i: I \rightarrow B, e \circ \underline{e}: E' \rightarrow B)$, the result of export retriction of M via $\underline{e}$, (resp. its syntactic completion) is correct.

Sketch of proof: We have the following syntactical and semantical diagrams:

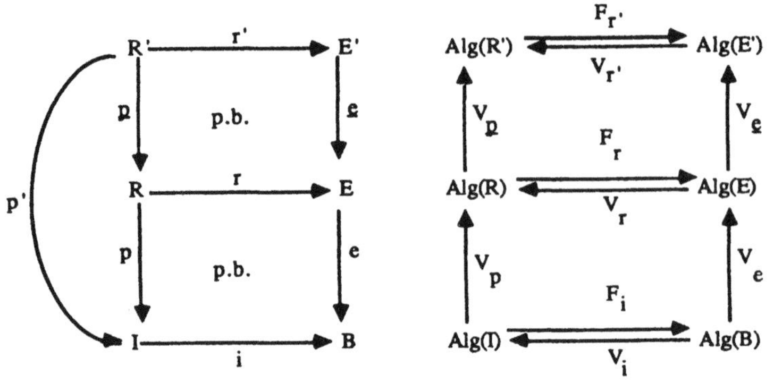

Figure 4.1

The statement of the proposition follows from the fact that the evaluation of export-terms of E' essentially is identical to the evaluation of export-terms of E in the same body algebra seen as an E- resp. E'-algebra. □

Proposition 4.2: Import Extension

Let $M = (i: I \rightarrow B, e: E \rightarrow B)$ be a correct module, $\underline{p}: I \rightarrow I'$ a specification morphism and $F_{\underline{p}}$ relative robust. Then the module $M' = (i': I' \rightarrow B', \underline{e} \circ e: E \rightarrow B')$, the result of import extension of M via $\underline{p}$, (resp. its syntactic completion) is correct, where B' is constructed as a pushout object of the specification morphisms $\underline{p}: I \rightarrow I'$ and $i: I \rightarrow B$.

Sketch of the proof: We have the following syntactical and semantical diagrams:

Figure 4.2

Application of corollary 3.7 tells us that $F_{i'}$ is consistent relative to later application of $V_p V_{\underline{p}}$. And export completeness of M' can directly be derived from export completeness of M. Thus, we get correctness of M' by application of proposition 3.5. □

Proposition 4.3 : Module Extension

Let M = (i: I→B, e: E→B) and M' = (i': I'→B', e': E'→B') be correct modules such that (e: E→B) = (p': R'→I'). Then the joint module M" = (i'∘i: I→B', e': E'→B), the result of module extension of M by M', (resp. its syntactic completion) is correct.

Sketch of proof: We have the following diagrams in <u>SPEC</u> and <u>CAT</u>:

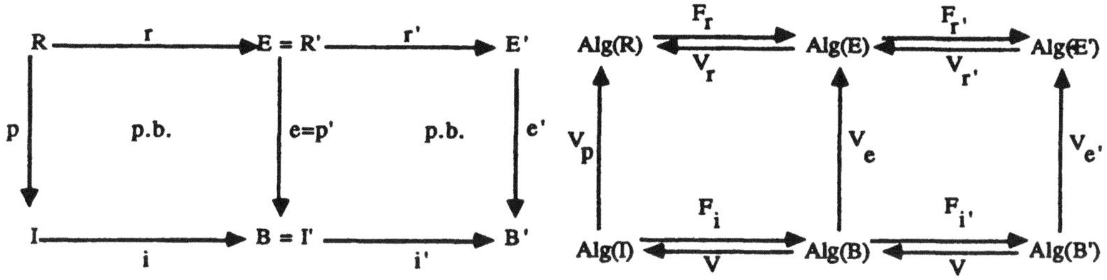

Figure 4.3

$F_{i''} = F_{i'}F_i$ is consistent relative to later application of V_p, since F_i is and consistency of $F_{i'}$ relative to later application of V_e implies its consistency relative to later application of $V_r V_e = V_p V_i$. That M" is export complete is a direct consequence of the export completeness of M and M'. Applying proposition 3.5, we have correctness of M". □

Theorem 4.4: Composition of Algebraic Modules

Let M = (i: I→B, e: E→B) and M '= (i': I'→B', e': E'→B') be correct algebraic modules and h : I→E' a specification morphism. If the functor $F_e \cdot F_h$ is robust relative to later application of V_p, then the composed module M" = (j∘i': I→B", g∘e: E→B") (resp. its syntactic completion) is correct, where B" is constructed as a pushout object of the specification morphsims i and e'∘h.

Proof: This proof is a straightforward combination of the results of propositions 4.1, 4.2, and 4.3. We have the following diagrams in <u>SPEC</u> and <u>CAT</u>:

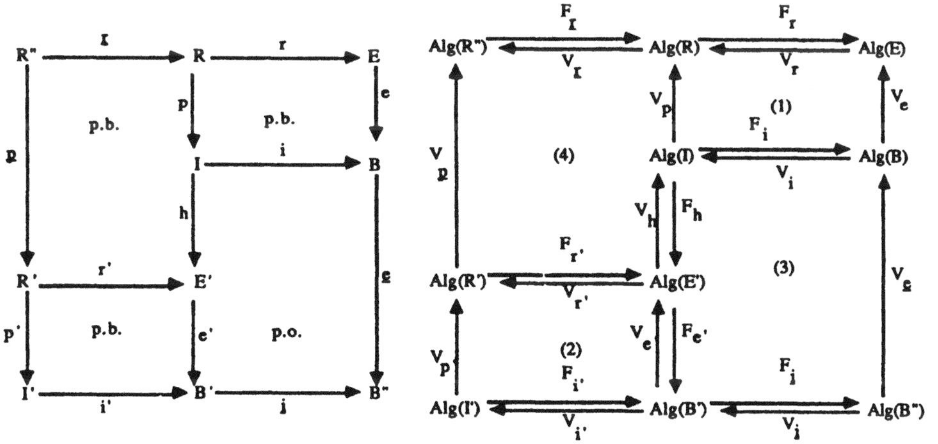

Figure 4.4

Application of proposition 4.2 yields that the combination of squares (1) and (3) is a correct module. Proposition 4.1 tells us that the squares (2) and (4) together form another correct module. Finally we get with proposition 4.3 that the combination of the previously constructed modules again is a correct module. □

5. Example: Composition of Modules

The following module provides an actual list structure that can satisfy the import requirements of the module "scheduler" of section 2.

```
module      list
reexport    sorts data
import      sorts bool
            opns  eq: data, data → bool
                  true, false: → bool
            eqns  eq(x,x) = true
                  eq(x,y) = eq(y,x)
                  eq(x,y) = true, eq(y,z) = true ⇒ eq(x,z) = true
export      sorts list
            opns  e: → list
                  app: data, list → list
                  rem: data, list → list
                  rest: list → list
            eqns  rest(e) = e
                  rest(app(d, e)) = e
                  rest(app(d, app(d',l))) = app(d, rest(app(d',l)))
                  rem(d, e) = e
body        opns  ite: bool, list, list → list
            eqns  ite(true, l, l') = l
                  ite(false, l, l') = l'
                  rem(d, app(d', l)) = ite(eq(d, d'), rem(d, l), app(d', rem(d, l)))
end module  list
```

This module, that encapsulate the implementation of the operation "rem", is correct, since its free functor from the class of import algebras to the class of body algebras is persistent on the reexport part, compare lemma 3.3.

A new module, we call it "actual scheduler", is generated if we compose "scheduler" with "list" by this import export match:

process ↦ data	empty ↦ e	remove ↦ rem
process queue ↦ list	add ↦ app	tail ↦ rest

This assignment of names defines uniquely a specification morphism h between import (scheduler) and export (list). The reexport part of the "actual scheduler" consists of the sort "process" since it has been actualized by the sort "data", "list" imports itself. The import is the import specification of "list", the export becomes the export specification of "scheduler" and the body specification is the union of the bodies of "list" and "scheduler" where sorts and operations are identified according to h. The "actual scheduler" is correct, i.e. its semantics does not change the semantics of the sort process, since the free functor $F_{e' \circ h}$ is robust relative to later application of V_p, where e' is the specification morphism from the export to the body of the module "list" and p' is the specification morphism from the reexport to the import of module "scheduler". This robustness is guaranteed because "list" contains neither equations of the sort "data" nor operations providing elements of sort "data".

Since "scheduler" is not correct on the whole class of import algebras (compare section 2), we additionally have to make sure that both functors $V_{e' \circ h} F_{e' \circ h}$ and $V_i F_i$ preserve the required subclass property (i is the morphism from import to body of "scheduler"). This is true because no equation on terms of sort "process queue" only generated by "empty" and "add" on data of sort "process" can be derived from the set of body equations of both modules.

6. Further Research

We introduced a new module concept for algebraic specifications that can cope with incomplete body specifications. Correctness of these modules has been defined in a way that extends the concepts of persistency to the level of modules. We investigated the conditions under which some basic constructions on correct modules yield correct modules.

Especially the side condition that has to be satisfied in the case of composition is not satisfactory. It would be better to have some local conditions that, if satisfied by each module, guarantee correctness of the composition and extend to the resulting module. A candidate for this is "robustness of the free construction from the export to the body relative to the reexport". If it can be achieved to prove in the extension lemma that not only consistency is extended from F_i to $F_{i'}$, but also robustness from F_h to $F_{h'}$ (compare figure 3.3), it would be easy to conclude that the composition of correct and robust modules yields a correct and robust module as well.

Besides that, further work has to be done in two areas. On one hand, we have to investigate larger and more application-oriented examples in order to check feasibility of the concepts introduced in this paper. On the other hand, the theory, we started to elaborate for the functorial semantics of the module concept, has to be carried over to some operational semantics for modules. This semantics will be based on the ideas of term rewriting and normal forms, but has to reflect module interfaces and subalgebra restriction in an appropriate way. Our theoretical research in this area concentrates on two aspects: How is the operational semantics related to the functorial one and which conditions and assumptions guarantee compositionality of the operational semantics? The property of compositionality may be very useful to increase efficiency of term rewriting systems for algebraic specifications. It may help to drastically reduce the set of rewrite rules that must be checked and applied in order to evaluate a normal form of an expression or subexpression.

7. References

[BCV 85] Bidoit, M., Choppy, C., Voisin, F.: The ASSPEGIQUE Specification Environment, in: Kreowski, H.-J. (ed.): Informatik Fachberichte Nr. 116, Springer, Berlin, 1985

[BEP 87] Blum, E.K., Ehrig, H., Parisi-Presicce, F.: Algebraic Specification of Modules and Their Basic Interconnections, JCSS 34 (1987), 293-339

[EFHLP 87] Ehrig, H., Fey, W., Hansen, H., Löwe, M., Parisi-Presicce, F.: Algebraic Theory of Modular Specification Development, Research Report No. 87-06, Techn. Univ. of Berlin, 1987

[EM 85] Ehrig, H., Mahr, B.: Fundamentals of Algebraic Specification 1, Springer, Berlin, 1985

[EM 88] Ehrig, H., Mahr, B.: Fundamentals of Algebraic Specification 2, Springer, Berlin, 1988 (to appear)

[EW 86] Ehrig, H., Weber, H.: Programming in the Large with Algebraic Module Specifications, Proc. IFIP Congress 1986, Dublin, 1986

[FGJM 85] Futatzugi, K., Goguen, J.A., Jouannaud, J.P., Meseguer, G.: Principles of OBJ2, in: Proc. of the 12. ACM Symposium on Principles of Programming Languages, New Orleans, Lousiana (1985), 52-66

[Gogo 87] Gogolla, M.: On Parametric Algebraic Specifications with Clean Error Handling, in: Proc. of the Int. Joint Conf. on Theory and Practice of Software Development, Pisa, Italy (LNCS 249), 1987

[Gogu 86] Goguen, J.A.: Reusing and Interconnecting Software Components, Computer 19, 2 (Feb. 1986), 16-28

[GM 87] Goguen, J.A., Meseguer, J.: Order Sorted Algebra I: Partial and Overloaded Operations, Errors and Inheritance, (to appear) SRI International Computer Science Lab, 198

[HL 88] Hansen, H., Löwe, M.: A New Module Concept for Algebraic Specifications, Internal Report, Techn. University of Berlin, 1988

[HMM 86] Harper, R., MacQueen, D., Milner, R.: Standard ML, LFCS Report Series No. ECS-LFCS-86-2, Department of Computer Science, Univ. of Edinburgh, 1986

[Hus 85] Hussmann, H.: Rapid Prototyping for Algebraic Specifications - RAP Systems User Manual, Univ. of Passau, Research Report MIP-8504, 1985

[Par 87] Parisi-Presicce, F.: Partial Composition and Recursion of Module Specifications, in: Proc. of the Int. Joint Conf. on Theory and Practice of Software Development, Pisa, Italy, (LNCS 249), 1987, 217-231

[Par 88] Parisi-Presicce, F.: Iteration of Module Specifications, to appear in: Proc. of CAAP'88, Nancy, France, March 1988

[Rei 85] Reichel, H.: Behavioural Program Specification, in: Proc. Category Theory and Computer Programming, Guildford, September 1985, LNCS 240, pp. 390-411

[SW 83] Sanella, D., Wirsing, M.: A Kernal Language for Algebraic Specification and Implementation, Internal Report No. CSR-131-83, Univ. of Edinburgh, 1-44; (extended abstract) also in: Proc. Int. Conf. on Foundations of Computing Theory, Borgholm, Sweden, Springer LNCS 158, 413-427

[Wir 85] Wirth, N.: Programming in Modula 2, Springer, Berlin, 1985

PROVING INDUCTIVE THEOREMS BASED ON TERM REWRITING SYSTEMS

Dieter Hofbauer & Ralf-Detlef Kutsche *

Abstract

Sufficient criteria for an equation to be in the inductive theory of a term rewriting system are given. Inspecting only special critical pairs, we need not require the underlying system to be confluent, not even on ground terms. We are able to deal with equations which - if viewed as rules - are possibly not terminating if added to the given rewrite system; we have to restrict, however, their use in the induction process. Modular use of lemmata, already known inductive theorems, is incorporated into the results. As examples we treat natural number arithmetic, sorting lists of natural numbers, and sorting lists over arbitrary data structures.

Dedicated to the 50th anniversary of Prof. Dirk Siefkes.

1. Introduction

Term rewriting systems (TRS's) are widely used as a basic concept for investigating the foundations of functional programming, the operational semantics of equational specifications and related topics. Thus tools for handling them are needed particularly.

The equations which are valid in all models of a given system can be obtained by means of ordinary rewriting (i.e. replacing equals by equals) ; in contrast, the set of equations which hold just in the initial model - the *inductive theory* - is not accessible as easy (in general it is not even recursively enumerable). Equations from the inductive theory are usually proved by some kind of explicite induction (e.g. on the term structure).

Based on the work of Musser [Mus80], Goguen [Gog80], Huet and Hullot [HH82], Jouannaud and Kounalis [JK86] and others, we show how to prove inductive equations under certain conditions by just applying rewrite steps between distinguished pairs of terms.

In the initial papers [Mus80] and [Gog80], inductive theorems are proved by checking *consistency* w.r.t. an equationally specified equality predicate (compare Kapur and Musser [KM87]). The need for such a predicate was removed by [HH82] through distinguishing *constructors* and *defined* operator symbols. This approach was further developed - based on the notion of *inductive completeness* - by Paul [Paul84] and H. Kirchner [Kir85] who allow equations between constructor terms. [JK86] show that such an explicite partition of the operation symbols is not necessary. Their premises include the demand for *quasi-reducibility*. Quasi-reducibility is decidable (Kapur et al. [KNZ87]; see [KNRZ87] for a summary of related results.). It localizes effectively the assumption of sufficient completeness (or related conditions, which in general are undecidable) in earlier works on *"inductionless induction"*. By their criteria, [JK86] are able to prove nonterminating rules if the underlying TRS is terminating and ground confluent on the corresponding equivalence classes (*E-termination* and *RE-confluence*). (For ground confluence in the context of completion see Küchlin [Küch87].)

Our results allow to weaken the premises of both (ground) confluence and (E-) termination. Only critical pairs of the underlying system *on* the rules to be proved have to be considered. (This was already noticed by Fribourg [Frib86]. However, his results require the underlying TRS to be ground confluent.)

*Technische Universität Berlin, Fachbereich 20 (Informatik), FR 6-2, Franklinstr. 28/29, D - 1000 Berlin 10
E-mail : (uucp) ...!mcvax!unido!tub!tub-tfs!dieter , ...!mcvax!unido!tub!tub-tfs!ralf

Nonterminating rules can be proved if they are used at most once as "induction hypothesis" for each of these special critical pairs. (The approach of Toyama [Toya86] even yields inductive proofs over nonterminating TRS's, but requires confluence.)
We do not explicitly treat completion procedures (in the sense of Knuth and Bendix [KB70]) in this paper. Nevertheless we use them as heuristics to find appropriate sets of rules to apply our theorems. Examples show that the process of adding (reduced) critical pairs to the actual set of rules may not terminate whereas the heuristics of *generalizing* these pairs succeeds.

2. Inductive Proofs

We assume the reader to be familiar with basic notations for *term rewriting systems* (TRS's) as given e.g. in [Hu80], and just recall some of them. A TRS consists of a *set of rules* $(l \rightarrow r)$ such that all variables in r already occur in l ; thus ground terms are only rewritten to ground terms. Given a TRS R , by $-_R\rightarrow$ we denote the induced *rewrite relation on terms*, by $-=_R\rightarrow$, $-^*_R\rightarrow$ and $\leftarrow^*_R\rightarrow$ its reflexive, reflexive and transitive, and equivalence closure, respectively. Composition of relations will be denoted by $\bullet$.
t/u is the *subterm* of t at *occurrence* u , $t[u \leftarrow s]$ the result of replacing the subterm of t at occurrence u by s . Let $(l \rightarrow r)$, $(l' \rightarrow r')$ be variable disjoint rules such that u is a non-variable occurrence in l' and η is the most general unifier of l and l'/u . Then $< l'[u \leftarrow r]\eta \, , r'\eta >$ is called *critical pair of* $(l \rightarrow r)$ *on* $(l' \rightarrow r')$. (A *critical pair of* a TRS R *on* a TRS R' is understood as a critical pair of a rule of R on a rule of R' .)
In the following we will assume that there are ground terms of every sort in the underlying signature.

Definition

Let R be a TRS. The set of equations
$\mathrm{ITh}(R) := \{ s=t \mid s\sigma \leftarrow^*_R\rightarrow t\sigma$ for all ground substitutions $\sigma \}$ is called *inductive theory* of R .

For convenience we often speak of a *set of rules* to be in the inductive theory of a TRS, always meaning the *associated set of equations*. The inductive theory is just the set of equations valid in the initial model of the underlying system (see e.g. [Pad87]). As an elementary fact about the inductive theory we will use implicitly for TRS's R, R' of the same signature : $\mathrm{ITh}(R) \subseteq \mathrm{ITh}(R \cup R')$.

Definition

Let R be a TRS. A term t is *ground-reducible under* R (in other papers called *quasi-reducible* or *inductively reducible*) iff $t\sigma$ is R-reducible for all ground substitutions σ , a TRS R' is *ground-reducible under* R iff the left-hand side of every rule in R' is ground-reducible under R .

The following theorems give sufficient criteria for proving inductive theorems of a rewrite system R . We think of R as being the set of *defining rules*, thus R should be terminating. We do not require R to be (ground) confluent, although in many examples it is.
Theorem 1 treats the case where R together with the rules to be proved is terminating. In theorem 2 we do not even assume R to be E-terminating (cf. [JK86]) for a set of rules E which possibly are not terminating in presence of R . However, we have to assume that E is applied as an "induction hypothesis" at most once (see premise (4)). Both theorems allow the use of a set L of inductive theorems of R, which can be rules from R itself and/or theorems already proved ("lemmata"). In place of the equivalence induced by a set L of rules, any subset of the inductive theory could be used. Nevertheless, our formulation of the theorems is fitted on their use in the examples. Finally, theorem 3 unifies and generalizes both theorems.

Theorem 1

Let R, I, L be TRS's such that

(1) $L \subseteq ITh(R)$

(2) $R \cup I$ is terminating

(3) I is ground-reducible under R

(4) for every critical pair <c,p> of R on I there exist terms c', p' such that

$$\begin{array}{ccc} c & & p \\ R\cup I \downarrow * & & * \downarrow R\cup I \\ c' & \leftarrow^*_L\rightarrow & p' \end{array}$$

Then $I \subseteq ITh(R)$.

Proof

By noetherian induction on $-_{R\cup I}\rightarrow$ we show for all groundterms s, t :

if $s -^*_{R\cup I}\rightarrow t$ then $s \leftarrow^*_R\rightarrow t$.

(Without loss of generality we assume all rules to be variable disjoint.)

If s = t we are done. Else assume

<u>case 1</u> $s -_R\rightarrow t' -^*_{R\cup I}\rightarrow t$ for some (ground) term t'.

Then $t' \leftarrow^*_R\rightarrow t$ by induction hypothesis, thus $s \leftarrow^*_R\rightarrow t$.

<u>case 2</u> $s -_I\rightarrow t' -^*_{R\cup I}\rightarrow t$ for some (ground) term t'.

Then $t' = s[u\leftarrow r\sigma]$ and $l\sigma = s/u$ for some rule $(l\rightarrow r) \in I$, occurrence u in s and ground substitution σ.

<u>claim</u> There exist an occurrence v in l and a rule $(l'\rightarrow r') \in R$ such that l/v is not a variable and $l\sigma/v$ is a (ground) instance of l'.

<u>proof</u> Let γ be a substitution such that $x\gamma$ is an R-normal-form of $x\sigma$ for all variables x. If there would be no such occurrence v, the groundterm $l\gamma$ would not be R-reducible, contradicting (3).

Let η be the most general unifier of l/v and l'. Then <c,p> where $c = l[v\leftarrow r']\eta$ and $p = r\eta$ is a critical pair of R on I, thus there are c', p' such that $c -^*_{R\cup I}\rightarrow c' \leftarrow^*_L\rightarrow p' \leftarrow^*_{R\cup I}- p$, according to (4). We have $l\sigma/v = l'\sigma$, therefore $\sigma = \eta\sigma'$ for a (ground) substitution σ'.

Thus $s -_R\rightarrow s[uv\leftarrow r'\sigma] = s[u\leftarrow l[v\leftarrow r']\sigma] = s[u\leftarrow (l[v\leftarrow r']\eta)\sigma'] = s[u\leftarrow c\sigma'] -^*_{R\cup I}\rightarrow s[u\leftarrow c'\sigma'] \leftarrow^*_L\rightarrow s[u\leftarrow p'\sigma'] \leftarrow^*_{R\cup I}- s[u\leftarrow p\sigma'] = s[u\leftarrow r\eta\sigma'] = s[u\leftarrow r\sigma] = t'$.

From the induction hypothesis we get $s[u\leftarrow c\sigma'] \leftarrow^*_R\rightarrow s[u\leftarrow c'\sigma']$, $t' \leftarrow^*_R\rightarrow s[u\leftarrow p'\sigma']$ and $t' \leftarrow^*_R\rightarrow t$. Restricted to ground terms, $\leftarrow^*_L\rightarrow$ is included in $\leftarrow^*_R\rightarrow$ (just consider a ground instance of an L-derivation sequence, using (1)), thus $s[u\leftarrow c'\sigma'] \leftarrow^*_R\rightarrow s[u\leftarrow p'\sigma']$.

Hence $s \leftarrow^*_R\rightarrow t$.

Example 1

We define axioms of addition w.r.t. the usual signature of natural numbers, containing 0, s, +, *, powers etc., as follows :

(a1)	0 + y	→	y	ADD
(a2)	s(x) + y	→	s(x+y)	

By applying theorem 1 we now prove the commutated versions of ADD

(a1')	x + 0	→	x	ADD '
(a2')	x + s(y)	→	s(x+y)	

and left-to-right-associativity

(ass_+) (x+y) + z → x + (y+z)

to be inductive theorems over R := ADD.

We choose L to be empty. Apparently there are two critical pairs in either case, (a1) and (a2) resp., *on* each of our candidates from the set $I := \{ass_+\} \cup ADD'$, which reduce to identical normal forms via standard reduction. Obviously R∪I is noetherian, and every ground-instance of I- left-hand-sides must contain an innermost + -symbol, to which at least one R-step is applicable, i.e. I is ground-reducible under R .

Theorem 2

Let R, E, L be TRS's such that

(1) $L \subseteq ITh(R)$

(2) R is terminating

(3) E is ground-reducible under R

(4) for every critical pair <c,p> of R on E there exist terms c', p' such that

$$\begin{array}{l} c \\ R \downarrow * \\ c' \ -=_E\rightarrow \ p' \ \leftarrow *_L \rightarrow \ p \end{array}$$

Then $E \subseteq ITh(R)$.

Proof

By noetherian induction on $-_R\rightarrow$ we show for all groundterms s, t :

if $s -_R\rightarrow \cdot -=_E\rightarrow t$ then $s \leftarrow *_R\rightarrow t$.

For the cases $s = t$ and $s -*_R\rightarrow \cdot -*_R\rightarrow \cdot -=_E\rightarrow t$ we proceed as in the proof of theorem 1. In the case $s -=_E\rightarrow t$ a "critical pairs lemma" just as in the above proof yields the existence of (ground) terms s', t_0 such that $s -_R\rightarrow s' -*_R\rightarrow \cdot -=_E\rightarrow t_0 \leftarrow *_L\rightarrow t$. From the induction hypothesis we get $s' \leftarrow *_R\rightarrow t_0$. Since t_0 and t are ground, we have $t_0 \leftarrow *_R\rightarrow t$ (see the remark above), thus $s \leftarrow *_R\rightarrow t$.

Remark

In premise (4) of theorem 2 we cannot replace $-=_E\rightarrow$ by $-*_E\rightarrow$ or $\leftarrow *_E\rightarrow$. Consider the following example: $R = L = \{ a\rightarrow b, b\rightarrow c \}$, $E = \{ b \rightarrow a, a\rightarrow d \}$ for constants a, b, c, d.

Here (1), (2), (3) and (4) - if modified in that way - hold, applying *two* E-steps to one of the critical pairs. However, $(a\rightarrow d) \notin ITh(R)$. (Cf. Fig. 13b in [Hu80].)

Example 2

Theorem 2 helps us to deal with non-terminating rules, such as commutativity

($comm_+$) $x + y \rightarrow y + x$.

Let $E = \{comm_+\}$, $L = ADD'$. The assumptions of theorem 2 are easily checked, reducing again both critical pairs from (a1) and (a2) *on* ($comm_+$) to identical normal form. Thus we have proved $(comm_+) \in ITh(ADD)$.

Both theorems will now be combined and further generalized. As an additional feature we allow the use of a set A of inductive theorems which together with R∪I are terminating. In contrast to L-steps, A-steps can be mixed with R∪I-steps, they even can precede the (optional) E-step.

Theorem 3 (Main Theorem)

Let R, I, E, A, L be TRS's such that

(1) $L \cup A \subseteq ITh(R)$

(2) $R \cup I \cup A$ is terminating

(3) $I \cup E$ is ground-reducible under R

(4) for every critical pair <c,p> of R on I :

$$\begin{array}{ccccc} c & & & & p \\ R\cup I\cup A \downarrow * & & & & * \downarrow R\cup I\cup A \\ \cdot & -=_E\rightarrow & \cdot & \leftarrow^*_L\rightarrow & \cdot \end{array}$$

(5) for every critical pair <c,p> of R on E :

$$\begin{array}{ccccc} c & & & & \\ R\cup I\cup A \downarrow * & & & & \\ \cdot & -=_E\rightarrow & \cdot & \leftarrow^*_L\rightarrow & p \end{array}$$

Then $I\cup E \subseteq ITh(R)$.

Proof

By combining the proofs of theorems 1 and 2 we show by noetherian induction on $-^*{}_{R\cup I\cup A}\rightarrow$ for all groundterms s, t : if $s\ -^*{}_{R\cup I\cup A}\rightarrow \cdot -=_E\rightarrow t$ then $s \leftarrow^*{}_R\rightarrow t$.

<u>case 1</u> $s = t$. Then $s \leftarrow^*{}_R\rightarrow t$.

<u>case 2</u> $s -_R\rightarrow t' -^*{}_{R\cup I\cup A}\rightarrow \cdot -=_E\rightarrow t$ for some (ground) term t'.

By induction hypothesis $t' \leftarrow^*{}_R\rightarrow t$, thus $s \leftarrow^*{}_R\rightarrow t$.

<u>case 3</u> $s -_I\rightarrow t' -^*{}_{R\cup I\cup A}\rightarrow \cdot -=_E\rightarrow t$ for some (ground) term t'.

As in the proof of theorem 1, using (4) we get terms s', t_0, t_1 such that

$s -_R\rightarrow s' -^*{}_{R\cup I\cup A}\rightarrow \cdot -=_E\rightarrow t_0 \leftarrow^*{}_L\rightarrow t_1 \leftarrow^*{}_{R\cup I\cup A}- t'$.

Then $t_0 \leftarrow^*{}_R\rightarrow t_1$ from (1), and by induction hypothesis $s' \leftarrow^*{}_R\rightarrow t_0$, $t' \leftarrow^*{}_R\rightarrow t_1$ and $t' \leftarrow^*{}_R\rightarrow t$, thus $s \leftarrow^*{}_R\rightarrow t$.

<u>case 4</u> $s -_A\rightarrow t' -^*{}_{R\cup I\cup A}\rightarrow \cdot -=_E\rightarrow t$ for some (ground) term t'.

Conclude as in case 2 using (1) to get $s \leftarrow^*{}_R\rightarrow t$.

<u>case 5</u> $s -=_E\rightarrow t$. Due to (5) there exist terms s', t_0 such that

$s -_R\rightarrow s' -^*{}_{R\cup I\cup A}\rightarrow \cdot -=_E\rightarrow t_0 \leftarrow^*{}_L\rightarrow t$. Then we get $s \leftarrow^*{}_R\rightarrow t$ as in case 3 .

3. Examples

The following examples illustrate the use of the above theorems. For lack of space, in this paper we do not give proofs for termination and ground-reducibility. In most cases termination can be proved by means of standard orderings like KBO, RPO or its variant lexicographic RPO; only a few rules require a more detailed analysis. Ground-reducibility, a decidable but PSPACE-hard property [KNRZ87], can be assured using easy to verify sufficient criteria. For details see the full version of this paper [HK88].

3.1. Natural numbers : standard arithmetic

As <u>basic specification of natural number arithmetic</u> we choose

(a1)	$0 + y$	$\rightarrow$	y	ARITH
(a2)	$s(x) + y$	$\rightarrow$	$s(x+y)$	
(m1)	$0 * y$	$\rightarrow$	0	
(m2)	$s(x) * y$	$\rightarrow$	$(x*y) + y$	
(p1)	x^0	$\rightarrow$	$s(0)$	
(p2)	$x^{s(n)}$	$\rightarrow$	$x * x^n$	

In section 2 we already have proved some <u>inductive properties of addition</u> over ARITH, using theorem 1 :

(a1')	$x + 0$	$\rightarrow$	x	ADD '
(a2')	$x + s(y)$	$\rightarrow$	$s(x+y)$	
(ass_+)	$(x+y) + z$	$\rightarrow$	$x + (y+z)$,	and using theorem 2 :
($comm_+$)	$x + y$	$\rightarrow$	$y + x$	

Also for <u>multiplication</u>, we can prove the <u>commutated versions</u> of our definitions, using theorem 1 with $L = \{ass_+\} \cup ADD'$:

(m1')	$x * 0$	$\rightarrow$	0	MULT '
(m2')	$x * s(y)$	$\rightarrow$	$x + (x*y)$	

Note that ARITH $\cup$ ADD ' $\cup$ MULT ' still is noetherian, also together with the (ass_+)-rule, but neither ($comm_+$) nor (ass_+) and its reverse (ass_+'), taken together, will terminate. Also note that the reverse of (a2')

(a2'r) $s(x+y) \rightarrow x + s(y)$

together with (a1') can be added safely to ARITH $\cup$ MULT ' without losing termination, but of course cannot be added in the presence of ADD ' .

As another example of theorem 1 we can show the <u>distributive law</u> to be in ITh(ARITH) :

(distr) $(x+y) * z \rightarrow x*z + y*z$

using $L = \{ ass_+, comm_+ \}$ and discussing two critical pairs. Trying the same procedure for

(distr') $x * (y+z) \rightarrow x*y + x*z$

we have to consider 4 critical pairs. Moreover we need MULT ' to reduce them all up to (ass_+,$comm_+$)-equivalence ; thus we obtain a weaker result : (distr') $\in$ ITh(ARITH $\cup$ MULT ').
Now theorem 3 helps to remedy this unsatisfactory situation : simply choose I and L as before, $E = \emptyset$, and now A = MULT ' (already verified inductive theorems), and the result appears as desired : (distr') $\in$ ITh(ARITH).

We shortly elaborate this computation to illustrate a simple use of theorem 3 :

(a1) *on* (distr') : $x * (0+z)$, a1 $\downarrow$ distr'

<u>$x * z$</u>	$x*0 + x*z$
	$\downarrow$ m1'
	$0 +$ <u>$x*z$</u>
	$\downarrow$ a1
	<u>$x * z$</u>

where the critical pair reduces to the same normal form.

and (a2) *on* (distr') : $x * (s(y)+z)$, a2 $\downarrow$ distr'

$x * s(y+z)$	$x*s(y) + x*z$
$\downarrow$ m2'	$\downarrow$ m2'
$x + x*(y+z)$	<u>$(x + x*y) + x*z$</u>
$\downarrow$ distr'	
<u>$x + (x*y + x*z)$</u>	

where the underlined terms are (ass_+)-equivalent.

The cases (m1) *on* (distr') and (m2) *on* (distr') cause no complications.
In the same manner, choosing A = {distr}, $L = E = \emptyset$, we prove <u>multiplicative associativity</u>

(ass_*) $(x*y) * z \rightarrow x * (y*z)$ $\in$ ITh(ARITH) .

The second variant of theorem 3 now gives us <u>multiplicative commutativity</u>

($comm_*$) $x * y \rightarrow y * x$ to be an inductive theorem over ARITH.

We choose $L = \{comm_+\}$, A = MULT ', $I = \emptyset$, and reduce both critical pairs resulting from the superpositions

(m1) *on* ($comm_*$) : $0 * y$, m1 $\downarrow$ $comm_*$

<u>0</u>	$y * 0$
	$\downarrow$ m1'
	<u>0</u>

which are again equal.

and (m2) *on* ($comm_*$) : $s(x) * y$, m2 $\downarrow$ $comm_*$

$x*y + y$	$y * s(x)$
$\downarrow$ $comm_*$	$\downarrow$ m2'
<u>$y*x + y$</u>	<u>$y + y*x$</u>

which are ($comm_+$)-equivalent.

We conclude our collection of standard arithmetic by <u>exponentiation rules</u> :

$$\begin{array}{llll} \text{(1pow)} & s(0)^{m} & \rightarrow & s(0) \\ \text{(expsum)} & x^{n+m} & \rightarrow & x^{n} * x^{m} \end{array}$$

Both rules are inductive over ARITH, and are appropriate lemmata to prove

$$\begin{array}{llll} \text{(powpow)} & (x^{n})^{m} & \rightarrow & x^{n*m} \end{array} .$$

Diagram

This diagram shows the structure of our proofs in ITh(ARITH), where an arrow indicates the use of appropriate "lemmata" (by means of the auxiliary sets A and L).

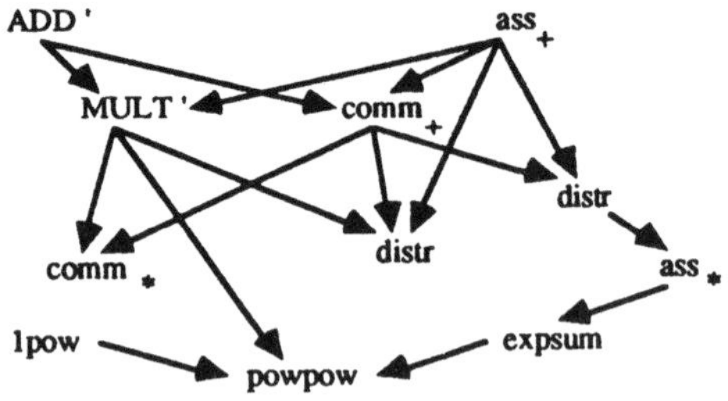

3.2. Natural numbers : binomial coefficients

Now we turn to a more interesting example, <u>binomial coefficients</u>. We define them by addition, in order to avoid fractions arising from the factorial definition :

$$\left.\begin{array}{llll} \text{(b1)} & \binom{0}{s(k)} & \rightarrow & 0 \\ \text{(b2)} & \binom{n}{0} & \rightarrow & s(0) \\ \text{(b3)} & \binom{s(n)}{s(k)} & \rightarrow & \binom{n}{s(k)} + \binom{n}{k} \end{array}\right\} \text{BIN}$$

We further include the sum of binomial coefficients with fixed n into our definition, writing $\sum_n^k$ for $\sum_{j=0}^{k} \binom{n}{j}$:

$$\left.\begin{array}{llll} \text{(s1)} & \sum_n^0 & \rightarrow & \binom{n}{0} \\ \text{(s2)} & \sum_n^{s(k)} & \rightarrow & \binom{n}{s(k)} + \sum_n^k \end{array}\right\} \text{SUM}$$

We put R := ARITH ∪ BIN ∪ SUM .

Our goal is to prove the theorem $\sum_{j=0}^{n} \binom{n}{j} = 2^n$, formalized as rule in our signature :

$$\begin{array}{llll} \text{(s3)} & \sum_n^n & \rightarrow & s(s(0))^{n} \end{array}$$

We again follow our main line to prove elementary properties first, for later use as lemmata . Start with

$$\text{(b4)} \quad \binom{n}{n+s(m)} \quad \rightarrow \quad 0$$, which easily turns out to be in ITh(R) , using A := (a2'r).

We separate this proof from the following of (b5) and (b6) ,

(b5)	$\binom{n}{n}$	→	s(0)
(b6)	$\binom{n}{s(n)}$	→	0

because (b6) requires some assistance, e.g. by (a1'r) and (a2'r), together with (b4), where neither (a1'r) nor {a2'r} ∪ ADD ' terminates. One succeeds in choosing L = ADD ' ∪ {b4}, which allows to apply the ADD ' -rules in either direction, mixed with one required (b4)-step.

Now we can proceed straightforward with some inductive properties of binomial sums, simultaneously proving

(s4)	$\sum_0^{s(k)}$	→	s(0)
(s5)	$\binom{n}{k} + \sum_{s(n)}^{k}$	→	$\sum_n^k + \sum_n^k$

and (s3) , which is our goal.

With L = {b4,b5,b6} ∪ ADD ' ∪ {ass_+,$comm_+$} we choose a non-terminating auxiliary set again, in additional we use E = ∅ and A = {b4,b5,b6} ∪ {ass_+}, and apply theorem 3. A problem occurs during the reduction of some critical pairs : One can happen to fail with that proof if one proceeds too far reducing with R ∪ I ∪ A . So it is necessary to check for L -equivalence during intermediate steps as well .

3.3. Lists : basic properties

We construct lists of arbitrary elements (e.g. numbers) by the empty list *nil* and the usual . -operation ("ladd", append an *element* x to a *list* u from the left, obtaining x.u). In additional we introduce "radd" and concatenation of two lists denoted by u_x and u ∘ v respectively.

(radd1)	nil_y	→	y.nil	LIST
(radd2)	(x.u)_y	→	x. (u_y)	
(conc1)	nil ∘ v	→	v	
(conc2)	(x.u) ∘ v	→	x. (u ∘ v)	

It is easy to prove by theorem 1

(conc1')	u ∘ nil	→	u	CONC '
(conc2')	u ∘ (x.v)	→	(u_x) ∘ v	

as additional axioms A ⊆ ITh(LIST). Using them to prove

(conc3) u ∘ (v_y) → (u ∘ v)_y ,

and adding this to A again we obtain

(ass) u ∘ (w ∘ v) → (u ∘ w) ∘ v ∈ ITh(LIST) .

Extending our system by a reverse-operation on lists and the rules

(rev1)	reverse(nil)	→	nil	REV
(rev2)	reverse(x.u)	→	reverse(u)_x	

we can prove a nice property of reverse, namely

(revrev) reverse(reverse(u)) → u ∈ ITh(LIST ∪ REV) .

During that proof the Knuth-Bendix completion procedure makes a "proposal for a possible lemma"

(rev3) reverse(u_x) → x.reverse(u) , which actually works if we add it to I .

3.4. Lists : sorting natural numbers

We keep the . -operation from the previous section,and add a <u>less-or-equal predicate on natural numbers</u>

(ord1)	0 ≤ 0	→	true	ORD
(ord2)	0 ≤ s(m)	→	true	
(ord3)	s(n) ≤ 0	→	false	
(ord4)	s(n) ≤ s(m)	→	n ≤ m	

in order to prove the <u>idempotency of sorting lists of natural numbers</u> :

(sortsort) sort(sort(u)) → sort(u)

We define <u>sorting</u> as follows :

(sort1)	sort(nil)	→	nil	SORT
(sort2)	sort(n.u)	→	insert(n, sort(u))	
(ins1)	insert(n, nil)	→	n.nil	
(ins2)	insert(n, m.u)	→	[n ≤ m, n, m.u]	
(n≤m)	[true, n, m.u]	→	n.m.u	
(n>m)	[false, n, m.u]	→	m.insert(n, u)	

Note that this if-then-else-specification by [, ,] assures that the ≤ –predicate is evaluated before further insertion is done.

Now we proceed as follows, always applying theorem 3. For better readability we present the results top-down :

(sortsort) sort(sort(u)) → sort(u) ∈ ITh(ORD ∪ SORT) holds,

if we can prove

(lemma1) sort(insert(n, u)) → insert(n, sort(u)) .

Here again we need a lemma to be an inductive theorem over ORD ∪ SORT

(lemma2) sort([n≤m, n, m.u]) → insert(n, insert(m, sort(u))) ,

but unfortunately here our method seems to fail. However, we can apply the ***technique of generalization*** and loosen the variable binding on the left hand side, thus proving an even "better" lemma instead :

(lemma2') sort([x≤y, n, m.u]) → insert(n, insert(m, sort(u))) .

In fact, now the proof succeeds with a small additional lemma which is easy to prove :

(lemma3) insert(0, u) → 0.u .

If we restructure the proof from bottom to top, placing the lemmata into the appropriate auxiliary sets in each step (i.e. into A in this case), or alternatively put all of them simultaneously into I, we get the desired result .

3.5. Arbitrary data structures

In section 3.4., we dealt with the example of sorting lists of natural numbers. As an example for inductive proofs in arbitrary data structures (cf. [BMS84]) the techniques of section 2 will now be applied to show inductive theorems for lists over an arbitrary domain D equipped with a (total) ordering ≤ on its elements. (E.g. D could be the real numbers with the ordinary less-or-equal predicate.) In order to handle the structure (D, ≤) in the framework of term rewriting systems we extend the signature for lists by new constants - for every element of D a distinct constant symbol - and a boolean valued binary operation symbol ≤ .

Throughout this section let R consist of the rules SORT from section 3.4. together with

$$\left.\begin{array}{l} \{ (d_1 \le d_2) \to \text{true} \mid \text{for all } d_1, d_2 \in D \text{ where } d_1 \le d_2 \text{ in } (D, \le) \} \cup \\ \{ (d_1 \le d_2) \to \text{false} \mid \text{for all } d_1, d_2 \in D \text{ where not } d_1 \le d_2 \text{ in } (D, \le) \} \end{array}\right\} \text{ORD}_D$$

Note that in the case of an infinite domain D we are dealing with infinite (for real numbers even uncountable) signatures and sets of rules without losing applicability of the theorems in section 2. In the following x, y, z, u will always denote variables whereas d_1, d_2, d_3 are constants.

Using theorem 3 (with A = L = Ø) we prove the idempotency of sort by choosing (sortsort), (lemma1) and (lemma2') as I, and letting E consist of the rules

insert(x, insert(y, u))	→	insert(y, insert(x, u))	
[x≤y, x, y.nil]	→	[y≤x, y, x.nil]	
insert(x, [y≤z, y, z.u])	→	insert(y, [x≤z, x, z.u])	
insert(d_1, [x≤d_2, x, d_2.u])	→	insert(x, d_1.d_2.u)	for all $d_1, d_2 \in D$ where $d_1 \le d_2$
insert(d_1, [x≤d_2, x, d_2.u])	→	insert(x, d_2.insert(d_1, u))	for all $d_1, d_2 \in D$ where $d_2 < d_1$

Consider as an example the set of critical pairs of ORD_D in a (particular) rule

insert(d_1, [x≤d_2, x, d_2.u]) → insert(x, d_2.insert(d_1, u)) where $d_1, d_2 \in D$, $d_2 < d_1$.

Superposition of a rule ($d_3 \le d_2 \to$ true) on this rule yields the critical pair < c, p > where

c = insert(d_1, [true, d_3, d_2.u]) and p = insert(d_3, d_2.insert(d_1, u)) .

Reducing c and p to R∪I - normal form, in both cases we get the term d_3.d_2.insert(d_1, u) . We use (among others) the rule ($d_1 \le d_3 \to$ false) which is in ORD_D due to the transitivity of ≤ on D .

Superposition of a rule ($d_3 \le d_2 \to$ false) on the above rule yields the critical pair < c, p > where

c = insert(d_1, [false, d_3, d_2.u]) and p = insert(d_3, d_2.insert(d_1, u)) .

Reducing them to normal forms yields

c' = d_2.insert(d_1, insert(d_3, u)) and p' = d_2.insert(d_3, insert(d_1, u)) respectively.

We succeed using the first rule of E to get the step $c' \overset{=}{\longrightarrow}_E p'$.

4. Conclusions

"Inductionless induction" - as it is sometimes called in the literature - is a very promising tool to cope with the problem of proving inductive properties of data types which are defined by term rewriting systems.

In the first part of this paper we have given refinements of well known sufficient criteria for establishing inductive theorems. We reduce the number of critical pairs to be computed, and omit the requirement of (ground) confluence, even allowing non-terminating rules to be proved.

In the second part we use these results to obtain a nice collection of examples of inductive theorems from very basic axioms which define abstract data types like numbers and lists. Admittedly, a very careful analysis of the axioms, and a clever choice of appropriate lemmata is necessary during the proofs, requiring human interaction rather than full automatization. Concepts and strategies for this problem are strongly needed. Nevertheless, the Knuth-Bendix completion process involved often results in proposals for "lemmata", and one often succeeds with the simple concept of generalization : do *not* prove what KBC needs, but prove a more general lemma.

References

[BMS84] Büchi, J.R., Mahr, B. and Siefkes, D. *Manual on REC - a language for use and cost analysis of recursion over arbitrary data structures*, Techn. Report 84-06, TU Berlin, FB 20 (1984).

[Frib86] Fribourg, L. *A strong restriction of the inductive completion procedure*, 13th ICALP,Lecture Notes in Comp. Sci., Vol.226 (1986), pp.105-15.

[Gog80] Goguen, J. *How to prove algebraic inductive hypotheses without induction, with applications to the correctness of data type implementation*, Lecture Notes in Comput. Sci., Vol.87, Springer-Verlag (1980), pp.356-73.

[Huet80] Huet, G. *Confluent reductions: abstract properties and applications to term rewriting systems*, J.ACM, Vol.27 (1980), pp.797-821.

[HH82] Huet, G. and Hullot, J.M. *Proofs by induction in equational theories with constructors*,J. Comp. and Syst. Sci., Vol.25 (1982), pp.239-66.

[HK88] Hofbauer, D. and Kutsche, R.-D. *Proving inductive theorems based on term rewriting systems* , Techn. Report 88-12, TU Berlin (1988).

[JK86] Jouannaud, J.P. and Kounalis, E. *Automatic proofs by induction in theories without constructors*, CNRS Rapport de Recherche No.295 (1986). Preliminary version in Proc. 1st LICS (1986).

[KM87] Kapur, D. and Musser, D.R. *Proof by consistency*, Artificial Intelligence, 31 (1987), pp.125-157.

[KNZ87] Kapur, D., Narendran, P. and Zhang, H. *On sufficient-completeness and related properties of term rewriting systems*, Acta Informatica, Vol.24 (1987), pp.395-415.

[KNRZ87] Kapur, D., Narendran, P., Rosenkrantz, D. and Zhang, H. *Sufficient-completeness, quasi-reducibility, and their complexity*, Bull. of the EATCS 33 (1987), pp.279-81.

[Kir85] Kirchner, H., *A general inductive completion algorithm and application to abstract data types*, Lecture Notes in Comp. Sci., Vol.170, Springer Verlag (1985), pp.282-302.

[Küch87] Küchlin, W. *Inductive completion by ground proof transformation*, Techn. Report No.87-08, Comp. and Inform. Sci., Univ. of Delaware, Newark, DE 19716 (1987).

[Mus80] Musser, D.R. *On proving inductive properties of abstract data types*, Proc. 7th ACM Symp. on Principles of prog. languages (1980), pp.154-62.

[Pad87] Padawitz, P. *Foundations of specification & programming with Horn clauses*, Habilitations-schrift, Universität Passau (1987), to appear as EATCS monograph.

[Paul84] Paul, E. *Proof by induction in equational theories with relations between constructors*, 9th Coll. on trees in algebra and programming, Ed. Courcelle,B., Cambridge Univ. Press (1984), pp.211-25.

[Toya86] Toyama, Y. *How to prove equivalence of term rewriting systems without induction*, Lecture Notes in Comp. Sci., Vol.230, Springer Verlag (1986), pp.118-27.

Jungle Evaluation for Efficient Term Rewriting *

Berthold Hoffmann and Detlef Plump **

Abstract
Jungles are acyclic hypergraphs that represent sets of terms over a signature so that equal subterms can be shared.
Term rewrite rules can be translated into jungle evaluation rules having the property that each evaluation step performs several term rewrite steps simultaneously.
It is shown that the translation preserves termination. Confluent and terminating term rewriting systems are translated into confluent and terminating jungle evaluation systems which are complete in the sense that each term normal form is represented by some jungle normal form.
Moreover, jungle evaluation can be speeded up by additional hypergraph rules that fold equal terms without loosing termination and confluence. Using these folding rules, even non-linear term rewriting systems can be modelled.

1 Introduction

Term rewriting is an interesting way of "computing by replacement" which is used in various areas of computer science: For the interpretation of functional and logical programming languages (cf. [Der 85], [O'D 87], [Pey 87]), for theorem proving (cf. [Hsi 85]), and for executing algebraic specifications of abstract data types and checking properties of specifications like consistency and completeness (cf. [EM 85], [FGJM 85], [Pad 83]).

In classical implementations terms are represented by *trees*, and rewriting is realized by *subtree replacement*. Unfortunately, this way of rewriting may be very expensive, both in time and space: The application of a rule may require large subterms to be copied, and for each copy of a term, the result must be computed anew.

In this paper, we investigate an improved model for implementing term rewriting where these sources of inefficiency are avoided:

- Terms are represented by acyclic hypergraphs rather than by trees. These hypergraphs are called *jungles*, a name coined in [Plu 86] and [HKP 88]. In jungles, multiple occurrences of terms can be shared.
- Rewriting is performed by hypergraph replacement, specified by *evaluation rules* according to the algebraic theory of graph grammars (see, e.g., [Ehr 79]). By applying these evaluation rules, new references to existing subterms are introduced instead of copying subterms completely.
- Additional hypergraph rules for *folding* multiple occurrences of subterms allow each term in a jungle to be represented uniquely, so that its result is computed at most once.

1.1 Example (Computation of Fibonacci Numbers)
Consider the term rewrite rules

$$\begin{aligned} \mathtt{fib(0)} &\rightarrow \mathtt{0} \\ \mathtt{fib(succ(0))} &\rightarrow \mathtt{succ(0)} \\ \mathtt{fib(succ(succ(x)))} &\rightarrow \mathtt{fib(succ(x))} + \mathtt{fib(x)} \end{aligned}$$

*This work is supported by the Commission of the European Communities under Contract 390 (PROSPECTRA Project) in the ESPRIT Programme.

** Fachbereich Mathematik und Informatik, Universität Bremen, Postfach 330 440, D-2800 Bremen 33. Usenet: {hof,det}%informatik.uni-Bremen.DE

specifying a function fib that computes fibonacci numbers, based on natural numbers with the constant 0, successor function succ, and addition +.

For computing the fibonacci number of 4, term rewriting starts as follows:

$$\begin{aligned} \texttt{fib}(\texttt{succ}^4(0)) &\rightarrow \texttt{fib}(\texttt{succ}^3(0)) + \texttt{fib}(\texttt{succ}^2(0)) \\ &\rightarrow \texttt{fib}(\texttt{succ}^2(0)) + \texttt{fib}(\texttt{succ}(0)) + \texttt{fib}(\texttt{succ}^2(0)) \ldots \end{aligned}$$

In both steps, the arguments of fib are copied. The resulting term contains two copies of $\texttt{fib}(\texttt{succ}^2(0))$; each of them must be rewritten anew. As a consequence, rewriting a term $\texttt{fib}(\texttt{succ}^n(0))$ to normal form requires time and space exponential in n.

Figure 1 shows the corresponding jungle evaluation steps, followed by one folding step. After the folding step, $\texttt{fib}(\texttt{succ}^2(0))$ and its subterms are represented just once in the jungle, and thus have to be evaluated at most once. By performing evaluation and folding steps in this order, the rewriting of a term $\texttt{fib}(\texttt{succ}^n(0))$ to normal form requires only a number of steps, and space linear in n (see section 4). □

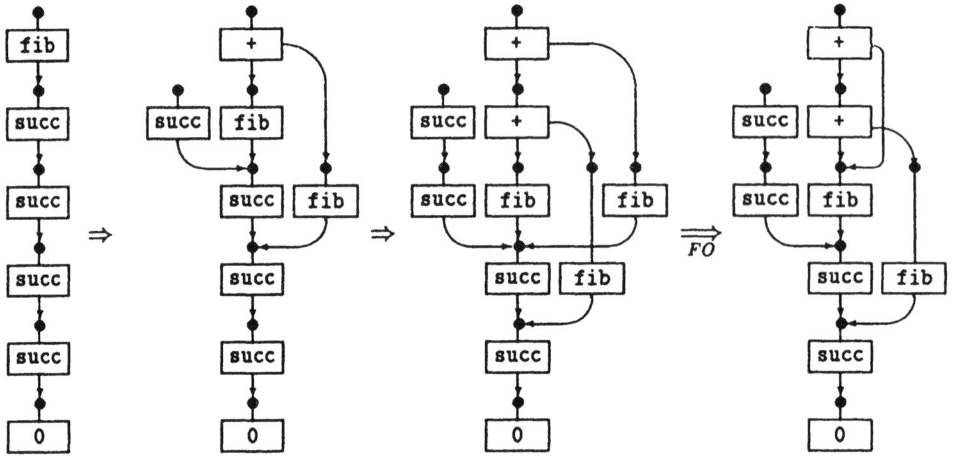

Figure 1: Jungle evaluation steps followed by a folding step

The first graph grammar approach to the evaluation of functional expressions has been undertaken by Ehrig, Rosen, and Padawitz ([ER 76], [Pad 82]). However, they use an extension of the algebraic theory of graph grammars having the drawback that most of the results for "standard graph grammars" are not applicable. The graph reduction approaches of Staples [Sta 80] and Barendregt et. al. [BEGKPS 87] use formalisms of graph rewriting for which no general theory is available.

In [Hof 83] and [Plu 86], the authors started an approach to model term rewriting by standard graph grammars. H-J. Kreowski, A. Habel, and D. Plump continued this work in [HKP 88], where hypergraphs are used instead of graphs, in order to make the technical treatment easier.

With this paper, we carry forth this research by using a slightly more general class of hypergraph grammars which allows for a translation of term rewrite rules into evaluation rules without need for "indirect pointers" or additional identity operations. We investigate general conditions for *termination* and *confluence* of jungle evaluation.

The paper is organised as follows:

In section 2, we recall the definition of jungles and state basic results about the relationship between jungles and terms. Rules for folding multiple representations of terms are introduced which allow to compute minimal jungle representations for sets of terms. In section 3, we construct jungle evaluation rules for modelling term rewriting. It is shown that these rules perform multiple parallel term rewrite steps and compute normal forms with respect to term rewriting. In particular, it is demonstrated that termination of term rewriting carries over to jungle evaluation. The corresponding result for confluence holds if the given term rewriting system is terminating. In section 4, we make some preliminary observations concerning the efficiency of our model of term rewriting and discuss briefly open problems and possible applications of our approach.

In the appendix, hypergraphs and hypergraph derivations are reviewed.

We assume familiarity with basic notions of term rewriting as signature, term, substitution, subterm replacement, etc. (see for instance [EM 85], [HO 80], [Klo 87]). All proofs omitted in this paper can be found in the technical report [HP 88].

Acknowledgements

We would like to thank H.-J. Kreowski and our colleagues from the PROSPECTRA team at Universität Bremen for providing valuable feedback. We are indebted to Clemens Lautemann who provided the key idea for the proof of the termination theorem 3.10. Special thanks are due to Annegret Habel for numerous fruitful discussions and many suggestions improving the presentation.

2 Jungles and Their Relationship to Terms

To realize sharing of common subterms, terms have to be represented by graph-like structures. We use the "Berlin-approach" of graph grammars (see e.g. [Ehr 79]) as a formal framework for the description of graph transformations. However, in the papers of Ehrig, Rosen, and Padawitz on function evaluation ([ER 76], [Pad 82]) this approach is modified in order to allow certain changes of node labels. Our intention is to avoid these modifications to have the results of the "Berlin approach" available.

We represent operation symbols as edges rather than as nodes. Under this requirement functional expressions are more adequately represented by hypergraphs than by graphs since an operation symbol of arity n can be represented as one hyperedge whereas in the graph case n+1 edges together with a node are needed.

The adaptation of notions and results of graph grammar theory to the hypergraph case is straightforward due to the one-to-one correspondence between bipartite graphs and hypergraphs. The basic notions concerning hypergraphs and hypergraph derivations are summarized in the appendix.

This chapter starts with the definition of jungles given in [HKP 88] and shows in which way jungles represent terms. For each jungle there is a ***fully collapsed jungle*** representing the same terms most efficiently: multiple occurrences of (sub-)terms are represented only once. This fully collapsed jungle is uniquely determined up to isomorphism and can be generated from the given jungle by application of ***folding rules***.

2.1 General Assumption

Let $SIG = (S, OP)$ be an arbitrary, but fixed signature. All hypergraphs considered in the following are hypergraphs over SIG, i.e., nodes are labeled with sorts from S and hyperedges are labeled with operation symbols from OP. □

2.2 Definition (Jungle)

A hypergraph G is a ***jungle*** (***over*** SIG) if

1. G is acyclic, [1]
2. $outdegree_G(v) \leq 1$ for each $v \in V_G$, [2]
3. the labeling of G is ***compatible*** with SIG, i.e., for each $e \in E_G$, $m_G(e) = op : s_1 \ldots s_k \to s$ implies $l_G^*(s_G(e)) = s$ and $l_G^*(t_G(e)) = s_1 \ldots s_k$.

Remarks

1. Each hyperedge in a jungle labeled with an operation symbol $op : s_1 \ldots s_k \to s$ has a single source node labeled with s and a sequence of k (not necessarily distinct) target nodes labelled with $s_1, \ldots, s_k$. The sequence of target nodes is empty if op is a constant symbol.

2. Each subhypergraph of a jungle is a jungle, too. Such a subhypergraph is called a ***subjungle***. □

2.3 Example

Let SIG contain a sort **nat** and operation symbols $0 : \to$ **nat**, **succ**, **fib** : **nat** $\to$ **nat**, + : **nat nat** $\to$ **nat**. Then the following hypergraph is a jungle over SIG.

[1] A hypergraph is said to be acyclic if its underlying bipartite graph is acyclic.

[2] The outdegree of a node v in a hypergraph G is defined to be the outdegree of v in the underlying bipartite graph. I.e., $outdegree_G(v) = \sum_{e \in E_G} \#(v, s_G(e))$ where $\#(v, s_G(e))$ denotes the number of occurrences of v in the string $s_G(e)$.

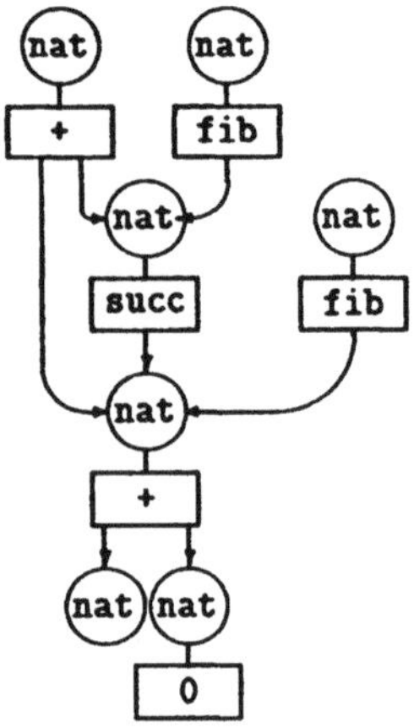

Nodes are drawn as circles and hyperedges as boxes, both with their labels written inside. A line without arrow-head connects a hyperdege with its unique source node, while arrows point to the target nodes (if there are any). The arrows are arranged from left to right in the order given by the target mapping. □

In addition to the graph-theoretic definition 2.2, the set of all jungles can be characterized by a set of generating hypergraph rules: each jungle can be constructed from the empty hypergraph by application of these rules. The interested reader may consult [HKP 88] where this characterization is used to establish a structural induction principle and an efficient parsing algorithm for jungles. Here we are not concerned with these topics but proceed with the relationship between jungles and terms.

Let us consider a jungle G where each node has an outgoing hyperedge. Then for each node v in G we can construct a term $term_G(v)$ by descending along the hyperedges and collecting the hyperedge labels, starting with the unique hyperedge outgoing from v. This procedure terminates since G is acyclic and yields a unique term since all nodes in G have outdegree one. Vice versa, given a variable-free term t it is straightforward to construct a jungle representing t (e.g. by using the well-known tree representation for terms).

However, since we want to translate term rewrite rules into jungle rewrite rules we also have to represent variables in jungles. The problem is that variables cannot be represented as labels because the application of hypergraph rules is based on hypergraph morphisms which preserve labels. Moreover, in the case of multiple occurrences of a variable x in the left-hand side of a jungle rewrite rule we would have to make sure that all instantiations of x are equal before we apply the rule.

Our solution is to use nodes without outgoing hyperedges as variables. Such a variable x can be mapped to any jungle node v by a hypergraph morphism provided that x and v are labeled with the same sort. Since variables are nodes rather than labels, each variable in a jungle occurs only once. (The consequences for the handling of non-left-linear term rewrite rules are discussed in chapter 3.)

2.4 Definition (Variables of a Jungle)
Let G be a jungle. Then $VAR_G = \{v \in V_G \mid outdegree_G(v) = 0\}$ is the set of all *variables* in G. □

2.5 Lemma (Terms Represented by a Jungle)
Let G be a jungle. Then

$$term_G(v) = \begin{cases} v & \text{if } v \in VAR_G, \\ m_G(e) & \text{if there is } e \in E_G \text{ with } s_G(e) = v \text{ and } t_G(e) = \lambda, \\ m_G(e)(term_G(v_1), \ldots, term_G(v_n)) & \text{if there is } e \in E_G \text{ with } s_G(e) = v \text{ and } t_G(e) = v_1 \ldots v_n \end{cases}$$

defines a function $term_G : V_G \rightarrow T_{SIG}(VAR_G)$. [3] □

Notation
The set $term_G(V_G)$ of all terms represented by a jungle G is denoted by $TERM_G$. □

[3] Given a variable set X, $T_{SIG}(X)$ denotes the set of all terms over X.

2.6 Example

Below we show the jungle G of example 2.3 together with its node names and list the terms represented by G. (We write terms with binary operators in infix notation.)

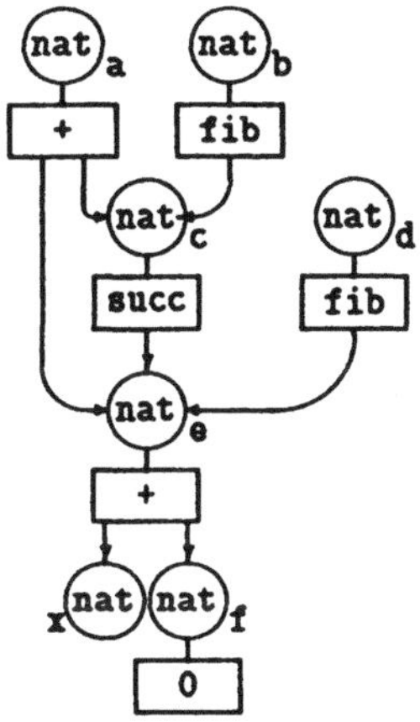

$$\begin{aligned}
term_G(\mathtt{a}) &= \mathtt{x+0+succ(x+0)}\\
term_G(\mathtt{b}) &= \mathtt{fib(succ(x+0))}\\
term_G(\mathtt{c}) &= \mathtt{succ(x+0)}\\
term_G(\mathtt{d}) &= \mathtt{fib(x+0)}\\
term_G(\mathtt{e}) &= \mathtt{x+0}\\
term_G(\mathtt{f}) &= \mathtt{0}\\
term_G(\mathtt{x}) &= \mathtt{x}
\end{aligned}$$

□

When comparing two different jungles it is natural to ask whether these jungles represent the same terms. The following definition provides the set of all jungles with an equivalence relation.

2.7 Definition (Equivalence)

Two jungles G and H are called *equivalent* if $TERM_G = TERM_H$. □

To represent a set T of terms by a jungle one may choose between equivalent jungles which represent T more or less redundantly. In fact each equivalence class of jungles representing a set of terms with at least one nonvariable term contains arbitrary large jungles. However, each equivalence class contains smallest jungles in which each two different nodes represent different terms.

2.8 Definition (Fully Collapsed Jungle)

A jungle G is said to be *fully collapsed* if for all $v_1, v_2 \in V_G$, $term_G(v_1) = term_G(v_2)$ implies $v_1 = v_2$ (i.e., if $term_G$ is an injective function). □

It turns out that fully collapsed jungles are, up to isomorphism, unique representatives of their equivalence classes.

2.9 Uniqueness Theorem

Equivalent fully collapsed jungles are isomorphic. □

Given a jungle, we want to generate the equivalent fully collapsed jungle by application of special hypergraph rules, called folding rules. To explain the idea behind these rules, consider a jungle G and two hyperedges e_1, e_2 in G which have the same labels and target nodes but different sources. Then $s_G(e_1)$ and $s_G(e_2)$ represent the same term. Identifying both source nodes and both edges yields a jungle equivalent to G.

2.10 Definition (Folding)

The set FO of *folding rules* contains the following hypergraph rules: [4]

For each constant symbol $c :\rightarrow s$:

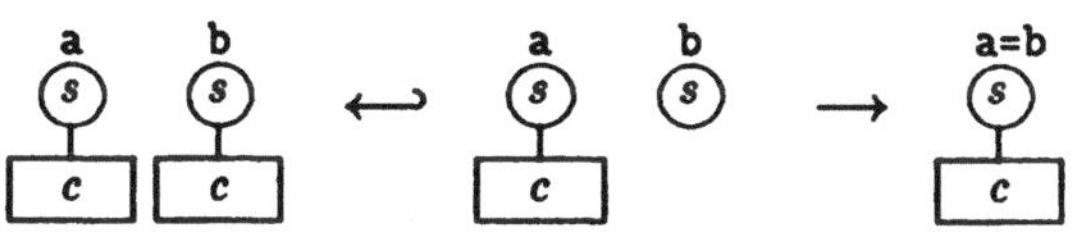

[4]The notation a=b expresses that the nodes a and b are identified by the right-hand morphism.

For each operation symbol $op : s_1 \dots s_k \to s$:

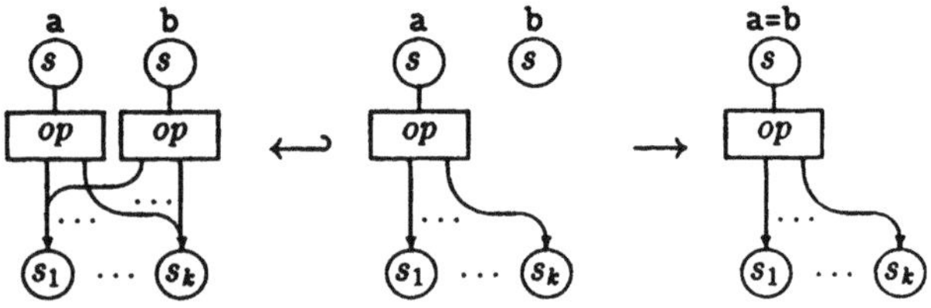

Let G, H be jungles. Then a direct derivation $G \underset{FO}{\Longrightarrow} H$ is called a *folding step* and a derivation $G \overset{*}{\underset{FO}{\Longrightarrow}} H$ is called a *folding*. □

2.11 Example (Folding)

Below we show the folding rule for the operation symbol `fib`. (Here and in the following examples we desist from indicating node labels. By default all nodes are assumed to be labeled with "nat").

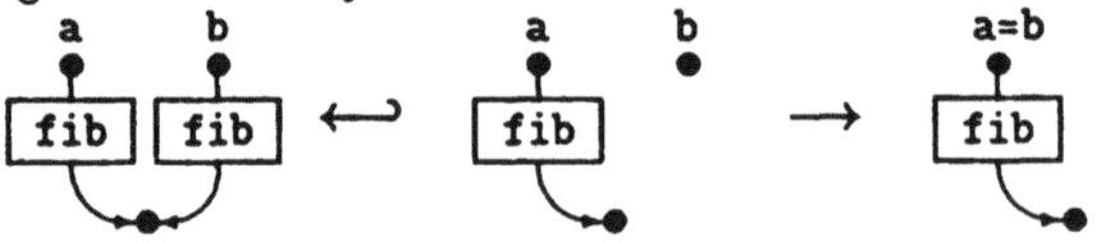

An application of this rule is shown in figure 2. The left and right occurrences of the folding rule are drawn with dashed hyperedges and unfilled nodes. □

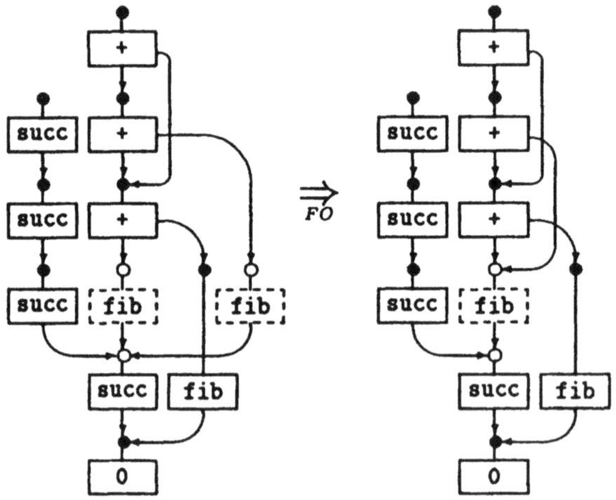

Figure 2: A folding step

By the definition of folding rules it is clear that applying a folding rule to a jungle always yields a jungle and, since each folding step deletes an edge, that each sequence of folding steps terminates. Moreover, given jungles G, H_1, H_2, and folding steps $H_1 \underset{FO}{\Longleftarrow} G \underset{FO}{\Longrightarrow} H_2$ there always is a jungle X such that $H_1 \overset{\lambda}{\underset{FO}{\Longrightarrow}} X \overset{\lambda}{\underset{FO}{\Longleftarrow}} H_2$.[5]

2.12 Theorem (Properties of Folding Steps)

1. $\underset{FO}{\Longrightarrow}$ preserves jungles and is terminating and strongly confluent.
2. A jungle G is fully collapsed if and only if there is no folding rule applicable to G. □

Given a jungle G, we know from the above theorem that the algorithm ***apply folding rules as long as possible*** terminates with a fully collapsed jungle H. Since folding preserves terms,[6] H is the unique fully collapsed jungle equivalent to G.

[5]Given hypergraphs G and H, we write $G \overset{\lambda}{\Rightarrow} H$ if $G \Rightarrow H$ or $G \cong H$.

[6]Here we already assume that folding steps do not rename variaoles. This is stated explicitly in assumption 3.6.

3 Jungle Evaluation

3.1 General Assumption
For the rest of this paper, *TR* denotes an arbitrary, but fixed term rewriting system[7] and $\rightarrow$ its rewrite relation. □

We want to construct a hypergraph rule $p = (L \leftarrow K \rightarrow R)$ that, when applied to some jungle G, performs rewriting according to a given term rewrite rule $l \rightarrow r$. How must p be defined?

The left-hand side L shall represent l in such a way that whenever there is a jungle L' representing l, there is a hypergraph morphism $L \rightarrow L'$ (since then p is applicable to each jungle containing L' as a subhypergraph). Therefore L has to be as little collapsed as possible.

The gluing hypergraph K contains those parts of L which are preserved by the application of p. We want to preserve all arguments of the topmost operation in l since these terms may be also the arguments of other operation symbols in G.

Finally, the right-hand side R has to represent r as well as the preserved subterms of l. We choose R to be fully collapsed since then R is the smallest jungle meeting this requirement.

3.2 Definition (Evaluation Rule and Evaluation Step)
Let $l \rightarrow r$ be a rewrite rule in *TR*. A hypergraph rule $(L \hookleftarrow K \xrightarrow{b} R)$[8] is called an *evaluation rule for* $l \rightarrow r$ if the following conditions are satisfied:

1. L is a variable-collapsed tree[9] with $term_L(root_L) = l$.
2. K is the subhypergraph of L obtained by removing the edge with source $root_L$.
3. R is a fully collapsed jungle such that $TERM_R$ contains all subterms of r and all proper subterms of l. [10]
4. For each $v \in V_K$, $term_R(b_V(v)) = \begin{cases} r & \text{if } v = root_L, \\ term_K(v) & \text{otherwise}. \end{cases}$

For each jungle G, a direct derivation $G \Rightarrow H$ through an evaluation rule is called an *evaluation step*. □

The application of evaluation rules to jungles always yields jungles again.

3.3 Theorem (Evaluation Steps Preserve Jungles)
Let G be a jungle and $G \Rightarrow H$ be an evaluation step. Then H is a jungle, too. □

3.4 Example (Jungle Evaluation)
Figure 3 below shows an evaluation rule for the recursive rewrite rule of `fib` given in the introduction.

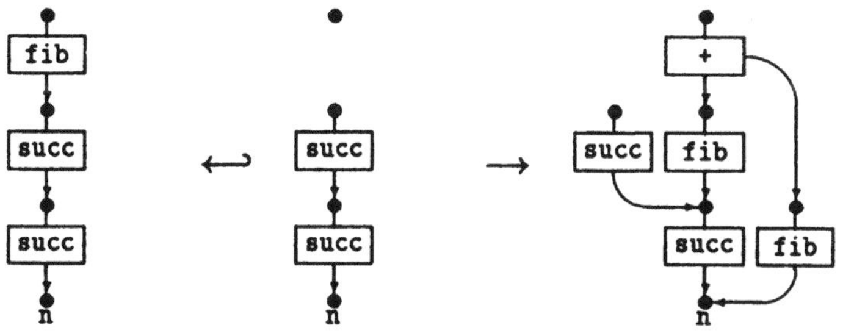

Figure 3: Evaluation rule for `fib(succ(succ(n)))`$\rightarrow$`fib(succ(n))+fib(n)`

In figure 4 the application of this evaluation rule to a jungle is depicted. (Figure 1 in the introduction shows two applications of the same rule.) The term at the root b of the occurrence of the rule is

[7] I.e., each rewrite rule $l \rightarrow r$ in *TR* consists of two terms l and r of equal sort such that l is not a variable and all variables in r occur already in l.

[8] "$L \hookleftarrow K$" denotes the inclusion of K into L.

[9] I.e., there is a unique node $root_L$ in L with $indegree_L(root_L) = 0$ and for each node v with $indegree_L(v) > 1$ we have $v \in VAR_L$.

[10] Where a subterm t of l is called *proper* if $t \neq l$.

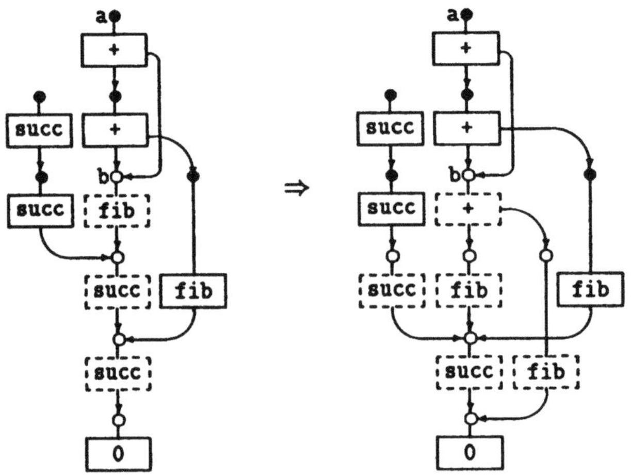

Figure 4: Application of the evaluation rule of figure 3 to a jungle

rewritten according to the underlying term rewrite rule. Terms at nodes which refer to b are rewritten as often as they refer to b (e.g., the term at a is rewritten twice). Terms at all other nodes are not changed. □

To be able to formally describe the effect of an evaluation step $G \Rightarrow H$ on the terms represented by G we define a node mapping from V_G to V_H.

3.5 Definition (Track)

1. Let $d : G \Rightarrow H$ be an evaluation step or a folding step, given by the diagram

$$\begin{array}{ccccc} L & \overset{b_1}{\leftarrow} & K & \overset{b_2}{\rightarrow} & R \\ \downarrow & & \downarrow & & \downarrow \\ G & \overset{c_1}{\leftarrow} & D & \overset{c_2}{\rightarrow} & H \end{array}$$

Since $b_{1,V}$ is bijective, pushout properties imply that $c_{1,V}$ is bijective, too. We define $track_d : V_G \rightarrow V_H$ by $track_d = c_{2,V} \circ c_{1,V}^{-1}$ where $c_{1,V}^{-1} : V_G \rightarrow V_D$ is the inverse of $c_{1,V}$.

2. Let $d : G \cong G_0 \Rightarrow G_1 \Rightarrow \ldots \Rightarrow G_n = H$ be a derivation given by an isomorphism $f : G \rightarrow G_0$ and evaluation or folding steps $d_i : G_i \Rightarrow G_{i+1}$ for $i = 0, \ldots, n-1$. Then $track_d$ is defined by $track_d = track_{d_{n-1}} \circ \ldots \circ track_{d_0} \circ f_V$. □

For each hypergraph derivation $G \overset{*}{\Rightarrow} H$ the hypergraph H is uniquely determined only up to isomorphism. In particular, the variables of a jungle may be renamed by derivations. To avoid a cumbersome handling with variable renamings we want to assume that derivations over evaluation and folding rules do not rename the variables in a jungle. This is not a severe restriction since for each such derivation $G \overset{*}{\Rightarrow} H$ one can rename the nodes of H to obtain an isomorphic jungle H' such that $G \overset{*}{\Rightarrow} H'$ satisfies the following assumption.

3.6 General Assumption

Let $d : G \overset{*}{\Rightarrow} H$ be a derivation over evaluation and folding rules. Then we assume $track_d(x) = x$ for each variable x in G. □

By the following soundness theorem, evaluation steps perform term rewriting on the terms represented by a jungle. In general, a single evaluation step $G \underset{p}{\Rightarrow} H$ corresponds to multiple applications of the rewrite rule underlying p to several terms in $TERM_G$.

3.7 Theorem (Soundness of Evaluation Steps)

Let p be an evaluation rule and $d : G \underset{p}{\Rightarrow} H$ be an evaluation step. Then for each $v \in V_G$,

$$term_G(v) \overset{n}{\rightarrow} term_H(track_d(v))^{11}$$

[11] $\overset{n}{\rightarrow}$ denotes the n-fold composition of $\rightarrow$.

where n is the number of paths[12] from v to the root of the occurrence of p in G. □

In the following we want to manipulate jungles by application of both evaluation rules and folding rules.

3.8 Definition (Jungle Reduction)

Let $d : G \Rightarrow H$ be an evaluation or a folding step. Then d is called a *reduction step*. A derivation consisting of reduction steps is said to be a *reduction*.

A jungle G is called *reduced* if there is no reduction step $G \Rightarrow H$. □

3.9 Theorem (Normalization of Reduction)

A jungle G is reduced if and only if G is fully collapsed and all terms in $TERM_G$ are normal forms[13]. □

We want to emphasize that folding is necessary for the above result. More precisely, the statement *a jungle G is evaluated if and only if all terms in $TERM_G$ are normal forms* does not hold as long as TR contains non-left-linear rewrite rules. As an example, consider the rewrite rule `f(x,x)` $\to$ `c` and a corresponding evaluation rule p. The rewrite rule is applicable to the term `f(c,c)` while p can be applied to a jungle representing `f(c,c)` only if both occurrences of `c` are represented by the same node. The reason is that both occurrences of the variable `x` are represented by a single node in the left-hand side of p.

It should be noted that in general jungle reduction cannot compute all normal forms of a term: Let `h` be an operation symbol and `c` $\to$ `d` and `c` $\to$ `e` be rewrite rules with constant symbols `c`, `d`, and `e`. Then the fully collapsed jungle representing `h(c,c)` can only be reduced to jungles representing `h(d,d)` or `h(e,e)` although `h(d,e)` and `h(e,d)` are further normal forms of `h(c,c)`.

In the following we look for conditions under which jungle reduction is terminating and confluent[14] It turns out that termination of term rewriting carries over to jungle reduction without restriction.

3.10 Termination Theorem

If term rewriting is terminating, then jungle reduction is terminating, too. □

It would be nice if confluence would carry over from term rewriting to jungle reduction as smoothly as termination does. However, this is not the case, as the following example shows.

Let `c` $\to$ `f(d)`, `c` $\to$ `f(e)`, and `f(d)` $\to$ `f(e)` be rewrite rules for an operation symbol `f` and constant symbols `c`, `d`, and `e`. Figure 5 demonstrates that term rewriting is confluent, whereas jungle reduction yields distinct normal forms. This is because the argument `d` is removed when applying the rewrite

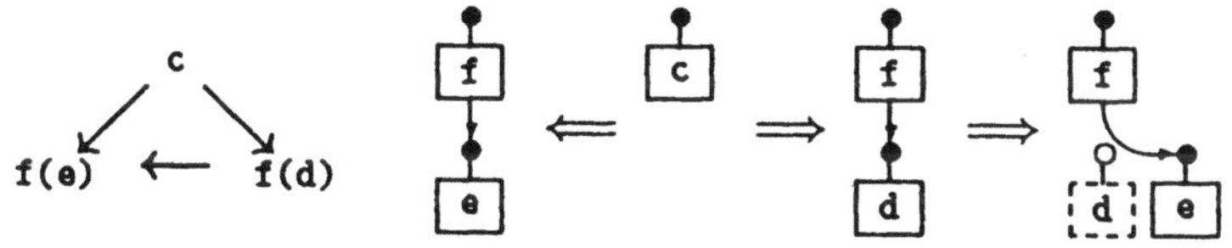

Figure 5: Non-confluence caused by garbage

rule `f(d)` $\to$ `f(e)`, while the corresponding subjungle (drawn with dashed box and unfilled circle) is preserved by the reduction. This subjungle is *garbage* with respect to the term rewriting performed.

In the following, we distinguish certain nodes in a jungle as "points of interest" and consider reductions that keep track of these points.

3.11 Definition (Pointed Jungle and Essential Jungle)

1. A *pointed jungle* is a pair (G, P_G) where G is a jungle and $P_G \subseteq V_G$. We may denote a pointed jungle only by its jungle component.

2. Let G be a pointed jungle. Then the *essential jungle* $G^\bullet$ is the subjungle of G consisting of all nodes and edges that are reachable from the nodes in P_G. □

Given a pointed jungle G, we consider the nodes and edges in $G - G^\bullet$ as *garbage*. It is possible to remove garbage from jungles explicitely by *garbage collecting rules* (which are the inverses of the jungle

---

[12] A *path* in a hypergraph is a path in the underlying bipartite graph.

[13] A term t is a *normal form* if no rewrite rule in TR is applicable to t.

[14] Let A be a set and $\to$ be a binary relation on A. $\to$ is called *terminating* if there is no infinite chain $a_1 \to a_2 \to a_3 \to \dots$ We say $\to$ is *confluent* if for all $a, a_1, a_2 \in A$, $a_1 \overset{*}{\leftarrow} a \overset{*}{\to} a_2$ implies that there is $a' \in A$ such that $a_1 \overset{*}{\to} a' \overset{*}{\leftarrow} a_2$.

generating rules defined in [HKP 88]). Here we just require that reductions preserve the distinguished nodes.

3.12 Definition (Pointed Reduction)
Let (G, P_G) and (H, P_H) be pointed jungles. Then a reduction $d : G \stackrel{*}{\Rightarrow} H$ is said to be *pointed* if $P_H = track_d(P_G)$. We write $G \stackrel{*}{\Rightarrow}\!\bullet\, H$ if d is pointed. □

With the notions just introduced, we are able to state that reduction is confluent "up to garbage" if term rewriting is terminating and confluent. Termination is needed because reduction is not confluent for some non-terminating but confluent term rewriting systems (cf. [HP 88] for an example).

3.13 Confluence Theorem
Let $\rightarrow$ be terminating and confluent and let $H_1 \bullet\!\stackrel{*}{\Leftarrow} G \stackrel{*}{\Rightarrow}\!\bullet\, H_2$ be pointed reductions. Then there are pointed reductions $H_1 \stackrel{*}{\Rightarrow}\!\bullet\, X_1$ and $H_2 \stackrel{*}{\Rightarrow}\!\bullet\, X_2$ such that $X_1^\bullet = X_2^\bullet$. □

4 Discussion and Outlook

The results of the previous sections can be used to show the correctness of the following procedure for computing normal forms of arbitrary terminating and confluent term rewriting systems.

4.1 Computing Term Normal Forms by Jungle Evaluation
Let TR be a terminating and confluent term rewriting system.

In order to compute the normal form of some term t, proceed as follows:

1. Choose a pointed jungle $(G, \{p\})$ with $term_G(p) = t$.
2. Construct a pointed reduction $d : G \stackrel{*}{\Rightarrow}\!\bullet\, H$ by applying evaluation and folding rules as long as possible.
3. Deliver $u = term_H(track_d(p))$ as a result.

By the termination theorem 3.10, step 2 terminates with H reduced, so u is a normal form according to the normalization theorem 3.9. Moreover, the soundness theorem 3.7 yields $t \stackrel{*}{\rightarrow} u$. Thus u is the unique normal form of t since $\rightarrow$ is confluent. □

4.2 Example (Efficiency of Evaluating Fibonacci Numbers)
Consider again the term rewrite rules for computing Fibonacci numbers shown in example 1.1.

We compare the efficiency of term rewriting and jungle reduction in terms of the steps needed to obtain normal forms. The Fibonacci number of 4 is computed in 9 term rewriting steps:

$$\begin{array}{lll}
\mathtt{fib(succ^4(0))} & \rightarrow & \mathtt{fib(succ^3(0)) + fib(succ^2(0))} \\
 & \rightarrow & \mathtt{fib(succ^2(0)) + fib(succ(0)) + fib(succ^2(0))} \\
 & \rightarrow & \mathtt{fib(succ(0)) + fib(0) + fib(succ(0)) + fib(succ^2(0))} \\
 & \rightarrow & \mathtt{fib(succ(0)) + fib(0) + fib(succ(0)) + fib(succ(0)) + fib(0)} \\
 & \stackrel{5}{\rightarrow} & \mathtt{succ(0) + 0 + succ(0) + succ(0) + 0}
\end{array}$$

In general the number of steps for rewriting a term $\mathtt{fib(succ^n(0))}$ grows exponentially with n, more precisely with ϕ^n, where $\phi = (1+\sqrt{5})/2 \approx 1.6180$ is the golden ratio. The size of the terms also grows exponentially.

A jungle reduction sequence corresponding to the above rewrite sequence is used as a running example throughout this paper. Figure 1 shows the first two evaluation steps and the first folding, while figure 4 shows the third evaluation step. Figure 2 shows the second folding step. Two applications of the non-recursive evaluation rules (which are not shown here) finish the reduction. Thus, only 5 evaluation steps and 2 folding steps are needed.

If, as in this example, the recursive `fib`-rule is always applied to the biggest argument and folding is done as early as possible, computing $\mathtt{fib(succ^n(0))}$ requires only $n+1$ evaluation steps and $n-1$ folding steps, i.e. the number of reduction steps grows only linearly with n. □

4.3 Outlook
The work presented here should be continued in various directions:

The procedure described in 4.1 is still rather general in that reduction steps are performed in a completely arbitrary order. This freedom can be used to develop strategies for constructing optimal reductions.

We can consider ***parallel jungle reduction.*** Since we have used a standard type of graph grammars where parallelism has been studied, it should be rather easy to proceed in that direction.

Also, we would like to investigate jungle rules which *cannot* be modelled by term rewriting, e.g. jungle rules with more than one root on their left-hand sides. The folding rules are a first step into that direction. Such rules can perform simultaneous replacement of several patterns which share variables.

Furthermore, we would like to obtain precise measures for the complexity of jungle reduction to compare it with the complexity of term rewriting. In particular, we are interested to find out how efficient folding rules can be implemented and how much folding gains in practical implementations.

Finally, we are confident that our approach can be used to model the interpretation of functional programming languages adequately. It seems as if graph rewriting is one of the key concepts for developing efficient interpreters and compilers for functional languages (cf. [Pey 87]).

5 Appendix: Hypergraphs and Hypergraph Derivations

This appendix summarizes all notions concerning hypergraphs and hypergraph derivations that are needed for our purposes.

We adopt the hypergraph definition given in [HKP 88]. However, we consider a kind of hypergraph replacement which corresponds to the general case of graph replacement in the algebraic approach of graph grammars (see [Ehr 79]). The notions of graph grammar theory (which can be found, e.g., in [Ehr 79] and [Kre 87]) are adapted to the hypergraph case.

5.1 Hypergraphs

1. Let $C = (C_V, C_E)$ be a pair of sets, called *node* resp. *edge labels.*

2. A *hypergraph* $G = (V_G, E_G, s_G, t_G, l_G, m_G)$ over C consists of a finite set V_G of *nodes*, a finite set E_G of *hyperedges* (or *edges* for short), two mappings $s_G : E_G \to V_G^*$ and $t_G : E_G \to V_G^*$, assigning a string of *source nodes* and a string of *target nodes* to each hyperedge, and two mappings $l_G : V_G \to C_V$ and $m_G : E_G \to C_E$, assigning labels to nodes and hyperedges.

3. Given an edge e in a hypergraph G, $NODES_G(e)$ denotes the set of all nodes occurring in $s_G(e)$ and $t_G(e)$.

4. Let G, H be hypergraphs. G is called a *subhypergraph* of H, denoted by $G \subseteq H$, if $V_G \subseteq V_H, E_G \subseteq E_H$ and s_G, t_G, l_G, and m_G are restrictions of the corresponding mappings of H.

Remark

Each hypergraph corresponds to a bipartite graph and vice versa: To transform a hypergraph into a bipartite graph (called the *underlying bipartite graph*), the hypergraph nodes become graph nodes of the one sort and the hyperedges become nodes of the other sort; for each hyperedge e with source string $s(e)$ and target string $t(e)$ one constructs $|s(e)| + |t(e)|$ edges. Conversely, given a bipartite graph, nodes of one sort are taken as hypergraph nodes and nodes of the other sort as hyperedges.

5.2 Hypergraph Morphisms

1. Let G, H be hypergraphs. A pair of mappings $f = (f_V : V_G \to V_H,$ $f_E : E_G \to E_H)$ is a *hypergraph morphism* from G to H, denoted by $f : G \to H$, if $s_H \circ f_E = f_V^* \circ s_G$, $t_H \circ f_E = f_V^* \circ t_G$, $l_H \circ f_V = l_G$, and $m_H \circ f_E = m_G$. f is called *injective* (*surjective, bijective*) if both f_V and f_E are injective (surjective, bijective).

2. A bijective hypergraph morphism is called an *isomorphism.* Two hypergraphs G and H are *isomorphic*, denoted by $G \cong H$, if there exists an isomorphism from G to H.

3. Let $f : G \to H$ be a hypergraph morphism and $U \subseteq G$. Then $f_V(V_U)$ and $f_E(E_U)$ determine a subhypergraph of H which is denoted by $f(U)$.

5.3 Hypergraph Pushouts

Let $b : K \to B$ and $d : K \to D$ be two hypergraph morphisms. Then a hypergraph H together with two hypergraph morphisms $h : B \to H$ and $c : D \to H$ is called a *pushout* (or *gluing*) *of b and d* if the following conditions are satisfied:

Commutativity

$h \circ b = c \circ d.$

Universal Property

For all hypergraphs H' and hypergraph morphisms $h' : B \to H'$ and $c' : D \to H'$ with

$h' \circ b = c' \circ d$ there exists a unique hypergraph morphism $f : H \to H'$ such that $f \circ h = h'$ and $f \circ c = c'$.

If H together with h and c is a pushout of b and d we also call the following diagram a pushout:

$$\begin{array}{ccc} K & \xrightarrow{b} & B \\ {\scriptstyle d}\downarrow & & \downarrow{\scriptstyle h} \\ D & \xrightarrow{c} & H \end{array}$$

Remarks

1. The hypergraph H is determined uniquely up to isomorphism by the pushout properties.

2. For the construction of hypergraph pushouts we refer to [HP 88] (see also [Ehr 79] for the graph case).

5.4 Hypergraph Rules

A ***hypergraph rule*** p is a pair of hypergraph morphisms of the form $p = (L \leftarrow K \rightarrow R)$. L, R, and K are called the ***left-hand side***, the ***right-hand side***, and the ***gluing hypergraph*** of p, respectively.

5.5 Hypergraph Derivations

1. Let G, H be hypergraphs, $p = (L \leftarrow K \rightarrow R)$ be a hypergraph rule, and $g : L \to G$ be a hypergraph morphism. Then G ***directly derives*** H ***through*** p ***and*** g, denoted by $G \underset{p,g}{\Longrightarrow} H$, if there are two pushouts of the following form:

$$\begin{array}{ccccc} L & \leftarrow & K & \rightarrow & R \\ {\scriptstyle g}\downarrow & & \downarrow & & \downarrow{\scriptstyle h} \\ G & \leftarrow & D & \rightarrow & H \end{array}$$

$g(L)$ and $h(R)$ are called the ***left*** resp. ***right occurrence*** of p.

2. Let G, H be hypergraphs and P be a set of hypergraph rules. Then a ***derivation from*** G ***to*** H ***over*** P, denoted by $G \overset{*}{\underset{P}{\Rightarrow}} H$, is a sequence of direct derivations of the form $G \cong G_0 \underset{p_1}{\Rightarrow} G_1 \underset{p_2}{\Rightarrow} \ldots \underset{p_n}{\Rightarrow} G_n = H$ where $p_1, \ldots, p_n \in P$.

5.6 Gluing Condition

Let G be a hypergraph and $p = (L \overset{b_1}{\leftarrow} K \overset{b_2}{\rightarrow} R)$ be a hypergraph rule. Given a hypergraph morphism $g : L \to G$ there exists a pushout

$$\begin{array}{ccc} L & \overset{b_1}{\leftarrow} & K \\ {\scriptstyle g}\downarrow & (\star) & \downarrow \\ G & \leftarrow & D \end{array}$$

if and only if the ***gluing condition*** is satisfied which consists of the following two conditions:

Contact Condition

For all $e \in E_G - E_{g(L)} : NODES_G(e) \cap V_{g(L)} \subseteq V_{g(b_1(K))}$.

Identification Condition

For all $x, y \in L$[15] $: x \neq y$ and $g(x) = g(y)$ implies $x, y \in b_1(K)$.

Remark

In general there may be several nonisomorphic hypergraphs D such that $(\star)$ becomes a pushout. However, if b_1 is injective, then D is determined uniquely up to isomorphism.

5.7 Hypergraph Derivations

Let G be a hypergraph, $p = (L \overset{b_1}{\leftarrow} K \overset{b_2}{\rightarrow} R)$ be a hypergraph rule and $g : L \to G$ be a hypergraph morphism such that the gluing condition is satisfied. Then the following construction yields a direct derivation $G \underset{p,g}{\Rightarrow} H$.

(1) $D = G - (g(L) - g(b_1(K)))$ is a subhypergraph of G. [16]

(2) $D \to G$ is the inclusion of D into G.

(3) $d : K \to D$ is defined for each $x \in K$ by $d(x) = g(b_1(x))$.

(4) H together with $h : R \to H$ and $c : D \to H$ is the pushout of $b_2 : K \to R$ and $d : K \to D$.

---

[15] "$x, y \in L$" is a short notation for "$x, y \in V_L$ or $x, y \in E_L$" and is used when nodes and edges have not to be distinguished.

[16] More precisely, D is the subhypergraph with $V_D = V_G - (V_{g(L)} - V_{g(b_1(K))})$ and $E_D = E_G - (E_{g(L)} - E_{g(b_1(K))})$.

References

[BEGKPS 87] H.P. Barendregt, M.C.J.D. van Eekelen, J.R.W Glauert, J.R. Kennaway, M.J. Plasmeijer, M.R. Sleep: *Term Graph Rewriting.* Proc. PARLE, Lecture Notes in Comp. Sci. 259, 141-158 (1987)

[Der 85] N. Dershowitz: *Computing with Rewrite Systems.* Information and Control 64, 122-157 (1985)

[Ehr 79] H. Ehrig: *Introduction to the Algebraic Theory of Graph Grammars.* Proc. 1st Graph Grammar Workshop, Lecture Notes in Comp. Sci. 73, 1-69 (1979)

[EM 85] H. Ehrig, B. Mahr: *Fundamentals of Algebraic Specification 1 – Equations and Initial Semantics.* Springer, Monographs in Computer Science, New York–Berlin–Heidelberg (1985)

[ER 76] H. Ehrig, B.K. Rosen: *Commutativity of Independent Transformations on Complex Objects.* Research Report RC 6251, IBM T.J. Watson Research Center, Yorktown Heights (1976)

[FGJM 85] K. Futatsugi, J. Goguen, J.P. Jouannoud, J. Meseguer: *Principles of OBJ2.* Proc. 1985 Symposium on Principles of Programming Languages, 52-66 (1985)

[HKP 88] A. Habel, H.-J. Kreowski, D. Plump: *Jungle Evaluation.* To appear in Proc. Fifth Workshop on Specification of Abstract Data Types (1988)

[Hof 83] B. Hoffmann: *Compiler Generation: From Language Descriptions to Abstract Compilers.* Dissertation, TU Berlin (1983)

[HO 80] G. Huet, D.C. Oppen: *Equations and Rewrite Rules, A Survey.* In R.V. Book (ed.): *Formal Language Theory: Perspectives and Open Problems.* Academic Press, 349-405 (1980)

[HP 88] B. Hoffmann, D. Plump: *Jungle Evaluation for Efficient Term Rewriting.* Technical Report, Universität Bremen, (1988)

[Hsi 85] J. Hsiang: *Refutational Theorem Proving using Term Rewriting Systems.* Artificial Intelligence 25, 255-300 (1985)

[Klo 87] J.W. Klop: *Term Rewriting Systems: A Tutorial.* EATCS Bulletin 32, 143-182, (1987)

[Kre 87] H.-J. Kreowski: *Is Parallelism Already Concurrency? - Part 1: Derivations in Graph Grammars.* Proc. 3rd Graph Grammar Workshop, Lecture Notes in Comp. Sci. 291, 343-360 (1987)

[O'D 87] M.J. O'Donnell: *Term Rewriting Implementation of Equational Logic Programming.* Proc. 2nd Conference on Rewriting Techniques and Applications, Lecture Notes in Comp. Sci. 256, 1-12 (1987)

[Pad 82] P. Padawitz: *Graph Grammars and Operational Semantics.* Theoretical Comp. Sci. 19, 117-141 (1982)

[Pad 83] P. Padawitz: *Correctness, Completeness, and Consistency of Equational Data Type Specifications.* Bericht Nr. 83-15, Technische Universität Berlin (1983)

[Pey 87] S.L. Peyton Jones: *The Implementation of Functional Programming Languages.* Prentice-Hall, Englewood Cliffs (1987)

[Plu 86] D. Plump: *Im Dschungel: Ein neuer Graph-Grammatik-Ansatz zur effizienten Auswertung rekursiv definierter Funktionen.* Diplomarbeit, Studiengang Informatik, Universität Bremen (1986)

[Sta 80] J. Staples: *Computation on Graph-like Expressions.* Theoretical Comp. Sci. 10, 171-185 (1980)

A NEW QUASI-REDUCIBILITY TESTING ALGORITHM AND ITS APPLICATION TO PROOFS BY INDUCTION

G.A.Kucherov[1)]

Introduction

In algebraic specifications of abstract data types we are mainly interested in term-generated (particularly initial) models. It means that every object of the intended algebra is associated with some ground term. This gives rise to a number of peculiarities of the use of term rewriting systems (TRSs) in implementation and analysis of abstract data types. While we interpret TRS as a set of computation rules on the set of abstract objects, it is important how the system acts on ground terms (rather than on all terms with variables). In particular, the notion of confluence transforms to ground-confluence (=confluence on the set of ground terms). Another specific notion is quasi-reducibility (=reducibility of all the ground instances) of a term with respect to a given TRS.

The utility of the notion of quasi-reducibility for abstract data types analysis was demonstrated in [8], where it was called inductive reducibility. In particular, close relationship between this notion and automatic proofs in initial algebra was shown. This approach was developed in [4], where the proposed algorithm generalizes the previously known methods of automatic inductive proofs [13,5,7]. The essence of the improvement was the use of ground-confluent TRSs instead of confluent ones. Quasi-reducibility is also directly related to sufficient-completeness property of algebraic specifications [14,9].

The decidability of quasi-reducibility for left-linear TRSs is implied by the result of [14], where the corresponding decision algorithm was presented. A slightly different algorithm was proposed in [8]. But both algorithms serve rather for principal purposes than for practical use. The decidability of quasi-reducibility for arbitrary TRSs was shown in [15,9].

In this paper we further investigate the concepts of quasi-reducibility and ground-confluence and their relation to proofs in initial algebras. Our presentation bases upon the concepts of cover and **complement** already described in the literature. Covers (more precisely more general notion of S-covering sets) were defined in [15]. In [10] covers were introduced to decide 'well-coveredness' property which implies quasi-reducibility (but not vice versa) and the corresponding algorithm was presented. Complements were introduced in [16] and were investigated in [2] in the framework of solving inequations. Some of the

---

1) Comput. Center, Siberian Div. of the Acad. Sci., Novosibirsk, 630090, USSR

ideas of [2] are closely related to ours.[1)]

Section 2 of this paper presents a number of results about covers and complements. Some of them are applied in section 3 to validate a new quasi-reducibility testing algorithm for left-linear TRSs. Correctness and completeness of the algorithm are proved. We show also how the problem for non-left-linear TRSs can be handled when a set of free constructors is predefined. Section 4 proves that ground-confluence and quasi-reducibility are really fundamental properties when proving equations in initial algebras. We show that ground-confluence is not generally semi-decidable even for noetherian TRSs. Finally, we reformulate Fribourg's algorithm, explain its correctness, and analyse some of its aspects. For the lack of space some of the proofs are omitted. They can be found in [11].

1. Basic notions and notations

We assume familiarity with foundations of term rewriting systems and equational theories [6]. Below we present a list of necessary notations (with possibly minimal comments).

For a one-sorted signature F $T_F(X)$ and T_F denote the set of terms over a set of variables X and the set of ground terms respectively. Occurences are considered as sequences of natural numbers in usual way. For $t \in T_F(X)$ depth(t) is defined as the maximal length of an occurence in t. For a set T of terms depth(T) stands for the maximal depth of a term of T. var(t) produces the set of variables of t. A variable $x \in var(t)$ is linear in t, if the corresponding variable occurence is unique, otherwise x is non-linear. A term containing only linear variables is linear. A substitution σ with the domain $\{x_1, \dots, x_k\}$ is written as $\langle x_1 \leftarrow t_1, \dots, x_k \leftarrow t_k \rangle$. SUB (resp. GSUB) denotes the set of substitutons (resp. ground substitutions) with finite domains. Composition of substitutions σ and ρ is defined as $(\sigma \circ \rho)(t) = \rho(\sigma(t))$.

We will be interested in the partial order relation $\leq$ on the set $T_F(X)$ defined as follows: $t \leq s$ iff $s = \sigma(t)$ for some $\sigma \in SUB$. It is known that two terms t,s are equivalent with respect to $\leq$ (i.e. $t \leq s$ and $s \leq t$) iff t can be derived from s by bijective variable renaming. This fact substanciates to assume sets of variables of "independent" terms to be always disjoint. By "independent" terms we mean those in which coinciding variables are not essentially the same. For example, subterms of a non-linear term are not generally independent.

A TRS R is left-linear, if the left-hand sides of all its rules are linear. $\longrightarrow_R^*$, $\longleftrightarrow_R^*$ are respectively transitive-reflexive and transitive-reflexive-symmetric closure of the reduction relation $\longrightarrow_R$

[1)] Just before the submission the author came across the paper [12] in which these notions also were considered generated by R. If for every $\rho\in GSUB$ $\rho(t)$ is reducible then t is called quasi-reducible. Ground confluence is defined as confluence on the set T_F. $t{\downarrow}_R$ denotes (any) R-normal form of t.

$=_E$ stands for the congruence relation on $T_F(X)$ induced by a set of equations E. If for a TRS R $=_E=\overset{*}{\longleftrightarrow}_R$, then R is said to model E. The initial algebra $I_F(E)$ is the quotient of T_F by $=_E$. For $t,s\in T_F(X)$ the relation $t=_I s$ holds if the equation t=s is valid in $I_F(E)$. It is known that $t=_I s$, iff for any $\rho\in GSUB$ $\rho(t)=_E\rho(s)$. If a TRS R models E, then the latter equation can be replaced by $\rho(t)\overset{*}{\longleftrightarrow}_R\rho(s)$, which is equivalent to $\rho(t){\downarrow}_R=\rho(s){\downarrow}_R$ if R is complete (or even ground complete) in addition.

A rule l >r superposes into a term t at the occurence $u\in O(t)\backslash O_{var}(t)$ if t/u and l are unifiable by a most general unifier б called superposition substitution. A set of superposition substitutions corresponding to superpositions of all the rules of R into t is denoted by $SS_R(t)$.

In our presentation we will need some basic facts of TRS theory (the Knuth-Bendix algorithm and its properties are among them).

2. Covers and complements

<u>Definition 2.1.</u> A set of terms $\Sigma\subseteq T_F(X)$ is called a cover of a set of ground terms $\Delta\subseteq T_F$, iff for any $m\in\Delta$ there exists $t\in\Sigma$ such that $t\leq m$. We say that Δ and its every element is covered with Σ. A set Σ is a complete cover if it covers the whole set T_F.

Trivial examples of a complete cover are a singleton set {x} (or any set including it) or the whole set T_F. As a nontrivial complete cover we will build a set $\Pi(k)$ (k is a natural number). For $t\in T_F(X)$ we define minvar(t) to produce the minimal length of a variable occurence in t. The definition of $\Pi(k)$ is the following.

$\Pi_{const}(k)=\{t\in T_F|\ depth(t)<k\}$,

$\Pi_{var}(k)=\{t\in T_F(X)|\ depth(t)=k,\ minvar(t)=k,\ t \text{ is linear}\}$,

$\Pi(k)=\Pi_{const}(k)\cup\Pi_{var}(k)$.

Since for all $t\in\Pi(k)$ $depth(t)\leq k$ then $\Pi(k)$ is finite.

<u>Proposition 2.1.</u> For any k $\Pi(k)$ is a complete cover.

The next useful concept is introduced by the following

<u>Definition 2.2</u> Let $\Sigma\subseteq T_F(X)$. A set $\bar{\Sigma}\subseteq T_F(X)$ is called a **complement** of Σ (with respect to a complete cover) if

1. $\Sigma\cup\bar{\Sigma}$ is a complete cover,
2. For any $s\in\Sigma, t\in\bar{\Sigma}$ s and t are not unifiable.

The second condition means that the sets of ground terms covered with Σ and $\bar{\Sigma}$ are disjoint.

We are interested in the existence and construction of a complement for different finite sets of terms. The following theorem gives a solution provided that only linear terms are under consideration.

<u>Theorem 2.1.</u> For any finite set Σ of linear terms a finite

complement $\bar{\Sigma}$ can be effectively constructed.

Proof. At first we note that if $t \in T_F(X)$ such that $var(t) \neq \emptyset$, $minvar(t) > depth(\Sigma)$, and t is unifiable with some $s \in \Sigma$ then $s < t$.

Consider $\Pi(k)$ for any $k > depth(\Sigma)$ and let $\bar{\Sigma} = \{t \in \Pi(k) \mid \forall s \in \Sigma \ s \not< t\}$. By the above statement condition 2 of definition 2.2 holds. Let us prove condition 1. Since $\Pi(k)$ is the complete cover then for any $m \in T_F$ $m = \rho(t)$ for some $t \in \Pi(k)$ and $\rho \in GSUB$. We are to prove that if $t \notin \bar{\Sigma}$ then m is covered with Σ. In this case $t = \sigma(s)$ for some $s \in \Sigma$ and $\sigma \in SUB$. Therefore $(\sigma \circ \rho)(s) = \rho(\sigma(s)) = \rho(t) = m$. The finiteness of $\bar{\Sigma}$ is implied by the finiteness of $\Pi(k)$. □

Note that the proof of the theorem gives a constructive way to compute a complement for a set of linear terms. Moreover, the depth of the complement is bounded with any $k > depth(\Sigma)$. However this complement is not minimal. Generally, it contains non-trivial subcovers of sets of the form $\{\rho(t) \mid \rho \in GSUB\}$ that can be replaced by a single term t.

Now we are going to investigate complements for the sets containing non-linear terms. This situation is much more complicated. As noted in [2] if F contains a binary symbol f then the set $\{f(x,x)\}$ has not a finite complement. We prove that it is the case for any set of non-linear terms. The starting point is the following

Theorem 2.2 Let $\Gamma \subseteq T_F(X)$ be finite and consist of non-linear terms. Then there does not exist a finite complement of Γ consisting of linear terms.

Corollary. For a finite set consisting of non-linear and ground terms there does not exist a finite complement consisting of linear terms. In particular, such a set cannot be a complete cover.

The next theorem gives conditions for a set containing non-linear terms to be a complete cover.

Theorem 2.3. Let Γ be a finite set of non-linear terms, Σ a finite set of linear terms. Then $\Gamma \cup \Sigma$ is a complete cover, iff $\Gamma^* \cup \Sigma$ is a complete cover, where Γ^* is a finite set of linear terms derived from Γ by substituting for non-linear variables by ground terms. In particular, if the terms of Γ do not contain linear variables then Γ^* consists of ground terms.

Corollary. For a finite set Γ of non-linear terms there does not exist a finite completement.

Theorem 2.3 states that non-linearity of terms is not relevant to the question of complete cover. More precise, if a set of terms is a complete cover, than it is still the case, if the non-linear terms are replaced by a certain set of linear terms. At the end of the next section we will show how this theorem can be applied to the so called theories with free constructors.

In the following we will need natural generalization of the concepts of cover and complement from terms to tuples of terms.

Definition 2.3. For two tuples $\alpha = (t_1, \ldots, t_n) \in (T_F(X))^n$, $\beta = (s_1, \ldots, s_n) \in (T_F(X))^n$ the relation $\alpha < \beta$ holds if for some $\sigma \in SUB$ $\sigma(t_1) = s_1$

for all $i \in (1,n)$.

By means of this definition the concepts of cover and complement are generalized straigthforwardly. All the results of this section freely extend to tuples of terms. In particular if Φ is a set of linear tuples (a tuple is linear if all its components are linear and have disjoint sets of variables) then the (generalization of) theorem 2.1 can be applied to Φ.

3. Quasi-reducibility testing algorithm for linear TRSs

In this section we use the notions of cover and complement to check quasi-reducibility for linear TRSs.

Let $t \in T_F(X)$ be any (linear or non-linear) term, $var(t)=\{x_1,\dots,x_n\}$ variables of t. A substitution $\sigma=\langle x_1 \leftarrow s_1,\dots,x_n \leftarrow s_n\rangle$ applying to t can be identified with the tuple $(s_1,\dots,s_n)$. So we can further extend the concepts of cover and complement to substitutions. For any substitutions $\sigma,\rho \in SUB$ with the same domain $\sigma<\rho$ is equivalent to $\rho=\sigma\circ\varrho$ for some $\varrho \in SUB$.

<u>Definition 3.1.</u> Let TRS R be given. A substitution $\sigma=\langle x_1 \leftarrow s_1,\dots,x_n \leftarrow s_n\rangle$ and a tuple $(s_1,\dots,s_n)$ are said to be reducible if for some $i \in (1,n)$ s_i is reducible. They are quasi-reducible if for every $\rho \in GSUB$, for some $i \in (1,n)$ $\rho(s_i)$ is reducible.

Note that quasi-reducibility of a linear tuple implies quasi-reducibility of some its component but it is not true in general.

Let $\rho=\langle x_1 \leftarrow m_1,\dots,x_n \leftarrow m_n\rangle \in GSUB$. If t is quasi-reducible then either ρ is reducible or for some occurence u $\rho(t/u)$ is reducible at root. The next theorem gives a finite set of substitutions covering all the ground substitutions of the second type.

<u>Theorem 3.1.</u> Given TRS R, t is quasi-reducible, iff the set $SS_R(t)$ covers all irreducible ground substitutions.

<u>Remark.</u> A superposition substitution of $SS_R(t)$ is considered here as applying to t, that is having the domain var(t) and possibly fictitious components $x_i \leftarrow x_i$.

<u>Proof.</u> 1) If-part. Once an irreducible ground substitution δ is covered with some σ of $SS_R(t)$, then $\delta=\sigma\circ\varrho$ for some $\varrho \in GSUB$, and $\delta(t)=\varrho(\sigma(t))$. But σ is a superposition substitution, hence $\sigma(t)$ is reducible and so is $\delta(t)$. If δ is reducible then $\delta(t)$ is a fortiori reducible.

2) Only-if part. Let $\delta \in GSUB$ be an irreducible ground substitution. Since $\delta(t)$ is reducible there exists a non-variable occurence u in t such that $\delta(t/u)$ is reducible at root. It means that $\delta(t/u)=\zeta(l)$ for some rule $l \to r$ of R and some $\zeta \in GSUB$. Therefore t/u and l are unifiable and there exists $\sigma \in SS_R(t)$ such that $\delta=\sigma\circ\varrho$ for some $\varrho \in GSUB$. □

As another pre-requisite for a quasi-reducibility testing algorithm we need the following technical

<u>Proposition 3.1.</u> Let t be any term and $var(t)=\{x_1,\dots,x_n\}$. Let l be

a linear term unifiable with t, i.e. $\delta(t)=\sigma(l)$ where δ and σ are most general unifiers and $\delta=\langle x_1\leftarrow s_1,\dots,x_n\leftarrow s_n\rangle$. Then the tuple $(s_1,\dots,s_n)$ is linear and $depth(\{s_1,\dots,s_n\})<depth(l)$.

Now we are in position to give a recursive quasi-reducibility testing algorithm for linear TRSs and to prove its correctness and completeness.

It is clear that only the left-hand sides of rewrite rules are relevant. Let L_0 denote a set of the left-hand sides of the initial TRS. Let L be a global (with respect to all recursive calls to the algorithm) set of terms initially assigned to L_0.

<u>Algorithm</u> CHECK(t)

1. <u>if</u> t is L-reducible <u>then</u> <u>return</u>(YES);
2. <u>if</u> t is a variable or irreducible ground term <u>then</u> <u>return</u>(NO);
3. Compute $SS_R(t)$ and the corresponding set of tuples Φ. According to theorem 2.1 compute a complement $\bar{\Phi}=\{\alpha_1,\dots,\alpha_k\}$ where $\alpha_i=(s_1^i,\dots,s_n^i)$;
4. $L:=L\cup\{t\}$;
5. <u>for</u> i=1 <u>to</u> k <u>do</u>

        <u>begin</u>
        <u>for</u> j=1 <u>to</u> n <u>do</u>
                <u>if</u> CHECK(s_j^i)=YES
                <u>then</u> <u>continue</u> external loop over i
        <u>return</u>(NO)
        <u>end</u>;

6. <u>return</u>(YES);

<u>Theorem 3.2.</u> The algorithm CHECK is correct (i.e. returns YES if t is quasi-reducible) and complete (i.e. always terminates).

<u>Proof.</u> Let us begin with some comments. The algorithm attempts to prove $\rho(t)$ to be reducible for any ground substitution (tuple) $\rho=(m_1,\dots,m_n)$. By theorem 3.1 $SS_R(t)$ should cover all irreducible ground substitutions, hence all the remaining ground substitutions should be reducible. By proposition 3.1 all the tuples $\beta\in\Phi$ are linear, therefore by theorem 2.1 $\bar{\Phi}$ is finite and consists of linear tuples. The algorithm turns recursively to test tuples of $\bar{\Phi}$ for quasi-reducibility, trying to find in each of them a quasi-reducible term, and adding to L a term t itself. The latter is a crucial step in the algorithm and corresponds to the inductive hypothesis about reducibility of $\rho(t)$.

Thus, to prove the correctness of CHECK we have to prove the correctness of step 4. The correctness becomes clear, if we think of the algorithm as carrying out a structural induction proof over a set of ground terms So step 5 is a standard structural induction step from reasoning about a term to reasoning about its subterms, and step 4 is the addition of the inductive hypothesis to the current set of premises.

Let $K=depth(L_0)$. By proposition 3.1 the depth of every term of the tuples of Φ are less than K. Then the depth of any term of the tuples of $\bar{\Phi}$ are not greater than K (see remarks after the proof of theorem 2.1). So K bounds the depth of terms under consideration. Since the number of such terms is finite and the size of L increases at each call to CHECK, the algorithm always terminates. □

<u>Example 3.1.</u> Let $F=\{a^0, h^1, f^2\}$, $L_0=\{h(f(x,y)), f(h(h(x)),y), h(a)\}$, $t=f(x,h(y))$. At the initial call to CHECK $\Phi_1=\{(x,f(y,z)), (h(h(x)),y), (x,a)\}$, $\bar{\Phi}_1=\{(a,h(x)), (h(a),h(x)), (h(f(x,y)),h(z)), (f(x,y),h(z))\}$. L is enlarged to $L_1=L\cup\{f(x,h(y))\}$ and then the algorithm turns to analyse the terms of the tuples of $\bar{\Phi}_1$. In the first tuple $(a,h(x))$ the first term is ground and irreducible, so the second one is supplied to the algorithm. For this argument the algorithm computes $\Phi_2=\{(a),(f(x,y))\}$, $\bar{\Phi}_2=\{h(x)\}$, then enlarges L to $L_2=L_1\cup\{h(x)\}$, and turns to analyse the only tuple of $\bar{\Phi}_2$. This tuple is quasi-reducible because the current set L contains $h(x)$. Returning to the analysis of $\bar{\Phi}_1$ the remaining tuples contain $h(x)$ as a subterm and therefore are easily proved to be quasi-reducible. Thus the term t is quasi-reducible.

The algorithm is, however, exponential because a complement may have an exponential size, but in contrast to algorithms [14,8] CHECK is purposeful. Its important property is that once quasi-reducibility of a term is proved, this information can be used at further calls to the algorithm. In other words, the set L can be global not only within a tree of recursive calls to CHECK but within all the calls to CHECK for a given TRS. Moreover, we can avoid keeping unnecessary information in L. If some term of L is reducible with respect to a term to be added, then it can be deleted from L. Thus, example 3.1 results in the set L consisting of the single term $h(x)$, for the other terms of L are reducible provided that $h(x)$ is reducible.

Now we show how theorem 2.3 can be applied to the theories with free constructors. Let $F=C\cup D$, where C is a set of constructors and D is a set of defined functions. Let a TRS R completely define D over C, i.e. for every $f\in D$ the term $f(x_1,\dots,x_n)$ is quasi-reducible. Let C be free in addition, i.e. no left-hand side of the rules of R belongs to $T_C(X)$. It means that the set of ground R-normal forms coincides with T_C. Thus, for $f\in D$ the set of the left-hand sides with the root symbol f should cover all the terms $f(m_1,\dots,m_n)$, where $m_i\in T_C$. Theorem 2.3 implies that the non-linear rules can be replaced by a finite number of linear rules and therefore the former are not relevant for a complete definition of D over C. We feel that the similar situation exists in general case, but we have not precise formulations so far.

4. Ground-confluence and its role in inductive proofs

Recently L.Fribourg [4] has designed a new TRS approach to automatic proofs by induction. Its main idea is that in automatic proofs by induction a fundamental TRS property is ground-confluence rather then confluence. A ground-confluent TRS can be constructed via the modified Knuth-Bendix algorithm computing a restricted subclass of critical pairs.

In this section we continue to develop Fribourg's approach. Throughout the section all considered TRSs are assumed to be noetherian.

Now we turn to the ground-confluence property. Fundamental character of this property in inductive proofs is shown by the following

<u>Proposition 4.1.</u> Let E be an equational theory, R a ground-confluent TRS which models E, t=s a theorem to be proved. Then $t=_I s$, iff

1. $R'=R\cup\{t\rightarrow s\}$ is ground-confluent, and
2. t is quasi-reducible with respect to R.

<u>Proof.</u> 1) If-part. Let $E'=E\cup\{t=s\}$. Since $t=_{E'}s$ then $\rho(t)=_{E'}\rho(s)$ for any ground substitution ρ, and $\rho(t)\downarrow_{R'}=\rho(s)\downarrow_{R'}$ because of the ground-confluence of R'. The quasi-reducibility of t with respect to R ensures that any ground R-normal form is a R'-normal form. But R' is ground-confluent and every ground term has only one R'-normal form. Thus $\rho(t)\downarrow_R=\rho(t)\downarrow_{R'}=\rho(s)\downarrow_{R'}=\rho(s)\downarrow_R$. Hence $\rho(t)\leftrightarrow^*_R\rho(s)$ and $t=_I s$.

2) Only-if part. $t=_I s$ means that the congruences $=_E$ (or $\leftrightarrow^*_R$) and $=_{E'}$ (or $\leftrightarrow^*_{R'}$) coincide on T_F. In particular,

for any ground R-normal forms m,n $\quad m=_E n$ iff $m=_{E'}n$ iff $m=n$ $\quad$ (*)

If t is not quasi-reducible with respect to R then for some ground substitution ϱ $\varrho(t)$ is a R-normal form. But $\varrho(s)\downarrow_R\neq\varrho(t)$, for otherwise $\varrho(t)\rightarrow_{R'}\varrho(s)\rightarrow^*_R\varrho(t)$ and the noetherity of R' is violated. So $\varrho(t)$ and $\varrho(s)\downarrow_R$ are distinct R-normal forms. But $\varrho(t)=_{E'}\varrho(s)\downarrow_R$ which contradicts to (*) and proves the quasi-reducibility of t.

If R' is not ground-confluent then some ground term m has two distinct R'-normal forms m',m''. It is clear that m',m'' are also R-normal forms and according to (*) $m'\neq_{E'}m''$. This contradiction proves the ground-confluence of R'. □

The proposition is similar to theorem 1 of [8] but the initial and resulting TRSs are required to be ground-confluent. Moreover, the proposition gives sufficient and necessary conditions for validity in the initial algebra. It is known that if we add a rule $t\rightarrow s$ to a confluent TRS R then the confluence can be completely destroyed even if $t\leftrightarrow^*_R s$. In the case of ground-confluence and validity in the initial algebra the situation is different: if $t=_I s$ then ground-confluence is invariant to the addition of $t\rightarrow s$ to a current TRS R.

Another conclusion from proposition 4.1 is that ground-confluence is not only undecidable but also non-semi-decidable even for noetherian TRSs. It is easy to build a noetherian ground-confluent TRS R specifying the calculation of addition, multiplication, and less predicate over natural numbers and the constants <u>true</u> and <u>false</u> The initial model is the standard model of natural numbers and by Matiyasevich theorem the validity of the equation $(p[\bar{x}]<q[\bar{x}])=_I$ <u>true</u> (where $p[\bar{x}],q[\bar{x}]$ are polynomials over $\bar{x}$) is not semi-decidable [3]. Therefore, the ground-confluence of R extended by the rule $(p[\bar{x}]<q[\bar{x}])\rightarrow$ <u>true</u> is not semi-decidable too.

Thus, to prove the validity of $t=_I s$ we have to prove both the quasi-reducibility of t and the ground-confluence of $R\cup\{t\rightarrow s\}$ provided that R is ground-confluent. To test quasi-reducibility we are going to use the algorithm CHECK designed in the previous section. This obliges us to assume R to be left-linear. (Note that the rule to be proved must not be left-linear.) To prove ground-confluence we are going to use Fribourg's idea about the restriction of the class of critical pairs to be computed. Now we reformulate Fribourg's algorithm in the recursive manner, explain its correctness and make some remarks about its properties. (Note that in the original paper [4] the correctness was proved only for theories with free constructors.)

Suppose the initial ground-confluent linear TRS R models E. Let R^* be a global TRS initially assigned to ø. In terms of [4] R and R^* contain a set of "defining axioms" and "rewrite lemmas" respectively. R' stands for a current $R \cup R^*$

<u>Algorithm</u> PROVE(t,s)

1. <u>if</u> $t\downarrow_{R'} = s\downarrow_{R'}$ <u>then</u> <u>return</u>(TRUE);
2. Orient the pair (t,s) into the rule $t \rightarrow s$
   (modulo mutual renaming of t and s);
3. $R^* := R^* \cup \{t \rightarrow s\}$;
4. <u>if</u> CHECK(t)=NO <u>then</u> <u>return</u>(FALSE);
5. Compute a subclass $\{(p_i, q_i)\}_{i=1}^{n}$ of critical pairs
   produced by superpositions of the rules of R into $t \rightarrow s$.
6. <u>for</u> i=1 <u>to</u> n <u>do</u>
   <u>if</u> PROVE(p_i, q_i)=FALSE <u>then</u> <u>return</u>(FALSE);
7. <u>return</u>(TRUE);

To prove the correctness we have to show that the validity of $t =_I s$ is equivalent to the validity of $p_i =_I q_i$ for all $i \in (1,n)$. Besides, the key point is to validate the use of the rule $t \rightarrow s$ itself in reducing p_i and q_i (at step 1 of recursive calls to PROVE). It is this step that corresponds to the use of induction hypothesis in structural induction proofs. A complete proof of the correctness is given by the following

<u>Theorem 4.1.</u> Let R be an initial ground-confluent TRS modelling E, t=s a theorem to be proved, $R' = R \cup \{t \rightarrow s\}$, and t be quasi-reducible with respect to R. Let $\{(p_I, q_i)\}_{i=1}^{n}$ be a set of critical pairs produced by superpositions of the rules of R into $t \rightarrow s$. Then $t =_I s$ iff for all $i \in (1,n)$ $p_i\downarrow_{R'} = q_i\downarrow_{R'}$.

The proof uses essentially the noetherian induction principle [6] because of the inductive nature of the algorithm itself. The theorem remains true if R' is enlarged by any rules provided that $\leftrightarrow^*_{R'} \subseteq \leftrightarrow^*_R$ on T_F (i.e. these rules are valid in the initial model).

In the rest of the paper we will dwell on two aspects of PROVE. The first one is that the check of quasi-reducibility (step 4) and the computation of critical pairs (step 5) are deeply interrelated. Both steps require unification of the left-hand sides of R with subterms of t, and computation of $SS_R(t)$. This allows us to avoid duplication of steps 4 and 5 in an obvious way.

The second point is that we can further restrict the class of critical pairs provided that the corresponding set of superposition substitutions covers all irreducible ground substitutions. In [4] there was proposed to superpose rules of R into t at some fixed occerence u (called "complete superposition occurence"). But we argue that this restriction is too strong. The appropriate subterm may not exist even in very simple cases. For example let us define a boolean type with the constant <u>false</u> and the negation function $\neg$ by the rule $\neg\neg\underline{false}\rightarrow\underline{false}$. Proving the theorem $\neg\neg x\rightarrow x$ we have two superpositions, and the corresponding set $SS_R(\neg\neg x)=\{\langle x\leftarrow\underline{false}\rangle, \langle x\leftarrow\neg\underline{false}\rangle\}$ obviously covers all irreducible ground substitutions. But the substitution $\langle x\leftarrow\underline{false}\rangle$ corresponds to the superposition at the root occerence while $\langle x\leftarrow\neg\underline{false}\rangle$ corresponds to the subterm $\neg x$. Thus there is not a subterm of $\neg\neg x$ with the required property. To sum up, one had better fix on a variable (or a set of variables, rather than a subterm) to restrict further the set of critical pairs.

References

[1] Boyer,R., Moore,J.S.: A computational logic. Academic Press 1979.

[2] Comon, H.: Sufficient completeness, term rewriting systems and "anti-unification". Lect. Notes Comput. Sci. 230, 1986, 128-140.

[3] Davis,M., Matijasevic,Y., Robinson,J.: Hilberts tenth problem: positive aspects of a negative solution. Mathematical Development Arising from Hilbert Problems, American Math. Society, Providence, RI, 1976, 323-378.

[4] Fribourg,L.: A strong restriction of the inductive completion procedure. Lect. Notes Comput. Sci. 226, 1986, 105-115.

[5] Goguen,J.A.: How to prove algebraic inductive hypothesis without induction, with application to the correctness of data type implementation. Lect. Notes Comput. Sci. 87, 1980, 356-373.

[6] Huet,G.: Confluent reductions: abstract properties and applications to term rewriting systems. Journ. ACM 27 (1980) No.4, 797-821.

[7] Huet,G.,Hullot,J.-M.: Proofs by induction in equational theories with constructors. Journ. Comput. System Sci. 25 (1982) No.2, 239-266.

[8] Jouannaud,J.-P., Kounalis,E.: Automatic proofs by induction in equational theories without constructors. Proc. Symp. Logic in Comput. Sci., Cambridge, Mass., June 16-18, 1986, 358-366.

[9] Kapur,D., Narendran,P., Zhang,H.: On sufficient-completeness and related properties of term rewriting systems. Acta Informatica 24 (1987) Fasc.4, 395-415.

[10] Kounalis, E.: Completeness in data type specifications. Lect. Notes Comput. Sci. 204, 1985, 348-362.

[11] Kucherov,G.A.: A quasi-reducibility testing algorithm for linear term rewriting systems. Computer systems 122, Novosibirsk 1987, 19-37. (in Russian)

[12] Lazrek,A., Lescanne,P., Thiel,J.-J.: Proving inductive equalities algorithms and implementation.INRIA, Rapports de Recherche, No.682, Juin 1987.

[13] Musser,D.: On proving inductive properetíes of abstract data types. Proc. 7th ACM Symp. on Princ. Progr. Lang., 1980, Las Vegas, 154-162.

[14] Nipkow,T., Weikum,G.: A decidability result about sufficient-completeness of axiomatically specified abstract data types. Lect. Notes Comput. Sci. 145, 1982, 257-267.

[15] Plaisted, D.A.: Semantic confluence tests and completion methods. Inf. and Cont. 65 (1985) 2, 182-215.

[16] Thiel, J.J.: Stop losing sleep over incomplete data type specifications. Proc. 11th ACM Symp. on Princ. Prog. Lang., 1984, Salt-Lake-City, 76-82.

FUZZY REASONING BASED ON f-HORN CLAUSE RULES

Dongbo Liu and Deyi Li

Department of Computer Science, Institute of
Computer Engineering, Beijing, China

ABSTRACT

Based on the first order predicate logic, in this paper, we present an approach to generalizing the syntax and semantics of ordinary Horn clause rules to establish the fuzzy proof theory. First of all, each Horn clause rule is associated with a numerical implication strength f, therefore we obtain f-Horn clause rules. Secondly, Herbrand interpretations can be generalized to fuzzy subsets of the Herbrand base in the sense of Zadeh. We show that as a result the procedural interpretation for Horn clause rules presented by R. A. Kowalski can be developed in much the same way for f-Horn clause rules. Hence, we obtain the fuzzy logic program system

1. Introduction

In the research of artifical intelligence, one has found the binary logic is not sufficient for the reasoning. There are plenty of states, which are fuzzy, between TRUE and FALSE. One can think integrally and parallelly by treating fuzzy phenomenon flexiblly.

Someone has implemented several approximate quantitative alternatives [4] , where the truth values are replaced by probabilities or other measures of uncertainty. However, probability is only suitable for treating random phenomenon, whereas randomness is quite different from fuzzness [2] .

In the qualitative reasoning, we have adopted the Horn clause subset of first order predicate logic. The Horn clause logic will be generalized from the qualitative case to the quantitative case according to Zadeh's fuzzy sets theory [1] in this paper. We first regard the truth value FALSE as the real number 0, TRUE as the real number 1, and extend the concept of truth value to include all real numbers in the interval [0,1] . Then we generalize two important concepts in Horn clause logic to the quantitative case. The first one is the implication in a rule consisting of a condition and a conclusion. The implication can be regarded as transferring truth value from the condition to the conclusion. We can associate each implication with a factor f, it is called implication strength. If the truth value of the condition is t, then the truth value of the conclusion is f x t. The second important concept is interpretation. We obtain fuzzy interpretations easily, provided we consider them as fuzzy subsets of the Herbrand base in the sense of Zadeh [1] . Therefore, they can be operated on in much the same way as the qualitative one, and we can directly transfer the existing results on the semantics of logic programs into the fuzzy logic reasoning.

Proof theory can be also generalized to the quantitative case. We have known that proofs using Horn clause rules are found by searching the and/or tree associated with the set of Horn clause rules. PROLOG interpreter performs this search depth first, from left to right. Obviously, and/or trees can be regarded as game trees for a suitably chosen game. When game trees are too large, the heuristic game values which are in the interval [0,1] have to be used. For searching this kind of game trees, there is a well-known algorithm called alpha-beta.

If we translate back from game trees with heuristic game values, then we obtain and/or trees with fuzzy truth values, this suggests that alpha-beta algorithm is suitable for answering questions to the sets of f-Horn clause rules.

In R. A. Kowalski's logic program system [5] , Horn clause rules are regarded as procedures. We obtain a fuzzy procedural interpretation for f-Horn clause rules, and a fuzzy logic program system by generalizing Kowalski's work [5] .

2. Syntax of finite sets of f-Horn clause rules

The implication of a Horn clause is called a Horn clause rules [6] . In the quantitative case, a f-Horn clause rule also has a conclusion and a condition, it is similar to Horn clause rule in form, and the difference is that each f-Horn clause rule has a numerical implication strength f, where $f \in (0,1]$. The formal specification of a f-Horn clause rule is

$$A \leftarrow (f)\text{-}B1 \ \& \ B2 \ \& \ \dots \ \& \ Bn, \qquad n \geq 0. \qquad (2.1)$$

suppose the truth value of Bi is tv(Bi), i=1,2, ... ,n, and the truth value of condition is t. When $n > 0$, $t = \mathrm{MIN}\{tv(Bi): u=1,2, \dots ,n\}$; when n=0, we define t=1. The truth value of conclusion $tv(A) = f \times t$.

When f = 1, quantitative f-Horn clause rules are corresponding to qualitative Horn clause rules in form. However, in both of the quantitative case and the qualitative case, the concepts of truth value are essentially distinct. For emphasizing this distinction, we use Sf to denote the set of f-Horn clause rules.

3. Semantics of finite sets of f-Horn clause rules

In this section, the semantics of Horn clause rules will be generalized, for the purpose of proposing the semantics of finite sets of f-Horn clause rules. Nowe we first extend some basic concepts of Horn clause logic.

Definition 3.1. Let Sf be a set of f-Horn claues rules, the Herbrand base H(Sf) of the Sf is the set of all ground atomic formulas that can be formed with the symbols contained in Sf.

Definition 3.2. Let Sf be a set of f-Horn clause rules, H(Sf) denote the Herbrand of Sf. The Herbrand interpretation I of Sf is defined as a mapping $H(Sf) \to [0,1]$.

In this case, a Herbrand interpretation is regarded as fuzzy as a subset of H(Sf) in the sense of Zadeh [1] . The mapping $H(Sf) \to [0,1]$ can be thought of as the membership function characterizing a fuzzy subset I of Sf. All Herbrand interpretations of a given set Sf can be specified by a functions meb taking as arguments a variable-free atomic formula A, and an interpretation I and having as result meb(a,I), the value of the membership function for I at the argument A.

In the quantitative case, when a set of f-Horn clause rules is true in a given interpretation I?

Definition 3.3. For a set Sf of f-Horn clause rules and its interpretation I,

(i) Sf is true in I iff every one of its f-Horn clause rules is true in I.

(ii) A f-Horn clause rule R in Sf is true in I iff every one of its ground instances is true in I.

(iii) A ground instance $A \leftarrow (f)-B1$ & $B2$ & ... & Bn of the R is true in I iff

$$\text{meb}(A,I) \geq f \times \text{MIN}\{\text{meb}(Bi,I): i=1,2, \dots, n\} \quad (3.1)$$

Here, we define $\text{MIN}\emptyset = 1$. In this definition, parts (i) and (ii) are the same as in the qualitative case. When f=1 and meb(Bi,I) = 0 or meb(Bi,I) = 1 (i=1,2, ... ,n), (iii) is also the same as in the qualitative case. So the qualitative case can be seen as the special case of it.

Definition 3.4. A Herbrand interpretation I such that a set Sf of f-Horn clause rules is true in I is called a Herbrand model of Sf.

Definition 3.5. For any set Sf of f-Horn clause rules (let its Herbrand base be H(Sf)), any $A \in H(Sf)$, and any $f \in (0,1]$, $Sf := \{A \leftarrow (f)-$ iff the right-hand side is true in every Herbrand model of Sf.

Note that the symbol ":=" means truth in all Herbrand models rather than in all models.

For a set Sf of f-Horn clause rules, it is clear that

$$Sf := \{A \leftarrow (f)-\} \text{ implies } Sf := \{A \leftarrow (f')-\}, \text{ for any } f,f' \in [0,1], \text{ and } f \geq f'. \quad (3.2)$$

Let Sf be a set of f-Horn clause rules, and M(Sf) denote the set of Herbrand models of Sf. $\cap M(Sf)$ is defined in the quantitative case by adopting

$$\text{meb}(A, \cap Is) = \text{INF}\{\text{meb}(A,I): I \in Is\}, \quad (3.3)$$

where Is is a set of Herbrand interpretations and INF is the greatest lower bound [1].

Theorem 3.6. Let Sf be a set of f-Horn clause rules, and $\cap M(Sf)$ denote the intersection of all Herbrand models of Sf. Therefore

$$\text{meb}(A, \cap M(Sf)) = \text{SUP}\{f: Sf := \{A \leftarrow (f)-\}\},$$

where SUP is the least upper bound.

PROOF. If Sf is a set of f-Horn clause rules. I is a Herbrand model of Sf, $A \in H(Sf)$, and $Sf := \{A \leftarrow (t)-\}$, then $\{A \leftarrow (f)-\}$ is true in I, and $\text{meb}(A,I) \geq f$, by definition 3.3. Therefore, $\text{meb}(A,I) \geq \text{SUP}\{f1: Sf := \{A \leftarrow (f1)-\}\}$, for any Herbrand model $I \in M(Sf)$, and $\text{meb}(A, \cap M(Sf)) \geq \text{SUP}\{f1: Sf := \{A \leftarrow (f1)-\}\}$. However, we have $Sf: A \leftarrow (f2)-$, where $f2 = \text{meb}(A, \cap M(Sf))$, for any Sf and all $A \in H(Sf)$. So the relation $\text{meb}(A, \cap M(Sf)) > \text{SUP}\{f: Sf := \{A \leftarrow (f)-\}\}$ is impossible. So far, the theorem has been proved.

For the qualitative case, fixpoint theory associates each set S of Horn clause rules with a mapping Ts from interpretations to interpretions, and it shows that fixpoints of Ts are models of S.
Now we follow the same way to establish the fixpoint semantics for a set Sf of f-Horn clause rules.

Definition 3.7. Let Sf be a set of f-Horn clause rules. For every $A \in H(Sf)$,

$$meb(A,Tsf(I))=SUP\{f \times \min\{meb(Bi,I):i=1,2, \ldots ,n\}: A \leftarrow (f)\text{-}B1 \ \& \ B2 \ \& \ \ldots \ \& \ Bn \text{ is a ground instance of a f-Horn clause rule in } Sf\}. \tag{3.4}$$

Definition 3.8. Let Sf be a set of f-Horn clause rules, and H(Sf) denote the Herbrand base of Sf. For two interpretations I1, I2 of Sf

$$I1 \subseteq I2 \text{ iff } meb(A,I1) \leq meb(A.I2) \quad \text{for all } A \in H(Sf). \tag{3.5}$$

Theorem 3.9. For any set Sf of f-Horn clause rule, Tsf is a monotone function.
PROOF. Let $I1 \subseteq H(Sf)$, $I2 \subseteq H(Sf)$, and $I1 \subseteq I2$. By definition 3.8, if $A \leftarrow (f)\text{-}B1$ & B2 & ... & Bn is the ground instance of a f-Horn clause rule in Sf, then $meb(Bi,I1) \leq meb(Bi, I2)$, i=1,2, ... , n. For any $A \in H(Sf)$, $meb(A,Tsf(I1)) \leq meb(A,Tsf(I2))$, by definition 3.7. Therefore $Tsf(I1) \subseteq Tsf(I2)$. It is said that Tsf is a monotone function.

The following theorem will associate models with fixpoints in the quantitative case.

Theorem 3.10. For any set Sf of f-Horn clause rules, and any $I \subseteq H(Sf)$,
$I \in M(Sf)$ iff $Tsf(I) \subseteq I$.

PROOF.(==>) If $I \in M(Sf)$, then Sf is true in I. For any ground instance $A \leftarrow (f)\text{-}B1$ & B2 & ... & Bn of a f-Horn clause rule in Sf we have

$$meb(A,I) \geq f \times MIN\{meb(Bi,I):i=1,2,\ldots,n\}.$$

by Definition 3.3. Hence

$$meb(A,I) \geq \sup\{f \times MIN\{meb(Bi,I):i=1,2,\ldots,n\}: A \leftarrow (f)\text{-}B1 \ \& \ B2 \ \& \ \ldots \ \& \ Bn \text{ is a ground instance of a f-Horn clause rule in } Sf\},$$

and $meb(A,I) \geq meb(A,Tsf(I))$ by Definition 3.7. Therefore $Tsf(I) \subseteq I$.

(<==) If $Tsf(I) \subseteq I$, then $meb(A,I) \geq meb(A,Tsf(I))$ for any $A \in H(Sf)$, by Definition 3.8. In addition, for any ground instance $A \leftarrow (f)\text{-}B1$ & B2 & ... & Bn of a f-Horn clause rule in Sf we have

$$meb(A,Tsf(I)) \geq f \times MIN\{meb(Bi,I):i=1,2,\ldots,n\},$$

by Definition 3.7. Hence

$$meb(A,I) \geq f \times MIN\{meb(Bi,I):i=1,2,\ldots,n\},$$

and this implies that Sf is true in I by Definition 3.3. It is said that

$I \in M(Sf)$.

Theorem 3.11. For any set Sf of f-Horn clause rules $\cap M(Sf)=lfp(Tsf)$.
PROOF. We have known $lfp(Tsf)=\cap\{I:Tsf(I)=I\}$. By Theorem 3.10., $Tsf(I) = I$ iff $I \in M(Sf)$. Hence $lfp(Tsf)=\cap\{I:Tsf(I)=I\}=\cap\{I:I\in M(Sf)\}=\cap M(Sf)$.

Theorem 3.12. For any set Sf of f-Horn clause rules, mapping Tsf is continuous, i.e.,

$$\cup\{Tsf(Ij):j\in N\} = Tsf(\cup\{Ij:j\in N\}),$$

for all chains $I1 \leq I2 \leq \ldots$ of Herbrand interpretation.
PROOF. For any atomic formula $A \in H(Sf)$, we have

$$\begin{aligned} &meb(A,Tsf(\cup\{Ij:j\in N\})) \\ &= SUP\{ f \times MIN\{meb(Bk,\cup\{Ij:j\in N\}):k=1,2,\ldots,n\} : \\ &\quad A \leftarrow (f)\text{-}B1 \ \&\ B2 \ \&\ \ldots \ \&\ Bn \text{ is a ground instance} \\ &\quad \text{of a f-Horn clause rule in } Sf\}. \end{aligned}$$

suppose $A \leftarrow (fa)\text{-}Ba1 \ \&\ Ba2 \ \&\ \ldots \ \&\ Bana$ is the ath ground instance of f-Horn clause rules in Sf having A as conclusion. The above expression can be shortened to

$$meb(A,Tsf(\cup\{Ij:j\in N\}))=SUP(fa \times MINk\ meb(Bak,\cup\{Ij:j\in N\})).$$

However, $meb(Bak,\cup\{Ij:j\in N\}) = SUPj\ meb(BAk,Ij)$, where j indexes the monotone chain $I1 \subseteq I2 \subseteq \ldots$ of Herbrand interpretations, therefore

$$\begin{aligned} &meb(A,Tsf(\cup\{Ij:j\in N\})) \\ &= SUPa\ SUPj\ fa \times MINk\ meb(Bak,Ij)=SUPa\ SUPj\ Vaj \ , \end{aligned}$$

where $Vaj=fa \times MINk\ meb(Bak,Ij)$. Using the same method we find

$$meb(A, \cup\{Tsf(Ij):j\in N\})=SUPj\ SUPa\ Vaj \ .$$

Now $SUPa\ SUPj\ Vaj=SUPj\ SUPa\ Vaj$ needs to be proved. The set consisting of all Vaj is bounded above, therefore it has a least upper bound, let it be V. Hence, $SUPa\ SUPj\ Vaj \geq V$. On the other hand, for all $a\in N$, we have $SUPj\ Vaj \leq V$, therefore $SUPa\ SUPj\ Vaj \leq V$. Hence, $SUPa\ SUPj\ Vaj = V$. Similarly, we show that $SUPj\ SUPa\ Vaj = V$. Therefore

$$\{Tsf(Ij):j\in N\}= Tsf(\cup\{Ij:j\in N\} .$$

It is easy to prove the important property of Tsf by the above Theorem 3.12.

Theorem 3.13. For any set Sf of f-Horn clause rules,

$$lfp(Tsf) =\cup\{Tsf^{n}(\emptyset):n\in N\},$$

where $\emptyset$ is a special interpretation such that $meb(A,\emptyset)=0$ for any $A\in H(Sf)$.

Now we present a theorem that can serve as foundation for the completeness result on the quantitative proof theory. A completeness result for a proof method is of the form : if an assertion is true, then it can be proved according to the method.
We assume $meb(A,\cap M(Sf)) = c$, and want to show that $A \leftarrow (c)\text{-}$ can be derived from Sf. By Theorem 3.12. and Theorem 3.14., we have $\cap M(Sf) =\cup\{Tsf^{n}(\emptyset):n\in N\}$.

But we try to draw the stronger conclusion from $meb(A,\cap M(Sf)) = c$ that there exists an $n\in N$ such that $meb(A,Tsf^n(\emptyset))=c$.
Here is one of the methods.

Lemma 3.15. For any finite set Sf of f-Horn clause rules, any $A\in H(Sf)$, and any real number $\varepsilon>0$,

$$meb(A,Tsf^n(\emptyset)):n\in N \text{ and } meb(A,Tsf^n(\emptyset))\geq\varepsilon\}$$

is finite.
PROOF. Let F(Sf) be the set of implications of f-Horn clause rules in Sf. Note that Sf is finite, therefore F(Sf) is a finite set. Let m be the greatest element of F(Sf) such that $m<1$. The real number $meb(A,Tsf^n(\emptyset))$ is a product of a sequence of elements of F. In this sequence, if q is the smallest integer such that m^q then at most q elements can be less than 1. The sequence can have any length, because 1 can occur in the sequence any number of time. So we conclude that the number of different products ($\geq\varepsilon$) of the sequences of elements of F(Sf) is not greater than $:F(Sf):^q$.

Theorem 3.15. For any finite set Sf of f-Horn clause rules, and any $A\in H(Sf)$, there exists an $n\in N$ such that $meb(A,\cap M(Sf))=meb(A,Tsf^n(\emptyset))$.
PROOF. If $\mu=meb(A,\cap M(Sf))=0$, then there exists $n=0\in N$ such that the expression is hold. Suppose $\mu>0$, then $\cap M(Sf)=lfp(Tsf)=\cup\{Tsf^n(\emptyset):n\in N\}$, by Theorem 3.11. and theorem 3.13. Hence

$$\begin{aligned} meb(A,\cap M(Sf)) &= SUP\{meb(A,Tsf^n(\emptyset)):n\in N\} \\ &= SUP\{meb(A,Tsf^n(\emptyset)):n\in N \text{ and } meb(A,Tsf^n(\emptyset))\geq\xi\}. \quad (3.6)\end{aligned}$$

for any $\xi<\mu$. If we choose such an ξ positive, according to $\mu>0$, then (3.6) is finite by Lemma 3.15. Hence the least upper bound is attained for an $n\in N$.

Note that the sets of f-Horn clause rules discussed in this section are finite, it is not superfluous condition.

4. Fuzzy proof theory

In this section we describe a fuzzy proof procedure precisely for f-Horn clause rules, and justify its results using the semantics results presented in the previous section.
As in the qualitative case, the quantitative proof procedure for f-Horn clause rules is also a search of an and/or tree. This tree determined by a set Sf of f-Horn clause rules and an initial atom G is defined as follows:

Definition 4.1.

(1) There are two kinds of nodes: and-nodes and or-nodes.
(2) Each or-node is labeled by a single atomic formula.
(3) Each and-node is labeled by a f-Horn clause rule in Sf and a substituion.
(4) The descendants of each or-node are all and-nodes, and the descendants of each and-node are all or -nodes.
(5) The root is an or-node labeled by G.
(6) For each f-Horn clause rule R in Sf with a left-hand side unifying with

the atomic formula A (with the most general substitution θ) in an or-node, there is an and-node descendant of the or-node labeled with R and θ. An and-node with no descendants is called a failure node.

(7) For each atomic formula B in the right-hand side of the f-Horn clause rule labeling an and-node, there is a descendant or-node labeled with B. An and-node with no descendants is called a success node.

(8) Each node is associated with a real number which is called the value of the node. The value of a success node is the implication of its associated f-Horn clause rule. The value of a nonterminal and-node is fxt, where f is the implication strength of the f-Horn clause rule labeling the and-node and t is the minimum of the values of its descendants. The value of a failure node is 0. The value of a nonterminal or-node is the maximum of the values of its descendants.

In the quantitative case, a proof tree is a subtree of an and/or tree defined as follows. The root of the proof tree is the root of the and/or tree. An or-node of the proof tree which also occurs in the and/or tree has one descendant in the proof tree which is one of the descendants of that node in the and/or tree. An and-node in the proof tree which also occues in the and/or tree has as descendants in the proof tree all of the descendants of that node in the and/or tree. Furthermore, All terminal nodes in a proof tree are success nodes. We assign values to proof tree nodes in the same way as we do to nodes in an and/or tree.

In the qualitative case, **correctness** of the (SLD-resolution)proof procedure says in the most elementary form:if $A \in H(Sf)$ is proved, then $A \in M(Sf)$. We can express correctness like this : results of the proof procedure are not more true than they are in the minimal model M(Sf). In the quantitative case limited to finite and/or trees, the form of the corresponding correctness is suggested.

Theorem 4.2. For any set Sf of f-Horn clause rules with a finite and/or tree and any $A \in H(Sf)$, the value of the root in the and/or tree with A as root is not greater than meb(A,$\cap$ M(Sf)).

PROOF. Note first that the value of the root in the and/or tree is the maximum of the values of the roots of its constituent proof trees. It can easily be verified that the value of the root of a proof tree with A as root is not greater than $Tsf^{n+1}(\emptyset)$, where n is the length of a longest path from the root to a terminal node. Here one unit of path length is from or-node to or-node along the path. It is proved by Theorem 3.16.

The following is completeness of the proof procedure.

Theorem 4.3. For any set Sf of f-Horn clause rules with a finite and/or tree and any $A \in H(Sf)$, the value v of the root in the and/or tree with A as root is at least meb(A,$\cap$ M(Sf)).

PROOF. By induction on n, we prove that $v \geq$ meb(A, $Tsf^{n}(\emptyset)$) for all $n \in N$. Then we conclude that

$v \geq SUP\{meb(A,Tsf^{n}(\emptyset)):n\in N\} = meb(A,\cup\{Tsf^{n}(\emptyset):n\in N\}) = meb(A,\cap M(Sf))$.

Now let's start the inductive proof of $v \geq meb(A,Tsf^{n}(\emptyset))$.

(1) for n=0, it is true.

(2) suppose it holds for $n=n_0$, then

$$meb(A,Tsf^{n_0+1}(\emptyset))=SUP\{ f \times MIN\{meb(Bk, Tsf^{n_0}(\emptyset)):k\in N\} : A\leftarrow (f)\text{-}B1 \ \&\ B2 \ \&\ \ldots \ \&\ Bn \text{ is a ground instance of a f-Horn clause rule in } Sf\}.$$

The set over which the superemum is taken is finite by Lemma 3.15. Therefore the superemum must be attained for ground instance A ←(f)-B1 & B2 & ... & Bn of a f-Horn clause rule R: A' ←(f)-B1' & B2' & ... & Bn' in Sf. Hence

$$meb(A, Tsf^{n_0+1}(\emptyset))=f \times MIN\{meb(Bk, Tsf^{n_0}(\emptyset)):k \in N\}. \qquad (4.1)$$

Let us consider the and/or tree for Sf having A as root. One of the descendants of the root must be the f-Horn clause rule. Because its left-hand side A' and A as ground instance, there is a most general substitution θ of A' and A. Hence one of the descendants of the root is the node (R,θ) labeled with R and θ. Its descendants are B1'θ, B2'θ,..., Bk'θ with values v1', v2', ... , vk' and having B1, B2, ..., Bk respectively as ground instances.
By the induction hypothesis, B1, B2, ..., Bk are roots of and/or tree having values v1, v2, ..., vk such that $vi \geq meb(Bi, Tsf^{n_0}(\emptyset))$, i=1,2,...,k. Because Bi'θ has Bi as instance, we must have vi' ≥ vi. For the value v of the entire and/or tree, with A as root, we have v=f x MIN{vi':i=1,2,...,k} and hence $v \geq f \times MIN\{meb(Bi, Tsf^{n_0}(\emptyset)):i=1,2, \ldots ,k\}$. We conclude $v \geq meb(A, Tsf^{n_0+1}(\emptyset))$ by (4.1), which completes the induction proof.

5. The Fuzzy Procedural Interpretation for f-Horn Clause Rules

In the logic program system presented by R. A. Kowalski [2,3] , a Horn clause rule is interpreted as a procedure. In the similar way, f-Horn clauses in Sf can be interpreted as follows:

(i) f-Goal ←(f)-A1 & A2 & ... & An (n ≥ 1), in which atoms are questions or goals relatedto the f needed tobe answered. In the fuzzy logic program system, the f-Goal will not be satisfied until the n subgoals are all satisfied.

(ii) f-Procedure A ←(f)-B1 & B2 & ... & Bn (n ≥ 0) is interpreted as a fuzzy rule of question answering. For given question ←(f)-A, when g > f, this rule for A gets us nowhere; when g ≤ f, subquestions ←(f')-B1, ←(f')-B2, ... , and ←(f')-Bn (n ≥ 0) need to be answered firstly, where f'=f/g.

(iii) f-Assertion A ←(f)-. It can be regarded as a special rule of fuzzy question answering. For question ←(g)-A, if g ≤ f, then it is satisfied; if g > f, then it is satisfied; if g f, then it is not unsatisfied. The question is answered directly with no subquestion derivation

A set Sf of f-Horn clause rules can be seen as a fuzzy logic program. It is intiated by an intial f-Goal, similar to the qualitative case, by using f-Procedure constantly, new f-Goals can be derived from old ones, so as to advance the computation procedure. Finally, it terminates with the derivation of the halt statement (it is derived from $\leftarrow$(f1)-A and A$\leftarrow$(f2)-, where f1 $\doteq$ f2).

6. Concluding remarks

The syntax and semantics of finite sets of f-Horn clause rules are quite similar to the qualitative case. Therefore many results of logic program on the semantics, especially, fixpoint semantics and model semantics can be transfered to the quantitative case. Even, sometimes, the proofs can be adopted nearly unchanged. Of course, a set of f-Horn clause rules is more powerful than the corresponding set of Horn clause rules in description because of the concept generlization of truth value, implication and so on.

This paper have established the complete formal specification of a finite set of f-Horn clause rules and its semantics interpretation. Furthermore, we presented the precise description of the fuzzy proof procedure, and the procedural interpretation for f-Horn clause rules. Hence we obtained the fuzzy logic program system.

To facilitate the implementation, f-Horn clause rules in a program must be well-defined. In addition, implication strengthes can not only be a constant or a variable, but also a function. So this is a good combination of fuzzy set theory and Horn clause logic.

At last, we have to warn the readers that, in the fuzzy proof theory, we required and/or trees to be finite because, in general, the Min-Max rule is only suitable for finte trees.

7. References

1. Zadeh, L. A., "Fuzzy Sets", Inform, and Control 8: 338-353 (1965)
2. Zadeh, L. A., "Fuzzy Sets as a Bases for a Theory of Possibility", FSS. 1 (1978): 3-28
3. van Emden, M. H. and Kowalski, R. A., "The Semantics of Logic as a Programming Language". J. Assoc. Comput. Mach. 23: 733-742 (1976)
4. Shortliffe, E., "Computer Based Medical Consultations: MYCIN". American Elsevier, New York, 1976
5. R. A. Kowalski, "Predicate Logic as Programming Language", Proc. IFIP 74, North Holland, Amsterdam, 1974, 569-574
6. Chang, C. L., and Lee, R. C. T., "Symbolic Logic and Mechanical Theorem Proving", New York, Academic Press, 1973.
7. Robinson, J. A., "A Machine-oriented Logic Based on the Resolution Principlé". J. ACM 12, 1 (Jan. 1965), 23-41
8. Winston, P. H., "Artifical Intelligence". Addison-Wesley, 1978
9. Dongbo Liu and Deyi Li, "Fuzzy PROLOG and Expert Systems", Min-Micro Computer System, 5 (1988), 26-31 (In Chinese)

BABEL: A Functional and Logic Programming Language based on constructor discipline and narrowing

Extended Abstract

Juan José Moreno-Navarro
Dpto. de Lenguajes y Sist.
Informáticos e Ing. del Software
Fac. de Informática
Univ. Politécnica
28043 Madrid
SPAIN

Mario Rodríguez-Artalejo
Dpto. de Informática y Automática
Fac. de C.C. Matemáticas
Univ. Complutense
28040 Madrid
SPAIN

That is why it was called Babel, because Yahve confused there all the world's tongues .....
Genesis 11,9

ABSTRACT

We present the experimental language BABEL, designed to achieve integration of functional and logic programming in a simple,flexible, and mathematically well founded way. The language relies on a constructor discipline and the use of narrowing to englobe rewriting and SLD-resolution. It is first-order, type-free and has a lazy operational semantics which supports infinite data structures and is sound w.r.t. a declarative semantics based on Scott domains. The paper includes BABEL's syntactical and semantical specification, some mathematical results on the semantics, and programming examples.

1. INTRODUCTION

During the last years, several attempts have been made to design declarative programming languages integrating the functional and logical paradigms; cfr.[1] , [2] . To achieve this integration is a highly desirable goal, then the resulting languages would fully exploit the facilities of logic (functions,relations and equality) and allow their users to keep them separate or to mix them in the way that best suits to a particular application.

We believe that any integrated functional and logic programming language should have a clean mathematical semantics, based on logic, as well as an operatinal semantics related to the logical one through soundness and completeness results. Equational logic,and some related notions such as rewriting (cfr. [3]) and narrowing (cfr, [4] , [5]) have been recognized as key notions to achieve this. In fact, narrowing can be used to incorporate functions and equations into Horn clause programs, via extended unification algorithms, as done in SLOG (Fribourg [6]), FUNLOG (Subrahmanyam and You [7]) and EQLOG (Goguen and Meseguer [8]). For a recent survey of extended unification algorithms, see Dincbas and van Hentenryck [9] . A different approach, followed by the designers of K-LEAF [10] , regards equality as an ordinary predicate and uses a form of SLD-resolution to simulate narrowing. This idea is further ellaborated in [11] , [12] .

The language BABEL we are going to present here uses narrowing to simulate SLD-resolution in the framework of a first order, type free functional language with constructor discipline(cfr. [13]). The idea of using narrowing in this way comes from Reddy [14] . In BABEL, boolean valued functions serve as predicates. This supplies two definite truth values and enables a quite flexible use of propositional conectives, in particular negation. BABEL expressions are evaluated through lazy narrowing and may yield several answers if they are not ground. This operational semantics is sound w.r.t. a declarative semantics based on complete Herbrand domains (cfr. [15]), and supports computation with infinite objects. Completeness results are in progress and will hopefully be published in the full version of this paper.

2. BABEL'S SYNTAX

We start with four disjoint sets of symbols:

Variables X, Y, Z ∈ VS Function Symbols f, g, h ∈ FS
Constructors c, d, e ∈ CS Predicate Symbols p, q, r ∈ PS

We assume that VS is countably infinite. In concrete examples, we shall use identifiers starting with an uppercase (resp. lowercase) letter for variables (resp. constructors, function symbols and predicate symbols). Symbols other than variables are assumed to have an associated arity.

Terms(s,t), expresions (M,N) and boolean expressions (B,C) are defined by:

t::=X /* variable */
| c /* constant */
| c($\overline{t}$) /* construction */

M::=t /* term */
| c($\overline{M}$) /* constructor applic. */
| f($\overline{M}$) /* function applic. */
| ($B \rightarrow M$) /* guarded expr. */
| ($B \rightarrow M \,\square\, N$) /* alternative expr. */

B::= **t** /* truth */
| **f** /* falsity */
| p($\overline{M}$) /* predicate applic. */
| $\neg B$ /* negation */
| (B,C) /* parallel conj. */
| (B;C) /* sequential conj. */
| (B|C) /* parallel disj. */
| (B/C) /* sequential disj. */
| ($B \rightarrow C$) /* guarded bool.expr.*/
| ($B \rightarrow C \,\square\, D$) /* alternative bool. expr.*/
| $M \equiv N$ /* strong equality */
| $M \not\equiv N$ /* strong inequality */

BABEL expressions evaluate to possibly infinite trees builded from constructors. The "strong equality" is interpreted as a continuous approximation of the identity (cfr. section 5).

For given expressions E_i (possibly boolean), we allow conditional expressions

$$B_1 \rightarrow E_1 \,\square\, B_2 \rightarrow E_2 \,\square \ldots \square\, B_k \rightarrow E_k \quad \text{resp.} \quad B_1 \rightarrow E_1 \,\square\, B_2 \rightarrow E_2 \,\square \ldots \square\, B_k \rightarrow E_k \,\square\, E_{k+1}$$

as syntactic sugar for

$$B_1 \rightarrow E_1 \,\square\, (B_2 \rightarrow E_2 \,\square \ldots \square (B_k \rightarrow E_k)..) \quad \text{resp.} \quad B_1 \rightarrow E_1 \square (B_2 \rightarrow E_2 \,\square\, .. \square (B_k \rightarrow E_k \,\square\, E_{k+1})..)$$

BABEL rules belong to six categories:

(F) Functional rules:

$f(\overline{t}(\overline{X})) := M(\overline{X})$

(GF) Guarded functional rules:

$f(\overline{t}(\overline{X})) := B(\overline{X},\overline{Y}) \rightarrow M(\overline{X})$.

(P) Predicative rules:

$p(\overline{t}(\overline{X})) := C(\overline{X})$

(GP) Guarded predicative rules:

$p(\overline{t}(\overline{X})) := B(\overline{X},\overline{Y}) \rightarrow C(\overline{X})$.

(R+) Positive relational rules:

$p(\overline{t}(\overline{X}))$.
$p(\overline{t}(\overline{X}))$:- $B(\overline{X},\overline{Y})$.

(R^-) Negative relational rules:

$\neg p(\overline{t}(\overline{X}))$.
$\neg p(\overline{t}(\overline{X}))$:- $B(\overline{X},\overline{Y})$.

Here, $\overline{X}$ and $\overline{Y}$ stand for tuples of variables, and $\overline{t}(\overline{X})$ stands for tupel of terms using variables from $\overline{X}$ only. Notice that only constructors may be used in $\overline{t}(\overline{X})$ ("constructor discipline" cfr. [13]). Similarly, $M(\overline{X})$ stands for an expression with variables from $\overline{X}$, etc.

Relational rules must be considered as syntactic sugar for guarded predicative rules of a particular form, as follows:

Sugared form	Unsugared form
$p(\overline{t}(\overline{X}))$.	$p(\overline{t}(\overline{X}))$:= **t**.
$p(\overline{t}(\overline{X}))$:- $B(\overline{X},\overline{Y})$.	$p(\overline{t}(\overline{X}))$:= $B(\overline{X},\overline{Y}) \rightarrow$ **t** .
$\neg p(\overline{t}(\overline{X}))$.	$p(\overline{t}(\overline{X}))$:= **f**.
$\neg p(\overline{t}(\overline{X}))$:- $B(\overline{X},\overline{Y})$	$p(\overline{t}(\overline{X}))$:= $B(\overline{X},\overline{Y}) \rightarrow$ **f**.

With this convention, all BABEL rules are inequational rules

(IR) ⟨Head⟩ := ⟨Body⟩

where ⟨Head⟩ is a "flat" form $f(\overline{t}(\overline{X}))$ or $p(\overline{t}(\overline{X}))$ - corresponding to some data pattern $\overline{t}(\overline{X})$ expressed via constructors - and ⟨Body⟩ is a (perhaps boolean) BABEL expression. The intended meaning of (IR) is that ⟨Head⟩ is at least so defined as ⟨Body⟩, for any possible value of the variables occurring in the rule.

To finish this section, let us introduce programs. A BABEL program is a finite set of

BABEL rules satisfying the following two decidable conditions:

(a) Left linearity: Each variable occurring in a rule has at most one occurrence at the head.

(b) No Superposition: Any two different rules for the same symbol do not superpose, unless they are both relational and (i) they have the same sign; or (ii) they have different signs, but mutually excluding bodies. Where:

(b.1) Two rules $\langle Head_i \rangle$:= $\langle Body_i \rangle$ (i = 1,2) are said to superpose iff $\langle Head_1 \rangle$, $\langle Head_2 \rangle$ are unifiable (in Robinson's classical sense; cfr. [16]).

(b.2) Two relational rules with different signs:

$p(\bar{t}(\bar{X}))$:- $B(\bar{X},\bar{Y})$. $\neg p(\bar{s}(\bar{U}))$:- $C(\bar{U},\bar{V})$.

(where $B(\bar{X},\bar{Y})$ and/or $C(\bar{U},\bar{V})$ may be empty) whose heads $p(\bar{t}(\bar{X}))$, $p(\bar{s}(\bar{U}))$are unifiable with most general unifier (m.g.u.) σ are said to have mutually excluding bodies iff the finite set $\{B(\bar{X},\bar{Y})\sigma, C(\bar{U},\bar{V})\sigma\}$ is propositionally unsatisfiable (see Definition 4 in section 5).

3. PROGRAMMING IN BABEL

In this section we present some simple BABEL programs. Their exact meaning and use will become clear in the two following sections.

Example 1: Appending lists.

This purely relational program shows that pure PROLOG can be mimicked in BABEL without any substantial notational changes.

Constructors:

```
[ ]        /* empty list */
[.|.]      /* list constructor */
```

Rules:

```
A1 : append([ ], L, R) :- R≡L.
A2 : append([X|R] ,L,[Y|S] ) :- X≡Y;  append(R,L,S).
```

Example 2: Testing binary trees for the equality of frontiers.

This program illustrates the interplay between functions and predicates.

Constructors:

```
[ ] , [.|.]        /* as before */
tree               /* constructor of trees */
leaf               /* constructor of leaves */
```

Rules:

```
EF1: equal-frontier(A,B) := equal-list(frontier(A),frontier(B)).
EL1: equal-list([ ],L) := L ≡ [ ]  .
EL2: equal-list( [X|R] ,[Y|S] ) := equal-atom(X,Y);equal-list(R,S).
/* Some rules for "equal-atom" */
FT1: frontier( leaf(X)) := [X] .
FT2: frontier(tree (leaf (X), B)) := [ X| frontier (B) ].
FT3: frontier(tree (tree(A,B),C)) := frontier(tree(A, tree(B,C))).
```

Example 3: The Alpine Club Puzzle.

In[17] we found the statement of the following puzzle, which was the subject of discussion of a few contributors to a PROLOG electronic mailing list:

"Tony, Mike and John belong to the Alpine Club. Every member of the Alpine Club is either a skier or a mountain climber or both. No mountain climber likes rain, and all skiers like snow. Mike dislikes whatever Tony likes and likes wathever Tony dislikes. Tony likes rain and snow. Is there a member of the Alpine Club who is a mountain climber, but not a skier?".

The following BABEL solution to the puzzle illustrates the more liberal approach to negation allowed by BABEL in comparison to PROLOG. Only nullary constructors (i.e. constants) are used.

```
AC1: alpinist (tony).     AC2: alpinist(mike).     AC3: alpinist(john).
SC:  climber(X) :- alpinist(X), ¬skier(X).
LK1: ¬likes(X,rain) :- climber(X).
LK2: ¬skier(X) :- ¬likes(X,snow).
LK3: ¬likes(mike,X) :- likes(tony,X).
LK4: likes(mike,X) :- ¬likes(tony,X).
LK5: likes(tony,rain).              LK6:  likes(tony,snow).
```

To solve the puzzle, BABEL has to reduce to t the boolean expression

```
G:  alpinist(X), climber(X), ¬skier(X).
```

Notice that rules LK3, LK4 do not violate the no superpositions rectriction, because the bodies are mutually excluding.

Example 4: Computing Hamming numbers.

Dijkstra [18] attributes to Hamming the problem of building the infinite ascending sequence of all positive numbers greater than one containing no prime factors other than 2, 3 and 5. The following BABEL solution illustrates the use of conditional expressions and infinite objects (streams).

```
Constructors
  [ ], [.|.]      /* as before */
  0               /* nullary constructor for zero */
  suc             /* succesor  */

Rules:
  NH:   nth-hamming(N,M) :-  nth-member(N, hamming-seq, M).
  NM1:  nth-member(1, [X|R] ,Y) :-  Y≡X.
  NM2:  nth-member(suc(suc(N)),[X|R] ,Y) :- nth-member(suc(N), R, Y).
  HS:   hamming-seq := merge-3(seq-prod(2, [1| hamming-seq] ),
                               seq-prod(3, [1| hamming-seq] ),
                               seq-prod(5, [1| hamming-seq] )).
  SP:   seq-prod(N, [X|R] ):=[ N*X | seq-prod(N,R)] .
  M3:   merge-3(L1,L2,L3) := merge-2(merge-2(L1,L2), L3).
  M2:   merge-2([X|R] ,[Y|S] ) :=  X<Y  → [X| merge-2(R, [Y|S] ) []
                                   Y<X  → [Y| merge-2([X|R] ,S)  []
                                          [X| merge-2(R,S)] .
  *1:   N*0 := 0.
  *2:   N*suc(M) := (N*M) + N.
  +1:   N + 0 := N.
  +2:   N + suc(M) := suc(N+M).
   1:   ¬0<0.
   2:   ¬suc(N)<0.
   3:   0 <suc(N).
   4:   suc(N) < suc(M):= N<M.
```

Notice the use of 1,2,3,5, as abreviations for suc(0), suc(suc(0)), etc., as well as the use of infix notation to increase legibility.

4. BABEL'S REDUCTION SEMANTICS

In this section we discuss BABEL's reduction mechanism, which is based on a lazy version of narrowing and defines the operational semantics of the language.

As is usual in the literature devoted to rewriting (cfr. e. g. [3]), we use finite strings u, v of integers to denote occurrences. Letters "E_i", "F_i" are used in the sequel to denote expressions (boolean or not). $E[u]$ and $E[u \leftarrow F]$ stand for the subexpression of E at occurrence u and for the result of syntactically replacing F for the subexpression of E at u, respectively. In particular, $E[\varepsilon]$ is E.

All along this section, when referring to BABEL rules we mean the unsugared forms.

Definition 1 (Applicable rules, pending rules, implicit rules).

(a) A rule applies to (resp. is pending for) E iff some variant of the rule which stands apart from E (i.e. has no variables occurring in E) applies to (resp. is pending for) E.

(b) A rule ⟨Head⟩ := ⟨Body⟩ standing apart from E applies to E iff ⟨Head⟩ and E are unifiable via some m.g.u. θ . If θ is some m.g.u. of {⟨Head⟩ , E} and σ is the restriction of θ to the variables occurring in E, we say that the rule applies to E via σ.

(c) A rule ⟨Head⟩ := ⟨Body⟩ standing apart from E is pending for E iff the attempt of unifying ⟨Head⟩ and E fails because of disagreement between a constructor and a defined function symbol occurring at corresponding places in ⟨Head⟩ and E. (Intuition: Some further evaluation inside E is demanded by the rule).

(d) The implicit rules for BABEL's primitives are the following ones:

Rules for conditional expressions

$(t \to X) := X.$ $\quad (t \to X \,\Box\, Y) := X.$ $\quad (f \to X \,\Box\, Y) := Y.$

Rules for the propositional connectives

(f;Y) := f.		(t/Y) := t.
(t;Y) := Y.		(f/Y) := Y.
(f,Y) := f.	¬t := f.	(t|Y) := t.
(t,Y) := Y.	¬f := t.	(f|Y) := Y.
(X,f) := f.		(X|t) := t.
(X,t) := X.		(X|f) := X.

Rules for equality and inequality

$c(X_1,\ldots,X_n) \equiv c(Y_1,\ldots,Y_n) \;:=\; X_1 \equiv Y_1,\ldots,X_n \equiv Y_n.$ /* t if n = 0 */

$c(X_1,\ldots,X_n) \equiv d(Y_1,\ldots,Y_m) \;:=\; f.$ /* for $c \neq d$; $n,m \geq 0$ */

$M \not\equiv N \;:=\; \neg(M \equiv N).$

Definition 2 (Redex occurrences)

The finite set $Red_\Pi[E]$ of redex-occurrences in a BABEL expression E with respect to a given BABEL program Π is defined by structural recursion.

$Red_\Pi[t] = \emptyset$ for any term t.

$Red_\Pi[c(M_1,\ldots,M_n)] = \bigcup_{i=1}^{n} i.Red_\Pi[M_i].$

$Red_\Pi[f(M_1,\ldots,M_n)]$ = if some Π-rule applies to $f(\overline{M})$
then if no Π-rule is pending for $f(\overline{M})$
then $\{\varepsilon\}$
else $\{\varepsilon\} \cup \bigcup_{i=1}^{n} i.\,Red_\Pi[M_i].$
else $\bigcup_{i=1}^{n} i.Red_\Pi[M_i].$

$Red_\Pi[p(M_1,\ldots,M_n)]$: Substitute "p" for "f" in the previous case.

Red $(B \to E)$ = if some implicit rule applies to $(B \to E)$
then $\{\varepsilon\}$
else $1.Red_\Pi[B]$.

Red $(B \to E \,\Box\, F)$ = if some implicit rule applies to $(B \to E \,\Box\, F)$
then $\{\varepsilon\}$
else $1.\,Red_\Pi[B]$.

$Red_\Pi[t] = Red_\Pi[f] = \emptyset$

$Red_\Pi[\neg B]$ = if some implicit rule applies to $\neg B$
then $\{\varepsilon\}$
else $1.\,Red_\Pi[B]$.

$Red_{\Pi}[(B \textbf{ par } C)]$ = **if** some implicit rule applies to (B **par** C)
then $\{\varepsilon\}$
else $1.Red_{\Pi}[B] \cup 2.Red_{\Pi}[C]$
where **par** stands for "," or "|"

$Red_{\Pi}[(B \textbf{ seq } C)]$ = **if** some implicit rule applies to (B **seq** C)
then $\{\varepsilon\}$
else $1.\ Red_{\Pi}[B]$
where **seq** stands for ";" or "/"

$Red_{\Pi}[M \equiv N]$ = **if** some implicit rule applies to $M \equiv N$
then $\{\varepsilon\}$
else $1.Red_{\Pi}[M] \cup 2.Red_{\Pi}[N]$

$Red_{\Pi}[M \not\equiv N]$ = **if** some implicit rule applies to $M \not\equiv N$
then $\{\varepsilon\}$
else $1.Red_{\Pi}[M] \cup 2.Red_{\Pi}[N]$

Each subexpression $E[u]$, $u \in Red_{\Pi}[E]$, is called a (lazy) redex. For any redex u there is some rule (perhaps an implicit one) that applies to $E[u]$.

Definition 3 (One-step reduction, reduction)

If $u \in Red_{\Pi}[E]$ and the rule ⟨Head⟩ := ⟨Body⟩ applies to $E[u]$ via σ, we write

$$E \vdash_{\Pi,\sigma} (E[u \leftarrow \langle Body \rangle])\,\sigma$$

and say that E reducesto $(E[u \leftarrow \langle Body \rangle])\,\sigma$ via by one step of lazy narrowing. More generally, we write

$$E \vdash^{*}_{\Pi,\sigma} E'$$

and say that E reduces to E' via by lazy narrowing iff there is some reduction sequence

$$E_0 = E \vdash_{\Pi,\sigma_1} E_1 \;\cdots\; E_{k-1} \vdash_{\Pi,\sigma_k} E_k = E'$$

and $\sigma = \sigma_1 \ldots \sigma_k$. We allow the limit case k=0, E' = E, $\sigma = \{\,\}$(the empty substitution).
For any reduction $E \vdash^{*}_{\sigma} t$ yielding a term, t is called the result and σ is called the answer. The reduction relation is nondeterministic, but the uniqueness of results is ensured if E includes no variables (see Corollary 1 in section 5).
We remark that our lazy narrowing reduction can be used to solve extended unification problems, by reducing expressions of the form $M \equiv N$. The resulting behaviour is similar to the "unification with lazy surderivation" in [9].
Now let us go back to the examples in section 3 to get a feeling of BABEL's computations and to see how BABEL's reduction includes both rewriting and SLD-resolution as particular cases.

Example 1: Appending lists.

The appending program admits multiple use, as in PROLOG. A SLD-resolution computation splitting the list [a,b,c] into two factors can be simulated through BABEL reduction as follows:

$$\text{append(U,V,[a,b,c])} \vdash^{A2}_{U:=[X1|R]} (X1 \equiv a\,;\ \text{append(R1,V,[b,c])}) \rightarrow t \vdash_{X1:=a}$$

$$(t;\ \text{append(R1,V,[b,c])}) \rightarrow t \vdash \text{append(R1,V,[b,c])} \rightarrow t \vdash^{A1}_{R1:=[\,]}$$

$$(V \equiv [b,c] \rightarrow t) \rightarrow t \vdash^{*}_{V:=[b,c]} (t \rightarrow t) \rightarrow t \vdash^{*} t$$

/* Answer: U = [a] , V = [b,c] */

Example 2: Testing binary trees for equality of frontiers.

When used to evaluate ground (i.e. variable free) expressions, the program behaves in a purely functional way, e.g.

equal-frontier(tree(tree(leaf(1),leaf(2)),leaf(3)),
tree(leaf(1),tree(leaf(3),leaf(2)))) $\vdash^{*}$ f

Under an appropriate redex selection strategy, BABEL's lazy reduction would achieve this result without evaluating the whole frontiers.
The same program is able to solve expressions with free variables;e.g.

$$\text{equal-frontier(tree(leaf(X),A),tree(B,leaf(Y)))} \vdash^{*}_{\substack{A1:=\text{leaf(Y)} \\ B:=\text{leaf(X)}}} t$$

Example 3: The Alpine Club puzzle.

A BABEL computation of the solution {X := mike} to the Alpine Club puzzle is as follows:

alpinist(X), climber(X), ¬ skier(X) $\vdash_{X:=\,mike}^{AC2}$ t, climber(mike), ¬ skier(mike) $\vdash_{LK2}^{*}$

climber(mike), ¬ (¬ likes(mike,snow) → f) $\vdash_{LK3}$

climber(mike), ¬ (¬ (likes(tony,snow) → f) → f) $\vdash_{LK6}$ climber(mike), ¬(¬(t → f) → f)

$\vdash^{*}$ climber(mike), t /* implicit rules */ $\vdash$ climber(mike) $\vdash_{SC}$

(alpinist(mike), ¬ skier(mike)) → t $\vdash_{AC2}$ (t, ¬ skier(mike)) → t

$\vdash^{*}$ (t,t) → t /* same reduction sequence as before */ $\vdash^{*}$ t /*implicit rules */

As in example 1, this mimics a PROLOG-like computation; but some (restricted) use of logical negation is possible.

Example 4: Computing Hamming numbers.

The functional behaviour of this program when computing "hamming-seq" under lazy evaluation is well known; the first n Hamming numbers can be computed in O(n) steps (cfr.[19]) The novelty in BABEL consists in the freedom to use logical variables. Some possible evaluations are:

5th Hamming number?	nth-hamming(5,M) $\vdash^{*}$	t	{M = 6}
Is 10 a Hamming number?	nth-hamming(N,10) $\vdash^{*}$	t	{N = 8}
Find Hamming numbers:	nth-hamming(N,M) $\vdash^{*}$	t	{N = 6, M = 8}

5. BABEL'S DECLARATIVE SEMANTICS

In this section, we define BABEL's declarative semantics, which is based on Scott's domains, and state some mathematical results whose proofs will be supplied in the full version of the paper.

We fix an arbitrary BABEL program with constructors CS, function symbols FS and predicate symbols PS for the rest of the section.

Definition 1 (Domains, Herbrand domains)

(a) Domains(in Scott's sense) are consistently complete algebraic CPOs; cfr.[20] . We need "consistence-completeness" to get a continuous interpretation of our "strong equality".

(b) B is the flat domain of the boolean values (t , f , ⊥).

(c) H is the Herbrand domain generated by CS. More precisely: The CPO completion of the set of finite terms constructed from CS $\cup \{\bot\}$, where ⊥ is a new nullary constructor, partially ordered by:

$s \sqsubseteq t$ iff t is the result of replacing some occurrences of ⊥ in s by other terms.

The elements in H can be identified with trees whose nodes are labelled by constructors (and ⊥). They serve as data structures for BABEL.

Definition 2 (Interpretations, Herbrand interpretations)

(a) An interpretation for Π is any algebra

$$I = \langle D, (c_I)_{c \in CS}, (f_I)_{f \in FS}, (p_I)_{p \in PS} \rangle$$

where D is a domain and $c_I: D^n \to D$, $f_I: D^n \to D$, $p_I: D^n \to D$ are monotonic and continuous mappings of the appropiate arities.

(b) I is called a Herbrand interpretation iff D = H and each c_I acts as a free constructor of trees (with "c" at the root).

Definition 3 (Models)

Let I be an interpretation for Π. I is a model of Π ($I \models \Pi$) iff I is a model of every rule in Π.

I is a model of a rule <Head> := <Body> ($I \models$ <Head>:= <Body>) iff for every environment ρ : VS $\longrightarrow$ D we have

$$[\![\langle Head\rangle]\!]_I(\rho) \sqsupseteq [\![\langle Body\rangle]\!]_I(\rho)$$

where "$\sqsupseteq$" is "$\sqsupseteq_D$" or "$\sqsupseteq_B$", as convenient.

Finally, the values $[\![E]\!]_I(\rho)$ of E in I under ρ is defined by structural recursion in the usual way; e.g.

$$[\![(B \rightarrow E)]\!]_I(\rho) = [\![B]\!]_I(\rho) \circledRightarrow [\![E]\!]_I(\rho)$$

$$[\![M \equiv N]\!]_I(\rho) = [\![M]\!]_I(\rho) \circledequiv [\![N]\!]_I(\rho) \text{ etc.}$$

We still have to define the semantic functions $\circledRightarrow$, $\circledequiv$, etc. For instance:

$$b \circledRightarrow d = \begin{cases} d \text{ if } b = t \\ \bot \text{ otherwise} \end{cases} \qquad d_1 \circledequiv d_2 = \begin{cases} t \text{ if } d_1 = d_2 \text{ is total and finite} \\ f \text{ if } \{d_1, d_2\} \text{ has no upper bound in D} \\ \bot \text{ otherwise} \end{cases}$$

It can be checked that all the semantic functions are continuous. They are even computable over Herbrand interpretations. Notice that $\circledequiv$ is only an approximation of the identity. It behaves as the identity for finite and total objects.

By adapting well known techniques (cfr. Lloyd [15]) it is possible to give a precise statement and proof of the following:

Theorem 1 (Minimal Herbrand model)

The collection $\mathfrak{I}_\Pi$ of all Herbrand interpretations for Π, partially ordered by

$$I \sqsubseteq J \text{ iff } f_I \sqsubseteq f_J \text{ for every } f \in FS \text{ and } p_I \sqsubseteq p_J \text{ for every } p \in PS$$

is a domain which admits an effective presentation (cfr. [21]).

The operator $T_\Pi : \mathfrak{I}_\Pi \longrightarrow \mathfrak{I}_\Pi$ mapping each I to the J given by

$$f_J(\bar{t}) := \bigsqcup \{ [\![\langle Body\rangle]\!]_I(\rho) \mid \text{There is some rule } f(\bar{s}) := \langle Body\rangle \text{ with } [\![\bar{s}]\!]_I(\rho) = \bar{t}$$

(and analogously for predicates) is well defined, continuous and computable, and has as its least fixpoint the least Herbrand model of Π w.r.t. $\mathfrak{I}_\Pi$'s ordering: I_Π.

We have also the following result:

Theorem 2 (Soundness of the reduction semantics)

Any computed answer holds in every model of Π, in the following sense:

$$E \vdash^{*}_{\Pi,\sigma} E' \implies \Pi \models (E\sigma \sqsupseteq |E'|)^{\forall}$$

where

(a) $|E'|$ has the recursive definition

$$|c| = c, \quad |c(M_1,\ldots,M_n)| = c(|M_1|,\ldots,|M_n|)$$

$$|t| = t, \quad |f| = f, \quad |E'| = \bot \text{ in any other case}$$

(b) "$\Pi \models (E\sigma \sqsupseteq |E'|)^{\forall}$" means that $[\![E\sigma]\!]_I(\rho) \sqsupseteq |E'|_I(\rho)$ holds for any model I of Π and any environment ρ.

Corollary 1 (Uniqueness of results)

Given two reductions $E \vdash^{*}_{\Pi} E'$ and $E \vdash^{*}_{\Pi} E''$ of any ground expression E, $|E'|$ and $|E''|$ must have the same constructor at any common occurrence.

Proof: $|E'|$ and $|E''|$ are both bounded by $[\![E]\!]_{I_\Pi}$

We finish with a definition which was announced in section 1:

Definition 4 (Propositional satisfiability)

A finite set $\mathcal{B}$ of BABEL boolean expressions is propositionally satisfiable iff there is some propositional interpretation I: At($\mathcal{B}$) $\longrightarrow$ B such that $[\![B]\!]_I = t$ for every B $\in \mathcal{B}$. Here, At($\mathcal{B}$) is the set of atomic subexpressions of expressions from $\mathcal{B}$, and $[\![B]\!]_I$ denotes the boolean value of B under I, according to the semantic functions of the connectives.

6. CONCLUSIONS AND FUTURE WORK

We believe to have provided a framework which allows to amalgamate functional and logic programming in a remarkably simple and flexible way. The functional basis, combined with the constructor discipline, seem to yield a reasonable form of lazy narrowing which can be seen as a generalization of SLD resolution. We feel that this is more confortable than using narrowing to add functions and equations to Horn Clause programs through semantic unification.

There is a still very primitive prototype of BABEL written in PROLOG by J.J. Moreno. It has been used to test some of the examples discussed in this paper. Unfortunately, it runs out of memory for most interesting examples involving infinite objects.

We conjecture that a completeness result reciprocal to Theorem 5.2 is true, and hope to complete a proof soon. We have found examples showing that completeness would fail if redex occurrences were chosen outermost; the condition about pending rules in Definition 4.2 has been designed to avoid this. Of course, infinite objects do preclude completeness of any innermost reduction strategy.

Some work on a version of BABEL with a polymorphic type discipline (cfr. Milner [22]) and higher order objects has already begun, and we plan to continue it in the near future.

REFERENCES

[1] De Groot D. and Lindstrom G. (eds.): Logic Programming: Functions, Relations and Equations, Prentice Hall, 1986.

[2] Bellia M. and Levi G. : The Relation between Logic and Functional Languages: A survey, J. Logic Programming 1986, 3 , pp. 217 - 236.

[3] Huet, G. and Open, D.C.: Equations and Rewrite Rules: A Survey, in : "Formal Language Theory: Perspectives and Open Problems", Academic Press, 1980, pp. 349-405.

[4] Fay,M.: First-Order Unification in an Equational Theory, Procs. of the 4th Workshop on Automated Deduction, Austin, Texas (1979), pp. 161-167.

[5] Hullot,J.M.: Canonical Forms and Unification. Procs. of the 5th Conference on Automated Deduction, Springer Verlag, LNCS 87 (1980), pp. 318-334.

[6] Fribourg, L. : SLOG: A logic programming language interpreter based on clausal superposition and rewriting. Procs. Symp. on Logic Programming Boston (1985), pp. 172 - 184.

[7] Subrahmanyam, P.A. and You, J.: FUNLOG: A computational model integrating logic programming and functional programming. In [1] , pp. 157 - 198.

[8] Goguen, J.A. and Meseguer, J. : EQLOG: Equality, Types and Generic modules for Logic Programming, in [1], pp. 295-363.

[9] Dincbas, M. and van Hentenryck, P.: Extended Unification Algorithms for the Integration of Functional Programming into Logic Programming, J. Logic Programming 1987:4, pp. 199-227.

[10] Levi, G., Bosco, P.G., Giovannetti, E.,Moiso, C. and Palamidessi,C.: A complete semantic characterization of K-LEAF, a logic language with partial functions.Procs. 4th Symp. on Logic Programming, San Francisco (1987), pp. 1-27.

[11] Bosco P.G. , Giovanetti E. and Moiso C. : Refined Strategies for Semantic Unification , Proc. TAPSOFT'87, LNCS 250, pp. 276-290.

[12] van Emden, M.H. and Yukawa, K. : Logic Programming with Equations, J. Logic Programming 1987:4, pp. 265-288.

[13] O'Donnell, M. J. : Equational Logic as a Programming Language. MIT Press , 1985.

[14] Reddy,U.S.: Narrowing as the Operational Semantics of Functional languages, Proc. 1985 Symp. on Logic Programming, IEEE Comp. Soc. Press, 1895, pp. 138-151.

[15] Lloyd, J. W. : Foundations of Logic Programming, 2nd edition, Springer Verlag,1987.

[16] Robinson, J.A.: A Machine Oriented Logic Based on the Resolution Principle,J. ACM 12, 1965, pp. 23-41

[17] Malachi, Y. Manna, Z. and Waldinger, R. : TABLOG: A New Approach to Logic Programming, en [1] , pp. 365-394.

[18] Dijkstra, E.W.: A Discipline of Programming, Prentice Hall, 1976.

[19] Bird, R. and Wadler, P.: Introduction to Functional Programming. Prentice Hall, 1988, pp. 188-189.

[20] Scott, D.S.: Domains for Denotational Semantics, Proceedings ICALP'82, LNCS 140, Springer Verlag, 1982, pp. 577-613.

[21] Larsen, K.G. and Winskel G. : Using Information Systems to solve Recursive Domain Equations Effectively, LNCS 173, Springer Verlag 1984, pp. 109-129.

[22] Milner, R.: A Theory of Type Polymorphism in Programming, J. Comp. and Syst. Sci. 17, 1978, pp. 348-375.

Refined Compilation of Pattern-Matching for Functional Languages*

Ph. Schnoebelen†

Abstract

This paper formally describes and studies an algorithm for compiling functions defined through pattern-matching. This algorithm improves on previous proposals by accepting an additional parameter: the domain over which the compiled function will be applied. This additional parameter simplifies the definition of the algorithm and suggests new applications.

The practical interest of this algorithm for the implementation of functional languages is demonstrated by several applications and/or extensions: conditional rewriting, equations between constructors, ...

Introduction

Today, functional programming languages are under active study, from both a theoretical and a practical points of view. As experience grows up, several features become (almost) universally agreed upon. Algebraic data types and functions defined by pattern-matching are two such features.

In order to have efficient implementations, we must compile pattern-matching definitions into "direct" expressions. Writing:

$$\begin{array}{rcl} fact(0) & \rightarrow & 1 \\ fact(n+1) & \rightarrow & (n+1) * fact(n) \end{array} \qquad (1)$$

is simpler, both conceptually and semantically, than writing:

$$fact(n) \stackrel{\text{def}}{=} \textbf{if } n = 0 \textbf{ then } 1 \textbf{ else } n * fact(n-1) \qquad (2)$$

but, for a machine, it is also less explicit. And though, (1) and (2) are equivalent, and a reasonably smart implementation should compile (1) into (2). Such compilation algorithms have been discovered independently by several implementors, and some descriptions can be found in the literature. Indeed, it was when faced with the problem of implementing FP2 [Jor86] that we began to think about such an algorithm and ended up writing one. We describe this algorithm in a way which allows an easy analysis of its behaviour and which fits several functional languages (including ML [Mil84], Miranda [Tur85], LML [Aug84] and, to some extent, OBJ2 [FGJM85]).

The algorithms described in [Aug85,Pey87] and, very briefly, in [Car84] are very similar and do transform (1) into (2). The algorithm we propose performs the same transformation but it may be parameterized by the domain over which the function should be compiled. If that parameter is set to the domain implicitly assumed in previous algorithms, we obtain identical results. But there are many

*This work has been supported, in part, by ESPRIT Project 415 and by CNRS Project C^3.

†Author's address: LIFIA-INPG, 46 Av. Félix Viallet, 38000 Grenoble, FRANCE. E-mail: schnoebelen@lifia.imag.fr

practical situations where it is useful to restrict the domain and such situations are handled by our algorithm. For example, we shall see in section 6 that, without any special adaptation, the algorithm gives more efficient code when we allow so called *equations between constructors*.

The paper is organized as follows : section 1 presents the notations that will be used throughout the paper. Section 2 formally defines the framework within which we are working, i.e. it describes the relevant features of the functional languages we compile, together with an (abstract) target language. Section 3 presents the compilation algorithm. Then section 4 describes a monotonicity property of the algorithm which suggests extensions to the functional language, and, in sections 5 and 6, we describe two such extensions: "subtypes" and "equations between constructors", while "conditional rewriting", another extension, is considered in section 7. As a rule, complete proofs may always be found in [Sch88].

1 Formal background - Notations

We suppose that the notions of terms, substitutions, ... are well-known (see e.g. [HO80]). In the following, we consider a set S of functions names given with their arity. $T_S(X)$ denotes the set of finite terms having function names in S and variables in X. $T_S(\emptyset)$ is written T_S. We write $Vars(t)$ for the set of variables occurring in a term t. When $Vars(t) = \emptyset$ we say that t is *ground*. t is *linear* if no variable occurs more than once in it.

We prefer to use the name "position", rather than "occurrence", for the list of integers which denotes (the path to) a subterm of a given term. These lists may be concatenated with the "." operation. There exists a standard prefix-ordering on positions: $\ll$ is the smallest transitive relation such that $\forall p \in N^*, \forall i \in N, p \ll p.i$.

Given a term $t \in T_S(X)$, we write $P(t)$ for the set of its positions. $P(t)$ is a finite subset of N^*. We extend these notions to tuples of terms: $p \in P(t_i)$ iff $i.p \in P(<t_1, \ldots t_n>)$. Of course ϵ (the empty list) also belongs to $P(t)$. Given a position $p \in P(t)$, we write t/p for the *subterm* occurring as position p in t and $t[p]$ for the *symbol* occurring at position p in t: $t[p] \in S \cup X$ (for completeness, we say that if t is a tuple then $t[\epsilon]$ is a special tupling function symbol, written $< \ldots >$). We write $NVP(t)$ for the set of non-variable positions of t, that is $\{p \in P(t) \mid t[p] \in S\}$.

For example, if $t = f(g(x,a),h(x))$, we have $P(t) = \{\epsilon, 1, 1.1, 1.2, 2, 2.1\}$. $t[2]$ is the symbol h while $t/2$ is the term $h(x)$. t is not linear as $t[1.1] = t[2.1] = x$. We have $NVP(t) = \{\epsilon, 1, 1.2, 2\}$.

There is a standard notion of *substitutions* as S-morphisms from $T_S(X)$ into itself. We write $\Sigma = \{\sigma, \rho, \ldots\}$ for the set of all substitutions and Σ_g for the set of all *grounding* substitutions, that is substitutions from $T_S(X)$ into T_S (we did *not* require that $Dom(\sigma) = \{x \in X \mid \sigma x \neq x\}$ be finite). Given a term $t \in T_S(X)$, we write $G(t)$ for the set of all its ground instances, that is $\{\sigma t \mid \sigma \in \Sigma_g\}$.

A term t_1 is said to *match* a term t_0, written $t_0 \preceq t_1$, if there exists $\sigma \in \Sigma$ such that $t_1 = \sigma t_0$. $G(t)$ is $\{t_1 \in T_S \mid t \preceq t_1\}$. If $t_0 \preceq t_1$ then for all $p \in NVP(t_0), t_1[p] = t_0[p]$. This suggests that we introduce the notion of *matching positions of a (ground) term t w.r.t.* t_0, written $MP_{t_0}(t)$:

Definition 1 *For all $t_0 \in T_S(X)$, for all $t \in T_S$, $MP_{t_0}(t)$ is inductively defined by:*

- $\epsilon \in MP_{t_0}(t)$ *iff* $t[\epsilon] = t_0[\epsilon]$
- $p.i \in MP_{t_0}(t)$ *(with $i \in N$) iff* $p \in MP_{t_0}(t)$ *and* $t[p.i] = t_0[p.i]$

MP_{t_0} also applies to *sets*: for $T \subseteq T_S$, $MP_{t_0}(T)$ is $\bigcap_{t \in T} MP_{t_0}(t)$. In what follows, we rather use

$NMP_{t_0}(T)$, the "non-matching positions of T w.r.t. t_0", defined as $NVP(t_0) - MP_{t_0}(T)$. $NMP_{t_0}(T)$ is $\bigcup_{t \in T} NMP_{t_0}(t)$. Now we may say that if a ground term t_1 matches t_0 then $MP_{t_0}(t_1) = NVP(t_0)$. Conversely, *if t_0 is linear*, $MP_{t_0}(t_1) = NVP(t_0)$ implies that $t_1 \succeq t_0$, so that, for linear t_0:

$$MP_{t_0}(t_1) = NVP(t_0) \Leftrightarrow t_1 \succeq t_0$$

2 Language framework

2.1 Source language

We consider a finite set $C = \{c, \ldots\}$ of function names (with known arity) and single them out as *constructors*. The domain of computation is the term algebra T_C (and from now on, Σ_g will denote the set of ground substitutions *having values in* T_C and $G(t)$ only contains the ground instances of t *that are in* T_C). For example if $C = \{O, Succ\}$ (with arity 0 and 1), T_C is the set $\{O, Succ(O), Succ(Succ(O)), \ldots\}$, usually called Nat. Languages such as ML and FP2 do have such a notion of type. Others (e.g. LML and Miranda) admit "infinite terms", which require lazy evaluation. For the sake of simplicity, our treatment does not consider heterogeneous (i.e. many-sorted) algebras.

Now that we have a domain T_C, we may define functions on it. We complete our signature by a set $F = \{f, g, \ldots\}$ of function names (with known arity) and write $S = \{s, \ldots\}$ for $C \cup F$. It is these functions that are defined by rewrite rules and will be compiled.

Definition 2 *A* n-ary rewrite rule *is a pair* $l \rightarrow r$ *where:*

- l *is a* linear *tuple*[1] *of* $T_C(X)^n$,
- $r \in T_S(X)$,
- $Vars(r) \subseteq Vars(l)$.

A n-ary function $f \in F$ is defined by a list of n-ary rewrite rules and we write $Rules(f) = (l_i \rightarrow r_i)_{i=1\ldots m}$. As a running example, we use the function $fibo$, defined on $T_C = Nat$, by the rules:

$$\begin{aligned} fibo(O) &\rightarrow 0 \\ fibo(Succ(O)) &\rightarrow 1 \\ fibo(Succ(Succ(n))) &\rightarrow fibo(n) + fibo(Succ(n)) \end{aligned} \tag{3}$$

Evaluating a term of T_S is rewriting it into a term of T_C. Figure 1 defines a function $Rewr$ from T_S to T_C. This gives an operational semantics to the language. $Rewr[t]$ may be error if we meet some term $f(t_1, \ldots t_n)$ where no rule of f applies to $< t_1, \ldots t_n >$. $Rewr[t]$ may also be undefined as the rewriting process may well never terminate. An easy consequence of this definition is:

$$t \in T_C \Leftrightarrow Rewr[t] = t \tag{4}$$

Remark 1 When several rules may apply, we always choose the first applicable one. Again, this is common practice in functional languages. One can always retain the equational semantics by requiring that the rules satisfy a confluence property.

Note also that an error (or exception) is raised when we try to apply a function f on some arguments $< t_1, \ldots t_n >$ which are not covered by any rewrite rule for f. There lies the distinction between constructors

[1]For technical reasons, it will be easier to write a rule $f(t_1, \ldots t_n) \rightarrow r$ as a rule $< t_1, \ldots t_n > \rightarrow r$.

For all $c \in C$, $f \in F$, $s \in S$ and $\sigma \in \Sigma_g$:

$Rewr : T_S \rightarrow T_C \cup \{\textbf{error}\}$

(S1) $$\frac{Rewr[t_i] = \textbf{error}}{Rewr[s(t_1, \ldots t_n)] = \textbf{error}}$$

(S2) $$\frac{\forall i = 1 \ldots n,\ Rewr[t_i] \neq \textbf{error}}{Rewr[c(t_1, \ldots t_n)] = c(Rewr[t_1], \ldots Rewr[t_n])}$$

(S3) $$\frac{\forall i = 1 \ldots n,\ Rewr[t_i] \neq \textbf{error}}{Rewr[f(t_1, \ldots t_n)] = Apply[Rules(f)] < Rewr[t_1], \ldots Rewr[t_n] >}$$

$Apply[R] : T_C^n \rightarrow T_C \cup \{\textbf{error}\}$

(S4) $$Apply[\emptyset] < t_1, \ldots t_n >= \textbf{error}$$

(S5) $$\frac{l_1 \not\leq < t_1, \ldots t_n >}{Apply[(l_i \rightarrow r_i)_{i=1 \ldots m}] < t_1, \ldots t_n >= Apply[(l_i \rightarrow r_i)_{i=2 \ldots m}] < t_1, \ldots t_n >}$$

(S6) $$\frac{\sigma l_1 = < t_1, \ldots t_n >}{Apply[(l_i \rightarrow r_i)_{i=1 \ldots m}] < t_1, \ldots t_n >= Rewr[\sigma r_1]}$$

Figure 1: Semantics of source language

and non-constructors. It would be possible to return $f(t_1, \ldots t_n)$ by considering f as a constructor and by allowing constructors to have rewrite rules: see section 6.

Finally, note that this definition exhibits *call by value*, it implies a "strict" semantics, i.e. the result of applying a function on an undefined argument is always undefined. Currently, lazy semantics are preferred, but it is a difficult semantic question to define what is meant by "pattern-matching" in lazy languages (see e.g. [Pey87,Lav87]). In practice it turns out that the actual implementations of lazy functional languages do not really treat these problems in the purest way and use "crude" pattern-matching, so that the algorithm we describe could very well be used in such implementations. **End of Remark**

2.2 Target language

Now we describe formally what we have chosen as a target language: $\mathcal{TL} = \{E, \ldots\}$. The following grammar define syntactic domains E, B and Te (Expressions, Boolean expressions and Terms) where x is any variable, c any constructor, s any n-ary function and p any position:

$$\begin{array}{rcl} E & ::= & Te \mid \textbf{if}\ B\ \textbf{then}\ E_1\ \textbf{else}\ E_2 \mid \textbf{no\_match} \\ B & ::= & \textbf{is.}c\textbf{?}(Te) \\ Te & ::= & x \mid \textbf{get.}p(Te) \mid s(Te_1, \ldots Te_n) \end{array}$$

The semantics of such expressions is given by the rules of Figure 2. $Eval_\rho$ evaluates an expression E

For all $x \in X$, $s \in S$, $p \in N^*$, $Te, B, E_1, E_2 \in \mathcal{TL}$ and $\rho \in \Sigma_g$:

$Eval_\rho : \mathcal{TL}_E \rightarrow T_S \cup \{\,\mathbf{error}\,\}$

$$\text{(T1)} \qquad Eval_\rho\ \mathbf{no\_match} = \mathbf{error}$$

$$\text{(T2)} \qquad Eval_\rho\, x = \rho(x)$$

$$\text{(T3)} \qquad Eval_\rho\, \mathrm{get}.p(Te) = (Eval_\rho\, Te)/p$$

$$\text{(T4)} \qquad \frac{Eval_\rho\, Te_i = \mathbf{error}}{Eval_\rho\, s(Te_1, \ldots Te_n) = \mathbf{error}}$$

$$\text{(T5)} \qquad \frac{\forall i = 1 \ldots n,\ Eval_\rho\, Te_i \neq \mathbf{error}}{Eval_\rho\, s(Te_1, \ldots Te_n) = Rewr[s(Eval_\rho\, Te_1, \ldots Eval_\rho\, Te_n)]}$$

$$\text{(T6)} \qquad \frac{Eval_\rho\, B = \mathbf{true}}{Eval_\rho\ \mathbf{if}\ B\ \mathbf{then}\ E_1\ \mathbf{else}\ E_2 = Eval_\rho\, E_1}$$

$$\text{(T7)} \qquad \frac{Eval_\rho\, B = \mathbf{false}}{Eval_\rho\ \mathbf{if}\ B\ \mathbf{then}\ E_1\ \mathbf{else}\ E_2 = Eval_\rho\, E_2}$$

$Eval_\rho : \mathcal{TL}_B \rightarrow \{\,\mathbf{true}\,,\,\mathbf{false}\,\}$

$$\text{(T8,9)} \qquad \frac{(Eval_\rho\, Te)[\epsilon] = c}{Eval_\rho\, \mathrm{is}.c?(Te) = \mathbf{true}} \qquad \frac{(Eval_\rho\, Te)[\epsilon] \neq c}{Eval_\rho\, \mathrm{is}.c?(Te) = \mathbf{false}}$$

Figure 2: Semantics of target language

in an environment ρ where ρ is a grounding substitution that assigns values to variables of X.

Remark 2 Of course, it is implicitly assumed that rules (T3), (T8) and (T9) return **error** if $Eval_\rho\, t$ does, but this cannot happen in the code we generate so we won't need the rules in the proofs. The evaluation strategy for the get.p and is.c? functions is call by value. The **if_then_else** construct has the usual non-strict semantics. **End of Remark**

The is.c? functions are usually referred to as *testor* functions, while the get.p functions correspond to the *selector* functions, in the abstract data type terminology. These concepts are now classical in programming methodology, even if [Wad87] points out that one advantage of pattern-matching definitions is to get rid of them. To improve readability, we usually write e.g. "**if** $x = 0$ **then** $CDR(l)$" rather than "**if** is.$O?(x)$ **then** get.$2(l)$" (when l is a list and x an integer).

Continuing our *fibo* example, we would like to compile (3) into:

$$\begin{aligned} fibo(x) \stackrel{\text{def}}{=}\ & \mathbf{if}\ \mathrm{is}.O?(x)\ \mathbf{then}\ 0 \\ & \mathbf{elseif}\ \mathrm{is}.O?(\mathrm{get}.1(x))\ \mathbf{then}\ 1 \\ & \mathbf{else}\ fibo(\mathrm{get}.1.1(x)) + fibo(\mathrm{get}.1(x)) \end{aligned}$$

which is our notation for the more readable:

$$fibo(x) \stackrel{\text{def}}{=} \textbf{if } x = 0 \textbf{ then } 0 \\ \textbf{elseif } x - 1 = 0 \textbf{ then } 1 \\ \textbf{else } fibo(x-2) + fibo(x-1) \tag{5}$$

where the test "**if** $x - 1 = 0$" may be replaced by "**if** $x = 1$" [Sch88].

The result is thus an **if_then_else** tree which leads to the right-hand side of the first applicable rule. We shall generate trees where all leaves are reachable which implies, as a corollary, that we will be able to check the *sufficient completeness* of the function definition[2]: for example, *fibo*'s definition is sufficiently complete over *Nat* because the expression we want to obtain has no **no_match** leaf.

We chose **if_then_else**'s for the target language instead of **case** instructions discriminating over the outer-most constructor of an expression (giving potentially better machine code, see [Aug85,Pey87]) because they greatly simplify the proofs of section 3 and the developments of section 4. Now, the algorithm can easily be rewritten so that it generates **case**'s.

3 The compilation algorithm and its correctness

In the description of the algorithm, we shall use a "choice" function:

Definition 3 *A* choice function *is a function that chooses a minimal (w.r.t. $\ll$) position into any (finite) subset of N^*.*

A choice function, choose, *is* regular *if $P' \subseteq P \wedge choose(P) = p \in P' \Rightarrow choose(P') = p$.*

Regular choice functions exist, e.g. taking the leftmost-outermost position, which corresponds to depth-first search, and which has been used in all our examples. A (better) heuristic choice function is given and formally justified in [Sch88].

Remark 3 Taking *any* choice function will give a correct algorithm. Nevertheless, in section 4 we shall require that the choice function be regular. **End of Remark**

We now describe a function "Compile_For" which compiles a n-ary function f from its list of defining rules and from a description of its possible arguments (i.e. its domain).

More precisely, Compile_For receives a list R of rewrite rules and a non-empty set T of n-tuples of ground terms of T_C^n. T may be infinite, and usually it is T_C^n. Compile_For returns an expression $E \in T\mathcal{L}$ (having $x_1, \ldots x_n$ as free variables) such that, for any $< t_1, \ldots t_n > \in T$, the result of applying f to $< t_1, \ldots t_n >$ is the result of evaluating E in a context where the x_i's are bound to the t_i's (see Theorem 2). Thus E is an explicit and effective definition for f and we can write $f(x_1, \ldots x_n) \stackrel{\text{def}}{=} E$, that is $f \stackrel{\text{def}}{=} \lambda x_1 \ldots x_n.E$. Note that E depends on T and that it yields a correct definition for f only if f is applied to arguments belonging to T.

Definition 4 Compile_For *is defined by the rules in Figure 3. Note that these rules apply in disjoint cases. Rules (C1') and (C2') define CompileRH, an auxiliary function used to build expressions corresponding to the right hand sides of the rules.*

Rule (C4) deserves some comments: in the general case where some tuples of T are matched by l_1 and some are not, we choose a minimal position p' in $NMP_{l_1}(T)$ (note that p' must have the form $k.p$). Such minimal positions exist because, as l_1 is linear, $NMP_{l_1}(T) = \emptyset$ would imply that all the tuples in T are matched by l_1. Let c be the constructor $l_1[p']$. We extract from T the tuples which have c at position p': this gives T'. Then $T' \neq \emptyset$, or else T would be disjoint from $G(l_1)$. Similarly, T' is strictly smaller than T, or else p' would be a matching position. Finally, E is simply defined as an **if_then_else** expression.

Note that the expression "$t[k.p] = c$" in rule (C4) (resp. "get.$p(x_k)$" in rule (C1')) is always applied to terms containing the position $k.p$ (resp. p).

[2] or more precisely its *convertibility*, i.e. that the rules cover all possible cases for the arguments. Note that this does not imply that the rewriting will terminate.

Compile R For : $2^{T_C^n} \rightarrow \mathcal{TL}$

(C1) Compile $\emptyset$ For T = **no_match**

(C2) *if* $T \subseteq G(l_1)$: Compile $(l_i \rightarrow r_i)_{i=1\ldots m}$ For $T = CompileRH_{l_1}\, r_1$

(C3) *if* $T \cap G(l_1) = \emptyset$:

Compile $(l_i \rightarrow r_i)_{i=1\ldots m}$ For T = Compile $(l_i \rightarrow r_i)_{i=2\ldots m}$ For T

(C4) *if* $T \not\subseteq G(l_1)$ *and* $T \cap G(l_1) \neq \emptyset$:

Compile $(l_i \rightarrow r_i)_{i=1\ldots m}$ For T = **if** is.c?(get.$p(x_k)$) **then** E_1 **else** E_2

$$\begin{aligned} \textit{where}\quad k.p &= choose(NMP_{l_1}(T)) \\ c &= l_1[k.p] \\ T' &= \{t \in T \mid t[k.p] = c\} \\ E_1 &= \text{Compile}\,(l_i \rightarrow r_i)_{i=1\ldots m}\ \text{For}\ T' \\ E_2 &= \text{Compile}\,(l_i \rightarrow r_i)_{i=2\ldots m}\ \text{For}\ T - T' \end{aligned}$$

$CompileRH_l : T_S(X) \rightarrow \mathcal{TL}$

(C1') *if* t *is some* $l/k.p$: $CompileRH_l\, t = \text{get}.p(x_k)$

(C2') *if* $s(t_1, \ldots t_n) \neq l/p$ *for all* $p \in P(l)$:

$$CompileRH_l\, s(t_1, \ldots t_n) = s(CompileRH_l\, t_1, \ldots CompileRH_l\, t_n)$$

Figure 3: Rules for Compile_For

Proposition 1 (Termination) *The algorithm for* Compile_For *always terminates.*

Proof. Rules (C1), (C2) and (C1') terminate at once. Rule (C2') clearly terminates and the only problem is with the recursive calls of Compile_For: given an invocation "Compile $(l_i \rightarrow r_i)_{i=1\ldots m}$ For T", let us write $n_i(T)$ for $\mid NMP_{l_i}(T) \mid$. It is easy to show that the quantity $m + \Sigma_{i \leq m} n_i(T)$ associated to any invocation of Compile_For is decreased by at least 1 in a recursive call. The only non trivial case is with the **then** arm in rule (C4): $T' \subseteq T$ implies $NMP_{l_i}(T') \subseteq NMP_{l_i}(T)$ and $n_i(T') \leq n_i(T)$ for all $i = 1 \ldots m$. Now, if $p' = choose(NMP_{l_1}(T))$, then $p' \notin NMP_{l_1}(T')$, which implies $n_1(T') < n_1(T)$ and then $\Sigma_{i \leq m} n_i(T') < \Sigma_{i \leq m} n_i(T)$. As m and $n_i(T)$ are finite and positive, the recursive calls must terminate. □

Theorem 2 (Correctness) *If* $t \in T \subseteq T_C^n$ *and* $\rho < x_1, \ldots x_n >= t$ *then:*

$$Apply[R]t = Eval_\rho\ \text{Compile}\ R\ \text{For}\ T$$

The proof is by induction over the recursive calls of Compile_For and *CompileRH*. It is given in [Sch88].

Remark 4 The compilation algorithm has to perform some operations (intersection, complementation, ...) on possibly infinite sets of terms. We do not have enough space to describe how these problems are handled in practice and simply refers to [Sch88] where we describe a solution using G-sets (a G-set is some $G(t)$, with $t \in T_C(X)$, or one of $G_1 \cup G_2$, $G_1 \cap G_2$, $G_1 - G_2$ where G_1 and G_2 are G-sets themselves) and relying on the fact that the emptiness of G-sets is decidable [Com88b]. **End of Remark**

4 A monotonicity property of the algorithm

This section studies how Compile R For T is modified when T is modified. Intuitively, when T gets smaller, Compile R For T should become smaller, i.e. be more efficient. We define a partial ordering $\sqsubseteq$ over $\mathcal{TL}_E$ with the axioms in Figure 4 (omitting reflexivity and transitivity). Clearly $\sqsubseteq$ is consistent with

For all $E, E_1, E_1', E_2, E_2', B \in \mathcal{TL}$:

$$\mathbf{no\_match} \sqsubseteq E$$

$$\frac{E_1 \sqsubseteq E_2}{E_1 \sqsubseteq \mathbf{if}\ B\ \mathbf{then}\ E_2\ \mathbf{else}\ E} \qquad \frac{E_1 \sqsubseteq E_2}{E_1 \sqsubseteq \mathbf{if}\ B\ \mathbf{then}\ E\ \mathbf{else}\ E_2}$$

$$\frac{E_1 \sqsubseteq E_2 \quad E_1' \sqsubseteq E_2'}{\mathbf{if}\ B\ \mathbf{then}\ E_1\ \mathbf{else}\ E_1' \sqsubseteq \mathbf{if}\ B\ \mathbf{then}\ E_2\ \mathbf{else}\ E_2'}$$

Figure 4: Axioms for $\sqsubseteq$

the intuitive notion of a given expression being simpler than another one. We have:

Theorem 3 (Monotonicity)

$$T_1 \subseteq T_2 \Rightarrow \text{Compile}\ R\ \text{For}\ T_1 \sqsubseteq \text{Compile}\ R\ \text{For}\ T_2$$

The proof assumes that Compile_For uses a *regular* choice function (see Remark 3). It is given in [Sch88]. The following two sections are direct applications of Theorem 3: we exhibit some situations that are important in the context of functional languages, and where it is possible to compile a function f over domains strictly smaller than the usual T_C^n, which results into smaller (but still correct) code.

5 Subtypes

A subtype is a subset of a type (the supertype). We will not dwell into language design considerations and just assume that when we are given a type τ, a function f_τ defined over τ, and a subtype τ' of τ. With Compile $Rules(f)$ For τ', we get a version $f_{\tau'}$ of f_τ, to be used when the arguments of f are known to belong to τ'. Over τ', f_τ and $f_{\tau'}$ give identical results but $f_{\tau'}$ is more efficient.

In this section we just give two examples with $Pos = \{Succ(O), Succ(Succ(O)), \ldots\}$, the set of strictly positive integers, and $Even = \{O, Succ(Succ(O)), \ldots\}$, the set of even integers. Both Pos and $Even$ are subsets of Nat, and we refer to [Sch88] for meaningful ways of defining and handling these sets. Over Pos and $Even$, $fibo$ compiles into:

$$fibo_{Pos}(x) \stackrel{\text{def}}{=} \mathbf{if}\ x = 1\ \mathbf{then}\ 1\ \mathbf{else}\ fibo_{Nat}(x-2) + fibo_{Nat}(x-1)$$

$$fibo_{Even}(x) \stackrel{\text{def}}{=} \mathbf{if}\ x = 0\ \mathbf{then}\ 0\ \mathbf{else}\ fibo_{Nat}(x-1) + fibo_{Nat}(x-2)$$

which are simpler than (5), $fibo$'s definition over Nat.

6 Equations between constructors

Some languages admit so-called *equations between constructors* (see e.g. [Tur85,Tho86] for Miranda). This greatly extends the scope of algebraic types by allowing the user to define types as unfree algebras. A standard example is the type of integer numbers, defined with constructors O, $Succ$ and $Pred$ together with the equations:

$$\begin{array}{rcl} Succ(Pred(n)) & \rightarrow & n \\ Pred(Succ(n)) & \rightarrow & n \end{array} \qquad (6)$$

In this section, we show how this fits into our framework. We first consider a general notion of congruence and later specialize it to equational congruences.

6.1 Ground normal forms

We assume that we are given a congruence, written $\equiv$, over T_C. Semantically, our domain of computation is $T_C/\equiv$ but, from an implementation point of view, it is better to use *representatives* of equivalence classes rather than classes themselves, and we assume that there exists a *normalization* function, $Norm$[3], that maps any term to its canonical representative, in such a way that:

$$t_1 \equiv t_2 \Leftrightarrow Norm[t_1] = Norm[t_2]$$

$T_C/\equiv$ is implemented as $\{Norm[t] \mid t \in T_C\}$ and we use GNF to denote this set of *ground normal forms*. The operational semantics of our rewriting language in such a framework is given by replacing rule (S2) of Figure 1 by a new (S2) rule given in Figure 5.

Remark 5 Note that rule (S6) is *not* changed, so that matching is not done "modulo $\equiv$". Furthermore $Norm$ needs not define a congruence. Again, this is standard practice. **End of Remark**

A main consequence of these definitions is that (4) does not hold any more: indeed, only terms t of GNF are such that $Rewr[t] = t$, so that finally $Rewr$ is a partial mapping from T_S into $GNF \cup \{\textbf{error}\}$ and $Apply[R]$ is from GNF^n into $GNF \cup \{\textbf{error}\}$.

For all $c \in C$:

$$\text{(S2)} \quad \frac{\forall i = 1 \ldots n,\ Rewr[t_i] \neq \textbf{error}}{Rewr[c(t_1, \ldots t_n)] = Norm[c(Rewr[t_1], \ldots Rewr[t_n])]}$$

Figure 5: Semantics of rewriting with $Norm$

The compilation algorithm needs not be changed: indeed, for functions $f \notin C$ we did not change the rules for *Apply* and the proof of Theorem 2 is still correct. Only, *it is safe to use the compilation algorithm over* GNF*, a subset of* T_C*, in order to get better code.* As an example, let us consider $C = \{O, Succ, Pred\}$ with $\equiv$ given by the rules in (6). We have:

$$GNF = \{O, Succ(O), Succ(Succ(O)), \ldots Pred(O), Pred(Pred(O)), \ldots\} \subseteq T_C$$

This set is usually called Int. Over Int, $fibo$ compiles into:

$$\begin{array}{l} fibo_{Int}(x) \stackrel{\text{def}}{=} \textbf{if } x = 0 \textbf{ then } x \\ \qquad \textbf{elseif } x > 0 \textbf{ then } \textbf{if } x - 1 = 0 \textbf{ then } 1 \\ \qquad\qquad\qquad\qquad \textbf{else } fibo_{Int}(x-2) + fibo_{Int}(x-1) \\ \qquad \textbf{else no\_match} \end{array}$$

which is more complicated that (5), $fibo$'s definition over Nat, because $Nat \subseteq Int$, but simpler than what we would have got with T_C (see [Sch88]).

6.2 Handling GNF

Unfortunately, it is not always possible to characterize GNF in such a way that we may compute intersections, ..., with it. Of course, there always remains the solution of using a larger set, up to T_C^n. In

[3] *Norm* can be given by a *ground canonical* rewrite system. These systems, introduced in [HR87], are able to handle e.g. commutative laws and seem to provide the right framework for our problems.

this section, we describe a special case where it is possible to effectively handle GNF: when $\equiv$ is given by a rewrite system over constructors[4].

In this situation, Figure 5 may be specialized into Figure 6. The new (S2) rule refers to some $Apply_c$

For all $c \in C$:

(S2) $$\frac{\forall i = 1 \ldots n,\ Rewr[t_i] \neq \mathbf{error}}{Rewr[c(t_1, \ldots t_n)] = Apply_c[Rules(c)] < Rewr[t_1], \ldots Rewr[t_n] >}$$

(S4') $$Apply_c[\emptyset] < t_1, \ldots t_n >= c(t_1, \ldots t_n)$$

Figure 6: Semantics of rewriting with constructors

function which behaves like the $Apply$ function of Figure 1 except when no rules apply, in which case no error is generated: simply the term is not rewritten any more. This is captured by the new rule (S4') (we omit the obvious rules (S5') and (S6') which complete the definition of $Apply_c$).

When $\equiv$ is given by rewrite rules, the problem of characterizing GNF has been given a complete solution in [Com88a,Com88b]: in the simpler case where the left-hand sides of the rules between constructors are linear, GNF may be described by a tree grammar, easily computable from the rules. The reader should really read [Com88a,Com88b] but he may get an intuition of the result by considering the grammar we obtain for Int's GNF set (where GNF_c denotes the set of all terms of GNF having c at their root):

$$\begin{array}{lcl} GNF & ::= & GNF_O \mid GNF_S \mid GNF_P \\ GNF_O & ::= & O \\ GNF_S & ::= & Succ(GNF_O) \mid Succ(GNF_S) \\ GNF_P & ::= & Pred(GNF_O) \mid Pred(GNF_P) \end{array}$$

7 Conditional rewriting

One can easily adapt our algorithm for conditional rewriting[5].

Definition 5 *A* conditional rewrite rule *is a triple $l : p \rightarrow r$ where:*

- *$l \rightarrow r$ is a rewrite rule,*
- *$p \in T_S(X)$ has sort "Boolean",*
- *$Vars(p) \subseteq Vars(l)$.*

For all $c \in C$, $f \in F$, $s \in S$ and $\sigma \in \Sigma_g$:

(S6) $$\frac{\sigma l_1 =< t_1, \ldots t_n > \quad Rewr[\sigma p_1] = \mathbf{true}}{Apply[(l_i : p_i \rightarrow r_i)_{i=1 \ldots m}] < t_1, \ldots t_n >= Rewr[\sigma r_1]}$$

(S7) $$\frac{\sigma l_1 =< t_1, \ldots t_n > \quad Rewr[\sigma p_1] = \mathbf{false}}{Apply[(l_i : p_i \rightarrow r_i)_{i=1 \ldots m}] < t_1, \ldots t_n >= Apply[(l_i : p_i \rightarrow r_i)_{i=2 \ldots m}] < t_1, \ldots t_n >}$$

(S8) $$\frac{\sigma l_1 =< t_1, \ldots t_n > \quad Rewr[\sigma p_1] = \mathbf{error}}{Apply[(l_i : p_i \rightarrow r_i)_{i=1 \ldots m}] < t_1, \ldots t_n >= \mathbf{error}}$$

Figure 7: Semantics of conditional rewriting

[4]Another advantage of using rewrite rules to define $\equiv$ is that the problem mentioned in Remark 5 may be avoided by requiring a confluence property between the rules defining the functions and the rules defining $\equiv$, in which case everything becomes transparent.

[5]Or rewrite rules with guards. This is a common feature of functional languages: see e.g. Miranda.

p is called the *condition*. When such rules are considered, the semantics of the source language is obtained by replacing rule (S6) of Figure 1 by the three rules given in Figure 7. This demonstrates how "non-conditional" rewriting is a special case of conditional rewriting: when p is always "**true**", rules (S7) and (S8) may be discarded and rule (S6) needs not evaluate $Rewr[\sigma p]$.

Similarly, the compilation algorithm just has to be slightly modified to accept conditional rules: we replace rule (C2) in Figure 3 by the rule given in Figure 8. In this framework Proposition 1 and Theorems 2 and 3 remain true, what we shall not prove here.

(C2) if $T \subseteq G(l_1)$:

$$\text{Compile}\,(l_i : p_i \rightarrow r_i)_{i=1\ldots m}\,\text{For}\,T = \textbf{if}\ B\ \textbf{then}\ E_1\ \textbf{else}\ E_2$$

$$\begin{aligned} \text{where}\quad B &= \mathit{CompileRH}_{l_1}\, p_1 \\ E_1 &= \mathit{CompileRH}_{l_1}\, r_1 \\ E_2 &= \text{Compile}\,(l_i : p_i \rightarrow r_i)_{i=2\ldots m}\,\text{For}\,T \end{aligned}$$

Figure 8: Compile_For with conditional rules

References

[Aug84] L. Augustsson. A compiler for Lazy ML. In *Proc. ACM Conf. Lisp and Functional Programming, Austin, Texas*, August 1984.

[Aug85] L. Augustsson. Compiling pattern matching. In *Functional Programming Languages and Computer Architecture, Nancy, LNCS 201*, Springer-Verlag, September 1985.

[Car84] L. Cardelli. Compiling a functional language. In *Proc. ACM Conf. Lisp and Functional Programming, Austin, Texas*, August 1984.

[Com88a] H. Comon. An effective method for handling initial algebras. In *Proc. 1st Workshop on Algebraic and Logic Programming, Gaussig*, 1988.

[Com88b] H. Comon. *Unification et Disunification: Théorie et Applications*. Thèse de Doctorat, I.N.P. de Grenoble, France, 1988.

[FGJM85] Futatsugi, Goguen, Jouannaud, and Meseguer. Principles of OBJ2. In *Proc. ACM Symp. Principles of Programming Languages*, 1985.

[HO80] G. Huet and D. Oppen. Equations and rewrite rules: a survey. In R. Book, editor, *Formal Language Theory: Perspectives and Open Problems*, pages 349–405, Academic Press, 1980.

[HR87] H. Hsiang and M. Rusinowitch. On word problems in equational theories. In *Proc. 14th ICALP, LNCS 267*, Springer-Verlag, 1987.

[Jor86] Ph. Jorrand. Term rewriting as a basis for the design of a functional and parallel programming language. A case study: the language FP2. In *Fundamentals of Artificial Intelligence, LNCS 232*, Springer-Verlag, 1986.

[Lav87] A. Laville. Lazy pattern matching in the ML language. In *Proc. 7th Conf. Found. of Software Technology and Theoretical Computer Science, Pune, INDIA, LNCS 287*, December 1987.

[Mil84] R. Milner. A proposal for Standard ML. In *Proc. ACM Conf. Lisp and Functional Programming, Austin, Texas*, August 1984.

[Pey87] S. Peyton-Jones. *The Implementation of Functional Programming Languages*. Prentice Hall Int., 1987.

[Sch88] Ph. Schnoebelen. *Refined Compilation of Pattern-Matching for Functional Languages*. Research Report Lifia 71 Imag 715, Univ. Grenoble, April 1988.

[Tho86] S. Thompson. Laws in Miranda. In *Proc. ACM Conf. Lisp and Functional Programming, Cambridge, Mass.*, August 1986.

[Tur85] D. A. Turner. Miranda: a non-strict functional language with polymorphic types. In *Functional Programming Languages and Computer Architecture, Nancy, LNCS 201*, Springer-Verlag, September 1985.

[Wad87] P. Wadler. A critique of Abelson and Sussman or Why calculating is better than scheming. *SIGPLAN Notices*, 22(3), March 1987.

NARROWING IN PARTIAL CONDITONAL EQUATIONAL THEORIES

Uwe Wolter

Sektion Mathematik, Technische Universität Magdeburg [1])

(Abstract)

We extend the "Conditional Narrowing Algorithm" of H. Hussmann (MIF-8502, Universität Passau 1985) to partial conditional equational theories. Without loss of generality it is assumed that the partial conditional equational theory IMP be canonical, i.e. for every conditional existence equation $(X : G \to t \overset{e}{=} t') \in IMP$ holds $Var(G)=X$, and that for every term $r \in Term(t,t') \setminus Term(G)$ there exists a conditional existence equation $(Y : H \to p \overset{e}{=} p') \in IMP$ and a substitution σ with $H\sigma = G$ and $p\sigma = r$.

Using canonical theories we can drop the explicit declaration of the sets X,Y of variables.
A set K of "ground objects" and a set F of "ground facts", i.e. a set F of existence equations on K, define a model $\underline{F}(IMP , F , K)$ of IMP freely generated by K and F. Given a set H of existence equations, the narrowing algorithm computes solutions of H in $\underline{F}(IMP , F , K)$.
We divide IMP into two parts:

$IMP_{equ} = \{(G \to r \overset{e}{=} r' \in IMP \, / \, r \neq r'\}$ and $IMP_{def} = \{(G \to t \overset{e}{=} t' \in IMP\}$.

In the partial case we have to replace the final unification step by an iterative resolution step using the conditional existence equations of IMP_{def}.

<u>Partial conditional Narrowing Algorithm (PCNA)</u>

<u>begin</u>: $H \to H$

<u>narrowing</u>:

$$\frac{R \to H\tau}{(G \, , \, R[u \leftarrow r'])\sigma \to H\tau\sigma}$$

where $(G \to r \overset{e}{=} r') \in IMP_{equ}$, $u \in O(R)$, R/u is not a variable, and σ is a most general ϕ-unifier of R/i and r

<u>partial unification</u>:

$$\frac{R \to H\tau}{(G \, , \, R \setminus (p \overset{e}{=} p'))\sigma \to H\tau\sigma}$$

where $(G \to t \overset{e}{=} t') \in IMP_{def}$, and σ is a most general ϕ-unifier of t,p, and p' for some $(p \overset{e}{=} p') \in R$.

<u>test of facts</u>:

$$\frac{R \to H\tau}{(R \setminus (p \overset{e}{=} p'))\sigma \to H\tau\sigma}$$

where $p \overset{e}{=} p' \in R$ and $p\sigma = r$, $p'\sigma = r'$ for some $(r \overset{e}{=} r') \in F$, or $p\sigma = p'\sigma = r$ for some $r \in K \cup Term(F)$.

If the PCNA yields " $\phi \to H\tau$ " we have found a single special solution τ of H, whereas " $Diag(Var(H\tau)) \to H\tau$ " describes a set of solutions.

---

[1]) On leave to: Organisations- und Rechenzentrum, Humboldt-Universität zu Berlin, PSF 1297, DDR Berlin 1086

SOLVING EQUATIONS IN AN EQUATIONAL LANGUAGE†

Jia-Huai You

Abstract

The problem of solving equations is a key and challenging problem in many formalisms of integrating logic and functional programming, while *narrowing* is now a widely used mechanism for generating solutions. In this paper, we continue to investigate the problem of solving equations in O'Donnell's equational language. We define a restricted equality theory which we argue adequately captures the notion of first-order functional programming and permits an efficient implementation. In particular, we show that there exists a special class of narrowing derivations which generates complete and minimal sets of solutions.

1. Introduction

In the last few years the notion of solving equations has been widely used in the context of logic programming with an equational flavor. See, for example, [DePl85, Frib85, GoMe84, Levi87, MaMR86, Redd85, RéKi85, YoSu86a]. The problem of solving equations can be characterized in terms of unification *modulo* an equational theory, which is often called E-unification [Siek84]. Many of proposed systems that combine functional, equational and logic programming adopt the classic equality theory, i.e., the one described by the axioms of reflexivity, transitivity, symmetry and substitutivity. In theorem proving, the handling of equality has long been suggested to use the E-unification method [Plot72]. The same approach has been carried over to logic programming with equality. As pointed out by Gallier and Raatz [GaRa86], most proposed systems embody a mechanism of E-unification, either implicitly or explicitly.

Equality, often referred to as the classic equality theory, is difficult. Although the theoretical formalism is elegant, there have been doubts about whether an efficient system, comparable to Prolog, can be actually built. **Looking back at the development of Prolog, one finds that the success of Prolog was largely due to two factors: the use of a restricted logic, i.e., definite Horn clause logic for which an efficient inference mechanism exists and was discovered (i.e., SLD-resolution), and the invention of efficient implementation** techniques by Warren and other researchers. The ultimate success of combining functional, equational and logic languages in various forms will depend on the same two factors.

It has been observed that the functions definable in many functional languages are expressible in terms of the classic equality theory [CaMc79, GoTa79, ODon85]. However, the fact that functions are a special case of equality is equally important, because the adoption of the classic equality theory has made it difficult to define a faithful and efficient operational semantics.

From the viewpoint of functional programming, the intended use of a first-order functional program is to manipulate data objects, which are composed of data constructors (including primitive data objects) and variables (which in turn denote whole classes of some data objects). If a function does not map a value to a data object, then the function is said to be undefined at the given value. A first-order term, even if it is in *normal form*, may not denote a meaningful data object; thus the language definition may disregard these terms as having any meaning. It is therefore semantically desirable to allow equality to be held only among meaningful terms. Consider, for example, the familiar functional definition of sum over natural numbers:

$$R = \{ +(0,x) \to x,\ +(s(x),y) \to s(+(x,y)) \}$$

If another rule, say $a \to b$, is also present in R, we then have, by the classic equality theory, $+(a,0) =_E +(b,0)$; even though both terms have no meaning in terms of functions and should only denote undefined. By the classic equality theory again, the substitution $\{x/b\}$ is a solution to the equation $+(a,0) = +(x,0)$, for it makes two sides equivalent even though neither of the terms $+(a,0)$ and $+(b,0)$ makes sense in terms of functions.

Notice that the above phenomenon, though closely related to the notion of *types* (or *sorts*), is essentially the problem of what should be considered as data objects. Consider the following system:

$$R = \{ g(a,x) \to f(b,x),\ f(a,x) \to g(c,x) \},$$

where neither f nor g is a mapping from terms to some meaningful data objects. Even if one uses many-sorted equational logic one is still forced to consider what should be the data objects that the functions manipulate.

---

† Department of Computing Science, University of Alberta, Edmonton, Alberta, Canada T6G 2H1. Work supported by the Natural Sciences and Engineering Research Council of Canada.

Based on the above arguments, we define in this paper an equality theory† in which data objects are *constructor terms*, i.e., the terms that are composed of constructors and variables. Two terms are equal if they are equivalent to a common constructor term. The problem of solving equations in this theory can be described in terms of unification modulo the theory: two terms A and B are unifiable modulo this equality theory if there exists a substitution σ, such that σA and σB are equivalent to a common constructor term.

The general problem of unification modulo this equality theory is still difficult. For example, similar to the case of E-unification, there exist theories for which complete and minimal sets of solutions may not always exist. We show, however, that complete and minimal sets of solutions do exist for theories expressible in O'Donnell's language [ODon85]. In addition, we show that complete and minimal sets of solutions can be effectively enumerated by a special class of narrowing derivations.

We are not the first to suggest such a framework of equational reasoning. Fribourg [Frib85] has considered Horn clause equational theories based on constructors, while Yukawa [Yuka87] and Levi *et al.* [Levi87] have advocated essentially the same framework. In all these frameworks, a complete semantics reformalization is needed in order to achieve the desired integration of functional, equational and logic programming. Our approach and objective are somehow different: we would like to build a system by the method of building-in theories without a semantics reformalization of the system for each additional feature (functional programming is only one of them [You86]). Theory Resolution [Stickel86] provides foundations for this approach in the general case, and the work by Jaffar *et al.* [JaLM86, JaLM87] addresses the preservation of semantic properties of traditional Horn clause logic programming in the framework of incorporating special theories. The restricted equality theory studied in this paper is simply another first order theory, which can be built into resolution by replacing unification modulo identity (i.e., conventional unification) with unification modulo the theory of functional programming. Technically, there is no need in our approach to reconstruct the domain of interpretation (such as the Herbrand Universe). All we consider is whether two terms are equivalent to a common constructor term; the domain of interpretation may or may not only contain constructor terms.

The paper is organized as follows. The next section provides basic notations used in this paper. The restricted equality theory will be discussed in more details in Section 3. It is shown in Section 4 that narrowing is complete, and complete and minimal sets of solutions always exist for equational programs expressible in O'Donnell's language. Section 5 describes a process of transformation of narrowing derivations, based upon which a special class of narrowing derivations is defined and shown to generate complete and minimal sets of solutions. Finally, some remarks and a discussion of related work conclude the paper.

2. Some Definitions

We assume the well-known concept of terms and algebra [HuOp80]. $T(F,V)$ denotes the set of terms generated from a set of function symbols F and an enumerable, disjoint set of variables V. Terms are denoted by capital letters A, B, C, etc., as well as by the lower case letters t, s. Variables are denoted by x, y, z, etc. Terms are viewed as *labeled trees* in the following way: a term A is a partial function from the set of sequences of positive integers, denoted by I^*, to $F \cup V$ such that its domain satisfies:

(i) $\varepsilon \in D(A)$

(ii) $u \in D(t_i)$ iff $i.u \in D(f(t_1, ..., t_i, ..., t_n))$ $1 \le i \le n$.

$D(A)$ is called a set of *occurrences* of A; $O(A)$ denotes the nonvariable subset of $D(A)$. The set of occurrences is partially ordered: $u \le v$ iff $\exists w$ $u.w = v$, and $u < v$ iff $u \le v$ & $u \ne v$. When $u < v$ we then say that u is *outer* to v and v is *inner* to u. u and w are said to be *independent*, denoted by $u <> w$ iff $u \not\ge w$ and $w \not\ge u$. The quotient u/v of two occurrences u and v is defined as: $u/v = w$ iff $u = v.w$.

$V(\Phi)$ denotes the set of variables occurring in an object Φ. $A[u \leftarrow B]$ is the term A in which the subterm at occurrence u has been replaced by the term B. The subterm of A at occurrence u is denoted by A/u. A *substitution* σ is a mapping from V to $T(F,V)$, extended to an endomorphism of $T(F,V)$. The composition of substitutions σ and θ is defined as a mapping: $(\sigma \bullet \theta)x = \sigma(\theta x)$. If σ is a substitution and A is a term, we write σA for the application of σ to A.

A *term rewriting system* is a set of directed equations $R = \{\alpha_i \to \beta_i \mid i \in N\}$ such that variables appearing in β_i must also appear in α_i. The definitions of *normal form, termination, confluence*, and *canonicality* are all standard [HuOp80]. A substitution is said to be *normalized* if each substitute wherein is in normal form. A rewrite rule $\alpha_i \to \beta_i$ is said to be *variable-dropping* if there is some variable appearing in α_i but not

† It might be a slight abuse of terminology by calling the theory an "equality" theory, for it is not really an equivalence relation in the usual sense. The choice of this terminology, however, should not affect our discussion in this paper since we are primarily interested in a semantics definition for functional programming.

in β_i. The reduction (or rewriting) relation $\rightarrow^R$ associated with R is the finest relation over $T(F, V)$, containing R and closed by substitution and replacement. From now on we will use $\rightarrow$ for $\rightarrow^R$. We denote by $\xrightarrow{*}$ the reflexive, transitive closure of $\rightarrow$. Equivalently, we say that a term A reduces to B at occurrence u using $\alpha_k \rightarrow \beta_k \in R$, and write $A \rightarrow B$ iff $\exists\eta\ \exists u \in O(A)\ A/u = \eta(\alpha_k)\ \&\ B = A[u \leftarrow \eta(\beta_k)]$. We sometime write $A \rightarrow_{[u,k]} B$ (or $A \rightarrow_u B$ for abbreviation) to indicate the rule and the occurrence associated with the reduction step.

Given a term rewriting system R, the set of function symbols F is divided into *defined functions*, which are those that appear as the leftmost functor in the left hand side of some rule, and *constructors*, which are the remaining symbols in F. A *constructor term* is either a variable, a (constant) constructor, or a term composed of constructors and variables only. We denote by $T(F_c, V)$ the set of *constructor terms*. A term is said to be a *function term* if it contains at least one defined function symbol. For example, with the rewrite system $R = \{ +(0,x) \rightarrow x,\ +(s(x),y) \rightarrow s(+(x,y)) \}$ and $F = \{+,\ 0,\ s\}$, we have one defined function symbol +, and two constructors 0 and s.

A term t *narrows* to s at a nonvariable subterm t/u using the k-th rewrite rule $\alpha_k \rightarrow \beta_k$ in R, denoted by $t \leadsto_{[u,k,\rho]} s$ (or $t \leadsto_\rho s$ for abbreviation), if ρ is the most general unifier of t/u and α_k and $s = \rho(t[u \leftarrow \beta_k])$. It is always assumed that the sets of variables in t and α_k are disjoint. We denote by $\overset{*}{\leadsto}$ the reflexive, transitive closure of $\leadsto$.

An equational program R in O'Donnell's language is a term rewriting system wherein (i) every rule must be *left-linear*, i.e., no variable may appear more than once in the left hand side; and (ii) there is no *critical pair* in R, i.e., for any two rules $\alpha_i \rightarrow \beta_i$ and $\alpha_j \rightarrow \beta_j$ in R, α_i/u and α_j, where $u \in O(\alpha_i)$, have no common instance, except when $u = \varepsilon$ and $i = j$. The second property is called *nonoverlapping, nonambiguous* or *superposition free* in the literature. These two properties are syntactically checkable and guarantee the confluence property without resorting to the termination property.

3. An Equality Theory for Functional Programming

Traditionally, an *equational theory* described by a term rewriting system R is the set of equations E obtained by replacing $\rightarrow$ by =; and E-equality, denoted by $=_E$, is defined as the finest congruence containing E, and closed under replacement and instantiation. It is known that the equality $=_E$ is the same as the symmetric closure of $\xrightarrow{*}$ [HuOp80]. We restrict our attention to a subset of $=_E$, called E_c-equality, denoted by $=_{E_c}$, with the following restriction: $A =_{E_c} B$ if and only if there exists a constructor term C such that $A =_E C$ and $B =_E C$. In other words, each "congruence" class in $=_{E_c}$ must contain a (unique) constructor term. Obviously, the set of axioms underlying this equality theory can be given based on a predicate, true of constructor terms. We will not dwell upon the details of these axioms here. From now on we will use the term *E_c-equational theory* instead of equational theory to reflect the fact that the underlying equality is E_c-equality.

The proof method based on the reduction mechanism in conventional equational reasoning can still be utilized for proving E_c-equality. It is easy to see that given a confluent term rewriting system R, for any terms A and B, $A =_{E_c} B$ if and only if $\exists C \in T(F_c, V)$ such that $A \xrightarrow{*} C$ and $B \xrightarrow{*} C$.

Note that a function term may not be E_c-equivalent to itself, i.e., identity does not apply everywhere. It is this property which eliminates the need of cross checking of identity, even for nonterminating term rewriting systems. Operationally, all that needs to be done is to reduce two terms to constructor terms and then check identity. If this cannot be done then two terms are not E_c-equivalent. E_c-equivalence is decidable for terminating term rewriting systems and semi-decidable for nonterminating systems.

Solving equations under this equality theory can be described in terms of E_c-unification. Two terms A and B are said to be E_c-unifiable iff there exists a substitution σ, such that $\sigma A =_{E_c} \sigma B$. Note that by definition substitutes in an E_c-unifier need not be constructor terms. To describe how E_c-unifiers compare with each other, we need to extend E-equality (not E_c-equality) to substitutions†: $\sigma =_E \theta$ iff $\forall x \in V\ \sigma x =_E \theta x$. Comparison of unifiers is defined as: $\sigma \leq_E \theta\ [W]$ iff $\exists\eta\ \eta\bullet\sigma =_E \theta\ [W]$, where $[W]$ is variable restriction. We then say that σ is more general than θ. We will often omit $[W]$ if no confusion arises.

† Unlike the usual denotational approach where a function maps ⊥ (bottom) to ⊥, E_c-equality does not care about whether subterms (i.e., arguments in function applications) make sense or not. Note that E-equality is used only in the metalanguage for the purpose of comparing substitutions. Our system does not attempt to prove E-equality.

A set of substitutions Σ is said to be a complete set of E_c-unifiers of two terms A and B iff every substitution in Σ is an E_c-unifier of A and B, and for any E_c-unifier θ of A and B (which may not be in Σ), there exists a substitution in Σ, which is more general than θ. In addition, Σ is said to be *minimal* iff no two unifiers in Σ can compare by $\leq_E$. Similar to the case of E-unification [FaHu83], complete and minimal sets of E_c-unifiers may not always exist. Consider the following term rewriting system modified from [FaHu83]:

$$R = \{ \ g(a,x) \rightarrow x, \ \ f(g(x,y)) \rightarrow f(y), \ \ f(a) \rightarrow b \ \}.$$

It can be shown that this system is a *closed linear* term rewriting system defined by O'Donnell [ODon77]. However, there exist an infinite number of E_c-unifiers for $f(z)$ and $f(a)$:

$\sigma_0 = \{ \ z/a \ \}$

$\sigma_1 = \{ \ z/g(x_1, a) \ \}$

$\sigma_2 = \{ \ z/g(x_1, g(x_2, a)) \ \}$

......

$\sigma_i = \{ \ z/g(x_1, g(x_2, \cdots g(x_i, a) \cdots)) \ \}$

......

with each one being more general than the previous one. It can be shown by induction that $\sigma_i f(z) =_{E_c} \sigma_i f(a) =_{E_c} b$, for all i, where b is a constructor term. Thus all of them are E_c-unifiers of $f(z)$ and $f(a)$.

Fortunately, complete and minimal sets of E_c-unifiers always exist for E_c-equational theories described by a left-linear, nonoverlapping term rewriting system. We will prove this claim in the next section.

4. Existence of Complete and Minimal Sets of E_c-Unifiers

In this section we first show that given a left-linear and nonoverlapping term rewriting system, for any E_c-unifier of two terms, there exists a more general E_c-unifier which is *normalized*, even in the case that R is nonterminating. This leads to two other results: (i) narrowing is complete for E_c-unification, and (ii) complete and minimal sets of E_c-unifiers always exist.

In the remainder of this paper, we will assume that the underlying term rewriting system R is always left-linear and nonoverlapping if not otherwise said. Following [Hull80], the binary constructor symbol H used in the rest of this paper is assumed not in F and is for the purpose of notationally combining two reduction/narrowing sequences.

Lemma 1. For any E_c-unifier τ of two terms A and B, there exists an E_c-unifier θ for A and B, such that θ is normalized and $\theta \leq_E \tau \ [V(A,B)]$ □

Proof: (Sketch)

Let $\tau = \{x_1/t_1, ..., x_n/t_n\}$. For each t_i in τ, if t'_i is the normal form of t_i, then replace t_i with t'_i. Let τ' denote the substitution so obtained. Clearly, τ' is an E_c-unifier of A and B and $\tau' =_E \tau \ [V(A,B)]$. By the confluence property and the definition of E_c-equality, there exist a constructor term and a reduction derivation

$$\tau' H(A,B) \overset{*}{\rightarrow} H(C,C).$$

Now if τ' is normalized the lemma is proven; otherwise assume $f(s_1, ..., s_m)$ is in τ' which cannot be normalized and which is reducible at ε by a rewrite rule defining f. Since C is a constructor term, an occurrence of $f(s_1, ..., s_m)$ in $\tau' H(A,B)$, or in any subsequently reduced term from it must eventually "disappear." This is possible only by (i) matching a variable which is *dropped* in the right hand side of a rule, and (ii) matching a subterm of the form $f(t_1, ..., t_m)$ in the inner part of the left hand side of some rule.

The second case is not possible since by the nonoverlapping property $f(t_1, ..., t_m)$ is not unifiable with the left hand side of any rule, and thus $f(s_1, ..., s_m)$, as an instance of $f(t_1, ..., t_m)$, must not be reducible at ε; this contradicts the assumption that $f(t_1, ..., t_m)$ is reducible at ε. For the first case, replacing $f(s_1, ..., s_m)$ by a variable will not change the fact that the reduction derivation leads to $H(C,C)$, and the only reductions that are deleted are those performed on $f(s_1, ..., s_m)$. We therefore replace each non-normalizable function term in τ' by a distinct variable. Let the substitution so obtained be τ''. Clearly, τ'' is normalized, $\tau'' < \tau'$, and a reduction derivation leading to $H(C,C)$ exists:

$$\tau'' H(A,B) \overset{*}{\rightarrow} H(C,C).$$

τ'' is therefore an E_c-unifier of A and B. From the way τ'' was constructed we have $\tau'' \leq \tau' =_E \tau \ [V(A,B)]$. □

Theorem 2. For any E_c-unifier τ of two terms A and B, there exist constructor terms C_1 and C_2 and a narrowing derivation:

$$\text{(i) } H(A,B) \overset{*}{\rightsquigarrow}_\sigma H(C_1,C_2)$$

such that C_1 and C_2 are unifiable, and if θ is m.g.u. of C_1 and C_2, then $\theta\bullet\sigma \leq_E \tau\ [V(A,B)]$.

Conversely, for any narrowing derivation (i), there exists a reduction derivation:

$$\text{(ii) } \theta\bullet\sigma H(A,B) \overset{*}{\to} H(C,C)$$

where $C = \theta C_1 = \theta C_2$. □

Proof: From Lemma 1 there exists a normalized E_c-unifier τ' of A and B, such that $\tau' \leq_E \tau\ [V(A,B)]$. The rest follows from Hullot's results [Hull80]. □

Lemma 3. For any E_c-unifiable terms A and B there exists a complete set Σ of E_c-unifiers for A and B, such that for any E_c-unifier $\tau = \{x_1/t_1, ..., x_n/t_n\}$ in Σ, if t_i is a nonvariable term then t_i is not unifiable with the left hand side of any rule in R. □

Proof: Consider a narrowing derivation that generates an E_c-unifier for A and B:

$$H(A,B) = A_0 \rightsquigarrow_{[u_0,k_0,\rho_0]} \cdots \rightsquigarrow_{[u_{n-1},k_{n-1},\rho_{n-1}]} A_n = H(C_1,C_2),$$

where C_1 and C_2 are unifiable by m.g.u. θ. Since R is left-linear, any substitute in any $\rho_i\ _{|V(A_i)}$ must be extracted from the left hand side of the corresponding rule. By the nonoverlapping property, no nonvariable substitute so obtained may be unifiable with the left hand side of any rule. This is also true of all substitutes in θ. It is easy to show that this is still true for any substitute in the composition $\theta\bullet\rho_{n-1}\bullet...\bullet\rho_0\ _{|V(A_0)}$. Since narrowing is complete (Theorem 2), we conclude the proof by letting Σ be the set of E_c-unifiers generated by all narrowing derivations issuing from $H(A,B)$. □

We are now in a position to give the minimality result.

Theorem 4. Complete and minimal sets of E_c-unifiers always exist for any E_c-unifiable terms. □

Proof: From Lemma 3, there exists a complete set Σ generated by narrowing such that for any E_c-unifier τ in Σ, none of the nonvariable substitutes therein can be unifiable with the left hand side of any rule in R. Based on this property it is easy to see that for any $\tau, \tau' \in \Sigma$, τ and τ' compare by $\leq_E$ iff τ and τ' compare by $\leq$. For any $\leq$ ordering, a lower bound always exists (in the extreme case it is a variable), which is unique up to variable renaming. Thus, the set of all lower bounds plus those uncompared E_c-unifiers form a complete and minimal set of E_c-unifiers. Note that complete and minimal sets of E_c-unifiers are unique up to variable renaming. □

Note that Theorem 4 only claimed the existence, not finiteness. For example, with the system

$$R = \{\ f(c(x)) \to f(x),\ f(d) \to e\ \},$$

we have an infinite number of uncompared E_c-unifiers for $f(x)$ and e, which are $\{x/d\}$, $\{x/c(d)\}$, $\{x/c(c(d))\}$, ... $\{x/c(...c(d)...)\}$.... It is this phenomenon that makes it impossible to use a *filtering* method to eliminate redundant solutions.

5. A Special Class of Narrowing Derivations

The narrowing method has been criticized for its high degree of non-determinism which often leads to generation of redundant solutions. The two most popular strategies adopted for reduction, the *innermost* and *outermost* strategies, do not work well for narrowing. The following example shows that neither innermost narrowing, which ignores narrowing steps at outer occurrences, nor outermost narrowing, which ignores narrowing steps at inner occurrences can be complete even for E_c-equality defined in this paper. Consider

$$R = \{\ f(y,a) \to true,\ f(c,b) \to true,\ g(b) \to c\ \}.$$

The system is left-linear and nonoverlapping. To unify the term $f(g(x),x)$ with $true$, innermost narrowing leads to

(i) $f(g(x),x) \rightsquigarrow_{\{x/b\}} f(c,b) \rightsquigarrow_{\{\}} true$,

while outermost narrowing yields

(ii) $f(g(x),x) \rightsquigarrow_{\{x/a\}} true$.

The innermost and outermost strategies generate *uncompared* results; therefore, neither is complete.

Fribourg [Frib85] showed the completeness of the innermost narrowing strategy under certain sufficient conditions. Dincbas and Hentenryck [DiHe87] investigated the termination and efficiency issues of several strategies, including innermost, outermost and a form of lazy strategy based on a procedural semantics of functional programming. Reddy [Redd85] outlined a lazy narrowing strategy, and the outermost strategy on flattened equational programs by Levi *et al.* [Levi87] is essentially equivalent to lazy narrowing for conditional rules. Outer narrowing, suggested in [You88], is operationally similar to lazy narrowing, but is complete for E-unification. (Note that lazy narrowing does not yield a complete procedure for E-unification.) All these inference methods have been shown to work for the case of constructor-based languages, where no defined function symbol can appear in the inner part of the left-hand side of a rule.

We describe a process of transformation of narrowing derivations, and the resulting derivations are called D-narrowing derivations for lack of precise terminology.† We show that every narrowing derivation is subsumed by a D-narrowing derivation. Moreover, D-narrowing derivations form a compete and minimal set of E_c-unifiers.

The main idea behind D-narrowing is that narrowing steps in a narrowing derivation may be *rearranged* in a certain way to yield a new narrowing derivation, a D-narrowing derivation, which produces a more general E_c-unifier. A D-narrowing derivation thus may serve as a representative of a group of narrowing derivations that generate *compared* E_c-unifiers. D-narrowing is a weaker form of outer narrowing [You88]. The two requirements for a narrowing derivation to be an outer narrowing derivation are (1) the *outer-before-inner* property and (2) the use of a computation rule (similar to that of Prolog), i.e., for any intermediate term in a narrowing derivation select only one redex among a set of *independent* redexes at which narrowing is possible.

The outer-before-inner property says that *no narrowing step at an outer occurrence may possibly be performed earlier in the derivation using the same rule and at the same occurrence*. For example, the narrowing derivation (i) above satisfies the outer-before-inner property, in that the rule used in the second narrowing step is not applicable when tried to narrow the given term, due to the function symbol conflict between c and g. If a narrowing derivation is not outer-before-inner, it may be transformed to one that possesses this property. Consider $R = \{\, f(a,b,x) \to d(x),\ \ g(a) \to c \,\}$. The derivation

$$f(y,z,g(y)) \leadsto_{\{y/a\}} f(a,z,c) \leadsto_{\{z/b\}} d(c)$$

is not outer-before-inner since the second narrowing step can be performed right at the beginning, using the same rule and at the same occurrence. "Moving" the step at the outer occurrence to the front will give us

$$f(y,z,g(y)) \leadsto_{\{y/a,\, z/b\}} d(g(a)) \leadsto_{\{\}} d(c).$$

This type of rearrangement, however, may not guarantee completeness when defined functions appear in the inner part of the left hand side of a rule. Consider the following nonoverlapping term rewriting system and narrowing derivation.

$$R = \{f(g(d)) \to true,\ \ g(c) \to g(d)\}$$
$$f(g(x)) \leadsto_{\{x/c\}} f(g(d)) \leadsto_{\{\}} true$$

The second narrowing step can be performed right at the beginning using the same rule and at the same occurrence, which will generate an uncompared E_c-unifier $\{x/d\}$. Notice that the problem is caused by the appearance of the defined function symbol g in the inner part of the first rewrite rule; therefore, the outer-before-inner property should not be required to hold strictly when such rewrite rules are present. Outer narrowing with the consideration of this exceptional case is then called D-narrowing.

To characterize this rearrangement process precisely, we need a method to keep track of "copies" of subterms that are carried over by reduction (and therefore by narrowing). Intuitively, when a reduction is performed, the occurrences of a variable in the right hand side of the rule denote the rearrangement of a subterm that is matched to an occurrence of the same variable in the left hand side. The notion of residue map has been used by O'Donnell to study a closure property [ODon77] (also see [HuLé79, Réty87]).

Definition. The residue map r with respect to a left-linear term rewriting system R is defined as:

for all $v \in D(A)$

$r[A \to_{[u,k]} B]v$

$= \{u.w.(v/v') \mid \exists v''\ \alpha_k(v'') \in V(\alpha_k)\ \&\ \alpha_k(v'') = \beta_k(w)\ \&\ u.v'' = v'\}$ if $v > u$

$= \{v\}$ if $(u <> v$ or $v < u)$

$= \varnothing$ otherwise

The residue map n for narrowing is defined as:

for all $v \in D(A)$

$n[A \leadsto_{[u,k,\rho]} B]v = r[\sigma A \to_{[u,k]} B]v$ □

For example, with the rewrite rule $f(c(x)) \to g(x,x)$ and reduction $f(c(a)) \to g(a,a)$, both redexes of a in $g(a,a)$ are residues of a in $f(c(a))$. Note that if v is independent of u, or v is outer to u, then the residue of v is itself and unique.

The following proposition describes a special case of the *closure property* that was studied by O'Donnell [ODon77]. The proposition can be shown from the nonoverlapping property. For notational

† D stands for *demand-driven*. However, the strategy is not completely demand-driven.

convenience, we may use $\xrightarrow{*}_{[U,K]}$ to denote a reduction sequence, where U denotes redexes and K the corresponding rule indexes. Similarly, $\overset{*}{\leadsto}_{[U,K,\sigma]}$ denotes a narrowing derivation, where U denotes redexes, K the corresponding rule indexes, and σ the composition of all unifiers along the narrowing derivation. For notational convenience, the substitution σ generated by a narrowing derivation $C_0 \overset{*}{\leadsto}_{[U,K,\sigma]} C_m$ is assumed to be already restricted to the variables in C_0 whenever it is referenced.

Proposition 5. Let A be a term. Then
$\forall u,v \in O(A)\ (v < u\ \&\ A \rightarrow_{[v,k]} B\ \&\ A \rightarrow_{[u,j]} C) \Rightarrow \exists D \in T(F,V)\ (B \xrightarrow{*}_{[U,J]} D\ \&\ C \rightarrow_{[v,k]} D)$
where $U = r[A \rightarrow_{[v,k]} B]u$ and the rule indexes in J are all j. □

We now extend Proposition 5 to narrowing.

Lemma 6. (Outer-Before-Inner Property)
Suppose $A \leadsto_{[u,j,\rho_1]} C \leadsto_{[v,k,\rho_2]} D$ where $v < u$. If A is also narrowable at v using the k-th rule which does not contain the defined function symbol $A(u)$ in its left hand side, then there exists a narrowing derivation $A \leadsto_{[v,k,\sigma_1]} B \overset{*}{\leadsto}_{[U,J,\sigma_2]} D'$ such that $\sigma_2 \cdot \sigma_1 \leq \rho_2 \cdot \rho_1$ and $\exists \tau\ \tau(D') = D$, where $U = n[A \leadsto_{[v,k,\sigma_1]} B]u$ and the rule indexes in J are all j. □

Proof: Use Proposition 5 and the definitions of residue maps r and n. The details are omitted. □

Note that in Lemma 6 if A is also narrowable at v using the k-th rule which *does* contain the defined function symbol $u(A)$ in its left hand side, then both narrowing steps upon A at u and v, respectively, must be enumerated in order to guarantee no solutions will be lost, for uncompared solutions may eventually be generated by these two different derivations. Another observation is that according to the nonoverlapping property, these two derivations cannot generate compared solutions.

The second requirement for a narrowing derivation to be a D-narrowing derivation is the use of a computation rule, i.e., for any intermediate term in a narrowing derivation select only one redex among a set of *independent* redexes at which narrowing is possible. The branching factor for a set of independent redexes in any term is therefore one. The reduction of the branching factor to one is guaranteed by a property similar to ***independence of computation rule*** in **Prolog [Lloy84].**

Lemma 7. (Independence of Computation Rule)
Suppose $A \leadsto_{[u,k,\rho_1]} C \leadsto_{[v,j,\rho_2]} D$, where v and u are independent redexes in A. Then there exists a narrowing derivation $A \leadsto_{[v,j,\sigma_1]} B \leadsto_{[u,k,\sigma_2]} D$, such that $\sigma_2 \cdot \sigma_1 = \rho_2 \cdot \rho_1$. □

Proof: Omitted. □

Now given a narrowing derivation, apply Lemmas 6 and 7 repeatedly to rearrange the given derivation. This is like a "sorting" process. It is easy to see that this process always terminates. We can now give the definition of D-narrowing.

Definition Given a computation rule, a narrowing derivation is said to be a D-narrowing derivation if no rearrangement can be made by either of the processes described in Lemmas 6 and 7. □

Theorem 8. (Completeness of D-Narrowing)
Given two terms A and B to be unified, for any narrowing derivation:

$$H(A,B) = A_0 \leadsto_{[u_0,k_0,\rho_0]} \cdots \leadsto_{[u_{n-1},k_{n-1},\rho_{n-1}]} A_n = H(C_1,C_2),$$

such that C_1 and C_2 are constructor terms and unify by m.g.u. ξ, there exists a D-narrowing derivation:

$$H(A,B) = B_0 \leadsto_{[v_0,j_0,\sigma_0]} \cdots \leadsto_{[v_{m-1},j_{m-1},\sigma_{m-1}]} B_m = H(D_1,D_2),$$

such that D_1 and D_2 are constructor terms and unifiable. If δ is m.g.u. of D_1 and D_2, then $\delta \cdot \sigma_{m-1} \cdot \ldots \cdot \sigma_0 \leq \xi \cdot \rho_{n-1} \cdot \ldots \cdot \rho_0$. □

Proof: By induction on the number of steps in the given narrowing derivation, using Lemmas 6 and 7. □

To establish the minimality property, we show that any two unrelated D-narrowing derivations (no one is a prefix of the other) generate uncompared E_c-unifiers.

Lemma 9. (Minimality of D-Narrowing)
Let A and B be two E_c-unifiable terms. Let C_1, C_2, D_1 and D_2 be constructor terms. Consider two unrelated D-narrowing derivations:

$$\text{(i)}\ H(A,B) = A_0 \leadsto_{[u_0,k_0,\rho_0]} \cdots \leadsto_{[u_{n-1},k_{n-1},\rho_{n-1}]} A_n = H(C_1,C_2)$$

(ii) $H(A,B) = B_0 \rightsquigarrow_{[v_0, j_0, \sigma_0]} \cdots \rightsquigarrow_{[v_{m-1}, j_{m-1}, \sigma_{m-1}]} B_m = H(D_1, D_2)$

where C_1 and C_2 unify by m.g.u. τ_1, and D_1 and D_2 by m.g.u. τ_2. Then $\tau_1 \cdot \rho_{n-1} \cdot \cdots \cdot \rho_0 {}_{|V(A,B)}$ and $\tau_2 \cdot \sigma_{m-1} \cdot \cdots \cdot \sigma_0 {}_{|V(A,B)}$ cannot compare by $\leq_E$. □

Proof: (Sketch)

Let $\eta_1 = \tau_1 \cdot \rho_{n-1} \cdot \cdots \cdot \rho_0$ and $\eta_2 = \tau_2 \cdot \sigma_{m-1} \cdot \cdots \cdot \sigma_0$. From Lemma 3 we know that η_1 and η_2 compare by $\leq_E$ iff η_1 and η_2 compare by $\leq$. Since (i) and (ii) are unrelated derivations, branching will eventually occur. Let $A_i \rightsquigarrow_{[u_i, k_i, \rho_i]} A_{i+1}$ and $B_i \rightsquigarrow_{[v_i, j_i, \sigma_i]} B_{i+1}$ denote the two steps corresponding to the very *first* branching, where $A_i = B_i$.

Now u_i and v_i are either independent or dependent. The first case is not possible because of the use of the same computation rule. For the second case, we have either $u_i = v_i$ or $u_i \neq v_i$. Consider the subcase that $u_i = v_i$. It can be shown that if there does not exist a substitution ζ such that $\zeta \cdot \rho_i = \zeta \cdot \sigma_i$, then η_1 and η_2 cannot compare by $\leq$. Since $u_i(A_i) = v_i(B_i)$, the k_i-th rule and the j_i-th rule must define the same function symbol. In this case, that $(\exists\zeta)(\zeta \cdot \rho_i = \zeta \cdot \sigma_i)$ must be true draws from the nonoverlapping property. Now consider the second subcase. Without loss of generality, assume $v_i < u_i$. By the outer-before-inner property, there is a sub-derivation issuing from A_{i+1}:

$$A_{i+1} \rightsquigarrow_{[u_{i+1}, k_{i+1}, \rho_{i+1}]} \cdots \rightsquigarrow A_l \rightsquigarrow_{[u_l, k_l, \rho_l]} A_{l+1} \qquad l < n$$

such that $u_l = v_i$, i.e., the defined function symbol $v_i(A_i)$ must eventually be replaced by using a rule defining $v_i(A_i)$ in the first derivation. To show that η_1 and η_2 cannot compare by $\leq$, all we need to show is that

$$(\exists\zeta) \;\; \zeta \cdot \sigma_i = \zeta \cdot \rho_l \cdot \cdots \cdot \rho_{i+1} \cdot \rho_i .$$

By induction, we can show that if this is not true then R must be overlapping, i.e., R must have critical pairs. □

Combining Theorem 8 and Lemma 9, we obtain the following result.

Theorem 10. For any E_c-unifiable terms A and B, the set of E_c-unifiers generated by D-narrowing derivations issuing from $H(A,B)$ is a complete and minimal set of E_c-unifiers for A and B. □

6. Final Remarks and Related Work

In [YoSu86] we showed that narrowing is incomplete in general for E-unification even in the case of left-linear, nonoverlapping term rewriting systems. In this paper, by adopting a more suitable (but tighter) semantics, narrowing becomes complete. An efficient implementation is made possible by the fact that there is a class of special narrowing derivations that guarantees no redundant solutions will be generated. The fact that two terms are E_c-unifiable if they can eventually be replaced by two unifiable constructor terms allows us to detect certain non-unifiability syntactically by analyzing the possible "chains" of constructors in a given term rewriting system. This becomes very close to Prolog where all functors are constructors. We believe that many program analysis techniques for Prolog, which are crucial for compiler optimization, can be directly or indirectly used for E_c-unification. Another observation worth mentioning is that E_c-unification automatically enforces "well typing" by the correctness of an E_c-unification procedure. Given two E_c-unifiable terms, all substitutes in an E_c-unifier must be of correct types that can be inferred from the given declaration-free functional program [MiRe85].

Several works are particularly relevant to ours in different aspects. Jaffar *et al.* [JaLM86, JaLM87] and Stickel [Stickel86] essentially established foundations for incorporating first-order theories other than the classic equality theory into SLD-resolution and resolution, respectively. The E_c-equality defined in this paper is just another special first-order theory, which can be incorporated into Horn clause logic programming by replacing conventional unification with E_c-unification. The completeness of the integrated system is a consequence of the completeness of the E_c-unification procedure. At the outset of this paper we argued that it can be more beneficial to just consider special theories for particular applications. In particular, for the purpose of embedding functional programming into Prolog, the E_c-equality theory is more suitable than E-equality since it permits an efficient operational semantics.

The general difficulty inherited in equational reasoning was well discussed by van Emden and Yukawa [vanEYu87]. As a matter of fact, for the purpose of introducing a Prolog-like operational semantics, the equality theory defined in [vanEYu87] is actually another restricted version of the classic equality theory, based on the notion of canonicality of terms. Our equality theory is more restricted in that two terms are equivalent if they are both equivalent to a constructor term. Fribourg [Frib85], Yukawa [Yuka87] and Levi *et al.* [Levi87] independently proposed similar semantic formalisms in which individuals in a model are constructor terms only. The E_c-equality defined in this paper is weaker than all of the above in that individuals in a model may not necessarily be constructor terms. The main theoretical result of this paper is the minimality result for left-linear, nonoverlapping systems, both semantically and operationally.

Besides narrowing, there has been another approach based on transformation of systems of equations, initially explored by Martelli and Montanari [MaMo82]. This approach is so powerful that it has been incorporated into an E-unification procedure, complete for *any* equational theories [GaSn87]. The E-unification algorithm described in [MaMR86] for canonical term rewriting systems is superior to the naive use of narrowing. As a matter of fact, the relationship between this algorithm and D-narrowing can be precisely established to show that D-narrowing derivations can be effectively enumerated by a restricted version of their algorithm.

In summary, two assumptions serve as the key in our approach towards the problem of solving equations: the intended equality is restricted to the one which more closely captures the intuitive notion of functional programming and the equational language is restricted to left-linear, nonoverlapping systems. These two assumptions permit us to address the minimality issue and develop a special narrowing strategy. It is our belief functional programming can be embedded into Prolog without sacrificing efficiency.

References

[CaMc79] Cartwright, R. and J. McCarthy, "Recursive programs as functions in a first order theory," in *Mathematical Studies of Information Processing*, E.K. Blum *et al.* (eds.) LNCS 75.

[DePl85] Dershowitz, H. and D. Plaisted, "Logic programming cum applicative programming," in *Proc. 1985 Symposium on Logic Programming*, Boston, Mass., July, 1985.

[DiHe87] Dincbas, M. and P. van Hentenryck, "Extended unification algorithms for the integration of functional programming into logic programming," *Journal of Logic Programming*, 1987.

[vanEYu87] van Emden, M.H. and K. Yukawa, "Logic programming with equality," *J. of Logic Programming*, 4:265-288, 1986.

[FaHu83] Fages, F and G. Huet, "Unification and matching in equational theories," in *Proc. CAAP '83, LNCS 159*, pp. 205-220, 1983.

[Frib85] Fribourg, L., "SLOG: A logic programming language interpreter based on clausal superposition and rewriting," in *Proc. 1985 Symposium on Logic Programming*, Boston, Mass.

[GaRa86] Gallier J.H. and S. Raatz, "SLD-Resolution methods for Horn clauses with equality based on E-unification E-unification procedure," in *Proc. of 1986 Symposium on Logic Programming, SLC, Utah, pp. 168-179, 1986.*

[GaSn87] Gallier J.H. and W. Snyder, "A general complete E-unification procedure," in *Proc. of RTA '87, LNCS 256*, pp. 216-227, 1987.

[GoTa79] Goguen, J.A. and J.J. Tardo, "An introduction to OBJ: a language for writing and testing formal algebraic program specifications," in *Specification of Reliable Software, IEEE*, pp 170-189, 1979.

[GoMe84] Goguen, J.A. and J. Meseguer, "Equality, types, modules and generics for logic programming," *Journal of Logic Programming*, Vol 2. 1984, pp 179-210.

[HuLé79] Huet, G. and J-J. Lévy, "Call by need computations in nonambiguous linear term rewriting systems," Technical Report, 359, INRIA, Le Chesnay, France, 1979.

[HuOp80] Huet, G. and D.C. Oppen, "Equations and rewrite rules: a survey," in *Formal Language Theory: Perspectives and Open Problems*, R.V. Book (ed.), pp. 349-405, Academic Press, New York, 1980.

[Hull80] Hullot, J.M., "Canonical forms and unification," in *Proc. 5th Conference on Automated Deduction*, pp. 318-334, 1980.

[JaLM86] Jaffar, J., J-L. Lassez and M. Maher, "A logic programming scheme," in *Logic Programming: Functions, Relations, and Equations*, D. DeGroot and G. Lindstrom (eds.), Prentice-Hall, 1986.

[JaLM87] Jaffar, J., J-L. Lassez and M. Maher, "Constraint logic programming," in *Proc. 14th POPL*, pp. 111-119, Munich, West Germany, 1987.

[Levi87] Levi, G., C. Palamidessi, P. Bosco, E. Giovannetti and C. Moiso, "A complete semantic characterization of K-LEAF", in *Proc. 1987 Symposium on Logic Programming*, pp. 318-327. San Francisco, California, September 1987.

[Lloy84] Lloyd, J.W., *Foundations of Logic Programming*, Springer-Verlag, New York, 1984.

[MaMo82] Martelli, A., and U. Montanari, "An efficient unification algorithm," *ACM Transaction on Programming Languages and Systems*, Vol. 4, No. 2, pp. 258-282, 1982.

[MaMR86] Martelli, A., C. Moiso and G.F. Rossi "An algorithm for unification in equational theories", in *Proc. 1986 International Symposium on Logic Programming*, pp. 180-186, Salt Lake City, Utah, 1986.

[MiRe85] Mishra, P. and U. Reddy, "Declaration-free type checking" in *Proc. of 12th POPL*, pp. 7-21, 1985.

[ODon77] O'Donnell, M., "Computing in systems described by equations," *Lecture notes in computer science*, vol. 58, Springer-Verlag, New York, 1977.

[ODon85] O'Donnell, M., "*Equational Logic as a Programming Language*," The MIT Press, Cambridge, Massachusetts, 1985.

[Plot72] Plotkin, G., "Building-in equational theories," in *Machine Intelligence* 7, pp. 73-90, Edinburgh University Press, 1972.

[Redd85] Reddy, U., ''Narrowing as the operational semantics of functional languages,'' in *Proc. 1985 Symposium on Logic Programming*, pp. 138-151, Boston, Mass., July, 1985.

[RéKi85] Réty P, C. Kirchner, H. Kirchner and P. Lescanne, ''NARROWER: a new algorithm and its application to logic programming,'' in *Proc. Rewriting Techniques and Applications*, also in *Lecture Notes in Computer Science* 202, pp. 141-157, 1985.

[Réty87] Réty P., ''Improving basic narrowing techniques,'' in *Proc. Rewriting Techniques and Applications, Lecture Notes in Computer Science* 256, pp. 228-241, 1987.

[Siek84] Siekmann, J., ''Universal unification,'' in *Proc. 7th International Conference on Automated Deduction*, pp. 1-42, Napa, California, May, 1984.

[Stickel86] Stickel, M., ''Automated deduction by theory resolution,'' *J. of Automated Reasoning*, Vol. 2 1986. Short version in *Proc. IJCAI-85*, pp. 1181-1186, Aug. 18-23, Los Angeles, CA, 1985.

[You86] You, J-H. ''Logic programming for artificial intelligence: an approach based on building-in theories,'' Dept. of Computing Science, Univ. of Alberta, (a proposal), 1986.

[YoSu86a] You, J-H. and P.A. Subrahmanyam, ''Equational logic programming: an extension to equational programming,'' in *Proc. ACM 13th POPL*, pp. 209-218, St. Petersburg, Florida, January, 1986.

[YoSu86b] You, J-H. and P.A. Subrahmanyam, ''A class of confluent term rewriting systems and unification,'' in *Journal of Automated Reasoning* Vol. 2 (1986) 391-418.

[You88] You, J-H. ''Outer narrowing for equational theories based on constructors,'' in *Proc. 15th ICALP*, Tampere, Finland, July, 1988.

[Yuka87] Yukawa, K., ''Amalgamating functional and relational programming through the use of equality axioms,'' Ph.D. Thesis, Dept. of Computer Science, Univ. of Waterloo, 1987.

NONSTANDARD LOGICS OF RECURSIVE PROGRAMS AND DENOTATIONAL SEMANTICS

Ana Pasztor *

1 Introduction

For a formal treatment of recursive programs, denotational semantics seems to have substantial advantages over operational semantics (cf. e.g. Apt-de Bakker[9], de Bakker[10], Brookes[12] and [13], Gordon[29], Halpern-Meyer-Trakhtenbrot[32], Meyer[44], Milne-Strachey[45], Mosses[47], Plotkin[52], Schmidt[67], Schwarz[68], Scott[69] and [70], Stoy[71], Tennent[73], Trakhtenbrot[75] and many, many more). In this connection it has been proposed by A. Meyer, A.J. Kfouri and other people (see e.g. Meyer[43]) to extend the so called Nonstandard Logics of Programs' approach (NLP henceforth and explained below) to recursive programs and denotational semantics in some natural way.

The present paper presents a possibility to provide NLP with a denotational semantics which handles recursive programs.

The key step in doing NLP is selecting a semantics (for programs) and then forming its KPU-absolute version (cf. below and Barwise[11], Manders[41], Pasztor[49], [51], Sain[56] and [65]). In the present paper we will take denotational semantics as our starting point, and search for its KPU-absolute version along the lines of the NLP traditions as outlined e.g. in Andreka-Nemeti- Sain[8]. (KPU stands for Kripke-Platek axiomatization of set theory with urelements–cf. Barwise[11] .)

KPU-absoluteness (absoluteness henceforth) is a set theoretic property of logics (cf. e.g. Sacks[55] and Barwise[11]). Desirability of absoluteness was already pointed out in Sacks[55] . The definition of this property is somewhat technical, hence to illuminate what absoluteness means, we will rather try to indicate what the lack of this property means. If a logic L is not absolute, then some formulas of L do not speak about the models of L, but about something else. A typical example of this is the case of the higher order logic L_2, described in Sain[56] . Namely, there is a formula ϕ in L_2, such that for any infinite model $\mathcal{M}$ of L_2 , we have $\mathcal{M} \models \phi$ iff CH (the continuum hypothesis) holds in the universe of set theory sometimes called the "real world" (in which $\mathcal{M}$ and ϕ live). So we may fix $\mathcal{M}$, fix ϕ and add or remove

*Florida International University, School of Computer Science, University Park, Miami, FL 33199

a few sets from this universe of set theory such that the meaning of ϕ in $\mathcal{M}$ changes. It is important to note that the addition (or removal) of the new sets happens in such a way, that neither ϕ, nor $\mathcal{M}$ (nor their transitive hulls for that matter) are changed. In other words, no new set is an element of ϕ or $\mathcal{M}$, nor is it an element of their elements etc. Despite of all these precautions, $\mathcal{M} \models \phi$ holds before the addition (removal) of the new sets, and $\mathcal{M} \models \neg\phi$ after it. If a logic has this strange property, then it is impossible to find a complete inference system for it, and this because of purely "administrative" reasons, which might not correspond to any of the inherent properties of the original situation we are modeling (in devising this logic etc.). This "administrative" incompleteness is only one of the undesirable side effects of non-absoluteness. In NLP, the purpose of constructing and studying absolute versions of logics is aimed at, among other things, the removal of these undesirable properties (in such a way however, that we are able to monitor and control the "price" we have to pay for this change).

The semantics investigated in Andreka-Nemeti-Sain[8] was operational (no recursive programs were allowed there) and therefore its absolute version was based on a 3-sorted first order logic. The absolute version of denotational semantics however, will turn out to be based on 2-sorted first order logic, or more concretely, on the usual nonstandard or weak second order logic–see Makowsky- Sain[39], [40] for terminology. (This version of second order logic is well known in the literature of logic, and its first thorough treatment was probably given in Henkin[35]. This is why many papers (e.g. Andreka-Nemeti-Sain[8]) call this a Henkin-type second order logic[1]. The point is that this Henkin type second order logic is related to the classical, or "standard" second order logic exactly the same way NLP is related to standard logics of programs.)

Let us return to denotational semantics and the search for its absolute version. Let D be the set of possible states. Then, in denotational semantics, the meanings of (possibly nondeterministic) programs in D are binary relations $R \subseteq D \times D$ over D. Therefore, it is natural to concentrate on D and on the relations over D. Here is what we mean. The NLP -method (originating with L. Henkin) for obtaining the absolute version of a semantics "$\models$" consists of the following. First we look at some fixed definition of "$\models$." For metamathematical reasons, this definition can be thought of as a set theoretical formula μ. Then we investigate the essentially unbounded quantifiers occuring in μ. For each of these quantifiers we investigate the kinds of mathematical objects (i.e. elements of our set theoretic universe mentioned earlier) they range over. Then we add the set(s) of these objects to the models of our semantics "$\models$" as extra sorts. After this we change μ accordingly, obtaining μ'. As a result, all quantifiers in μ' are bounded. Then, by a result in Barwise[11], μ' is absolute. (The essence is, that everything formulas of the language speak about is now made explicit in the models, so that they really speak about the models, and not about something else.) The finishing touch is writing down axioms in the extended language of

[1] In pure proof theory, this is the only kind of second order logic (for certain rather natural reasons). Hence many proof theorists do not use any of the adjectives "weak" or "nonstandard," or "Henkin-type" discussed above.

the new many-sorted semantics "$\models$," to govern the behaviour of the extra sorts. (These axioms in the case of second order logic, for example, are usually the so called comprehension axioms.)

In denotational semantics, the meaning of a recursive program p in D is defined as the least fixed point of a functional F_p from $P(D \times D)$ to $P(D \times D)$. The (set theoretic or metamathematical) formula ϕ defining the meaning R of p in D reads as $\phi = [F_p(R) = R \wedge (\forall X \in P(D \times D))(F_p(X) = X \rightarrow R \subseteq X)]$ and clearly contains one unbounded quantifier ranging over binary relations on D. As a consequence, we will add a new sort and we will use 2-sorted models $< D, S, \in >$, where S is a set of binary relations on D and $\in$ is the usual membership relation. (On the other hand, we do not need objects of other than these sorts–unlike in e.g. Andreka-Nemeti- Sain[8], where three sorts are needed.) By writing down the comprehension axioms (to be able to make use of the new sort), we will arrive at weak (or nonstandard) second order logic, D being the first order sort, and S the second order sort (cf. Makowsky-Sain[39], [40] for details). So we obtain second order logic as a natural "framework logic," or "underlying logic" (in the sense of e.g. Sain[58], [62] or [65]) for a branch of NLP.

Weak second order logic has already been used in NLP (as a framework logic) in e.g. Makowsky-Sain[39],[40] and Sain[65] explicitly, and in Csirmaz[16], [17], Gonzales-Artalejo [27], Leivant[37], [38] and Sain[63] implicitly. More detailed references in this connection can be found in Makowsky-Sain[40]. Other papers representing the NLP-approach are e.g. Andreka[1], Andreka-Nemeti[2], Andreka-Nemeti-Sain[3][8], Csirmaz[18]- [22], Csirmaz-Paris[23], Hajek[30],[31], Nemeti[48], Pasztor[49]-[51], Richter - Szabo[54], Sain[57]-[62] and [64] etc., etc.

We also would like to note that recursive programs have already been treated in the NLP-approach in e.g. Cartwright[14], Cartwright-McCarthy[15], and, to a certain extent, Gergely-Ury[25]. However, these treatments were not aimed at realizing all the main kinds of applications of NLP outlined in e.g., Sain[61], Kfoury-Park[36] or Pasztor[49]-[51].

To avoid misunderstandings, we would like to conclude this Introduction by noting that the two terms "(KPU-) absolute logics of programs" and "nonstandard logics of programs" are two different names for the same thing. The mathematical content of this statement consists of facts 1. and 2. below.

1. Every nonstandard logic of programs is KPU-absolute.

2. Given any logic of programs, if it is KPU-absolute, then, by the main results of Manders[41] and Sain[56], this logic shares those features of the existing NLP's which are considered important or relevant by the main proponents of NLP. (E.g. the new logic might not be based on 3-sorted first order logic, but instead on 5-sorted or 1-sorted or on some completely different kind of logic–e.g. modal logic with Kripke models. It may not have an explicit time scale at all. These differences will not affect its belonging to NLP, since these features are considered to be irrelevant in NLP. Some relevant features would be the ability to compare "realistic" existing program verification methods,

create new, more powerful and still "realistic" ones, completeness etc.–to mention only a very few.)

2 The Framework Logic FL

In this section we define the framework logic FL of our absolute logic DL^{rec} of recursive programs. Technically, FL $=< FL_{ds} : d$ is a 1-sorted type $>$ and FL_{ds} is a classical, 2-sorted, first order language of type ds with the axioms of comprehension and extensionality as logical axioms. Every type ds is a 2-sorted expansion of the 1- sorted type d (of data). So the models of FL_{ds} are of the form $< \underline{D}, S, \in>$, where $\underline{D}$ is a model of type d with universe D, $S \subseteq P(D \times D)$ (the powerset of $D \times D$), and $\in$ is the "element of" relation ($\subseteq D \times D \times S$), and furthermore these models satisfy the comprehension axioms.

Definition 1: (1-sorted type d, model of type d)

(i) A 1-sorted similarity type (or signature–for short type henceforth) d is a pair $< d_0, d_1 >$ of functions, where: the elements of the domain D_0 of d_0 are called function symbols of type d and for every $f \in D_0$, $d_0(f) \in \omega$ is called the arity of f; similarly, the elements of the domain D_1 of d_1 are called relation symbols of type d, and for every $r \in D_1$, $d_1(r) \in \omega - \{0\}$ is called the arity of r.

(ii) Given a 1-sorted type d, a model $\underline{D}$ of type d is a pair $< D, D^0 >$, D being a nonempty set (called the universe of $\underline{D}$) and D^0 being a function: to each function symbol f of type d and of arity n, D^0 assigns a function $f^{\underline{D}} : D^n \to D$, and to each relation symbol r of type d and of arity $n(\neq 0)$, D^0 assigns a relation $r^{\underline{D}} \subseteq D^n$.

□ Definition 1

Definition 2: (many-sorted type m, model of type m)

(i) A many-sorted type m is a triple $< m_0, m_1, m_2 >$, where m_2 is a set, whose elements are called sorts, and m_0 and m_1 are functions: the elements of the domain M_0 of m_0 are called function symbols and for every $f \in M_0$, $m_0(f) \in m_2^* - \{\lambda\}$ (i.e. is a nonempty, finite string of sorts) and is called the arity of f; similarly, the elements of the domain M_1 of m_1 are called relation symbols, and for every $r \in M_1$, $m_1(r) \in m_2^* - \{\lambda\}$ and is called the arity of r.

(ii) Given a many-sorted type $m =< m_0, m_1, m_2 >$, a model of type m is a pair $\mathcal{M} =< \mathcal{M}^0, \mathcal{M}^1 >$ of functions, where $\mathcal{M}^0$ assigns to every sort $s \in m_2$ a nonempty set $\mathcal{M}_s$, called the universe of sort s of $\mathcal{M}$, and $\mathcal{M}^1$ assigns to every function symbol f of arity $s_0 s_1 \dots s_n$ a function $f^{\mathcal{M}} : \mathcal{M}_{s_0} \times \mathcal{M}_{s_1} \times \cdots \times \mathcal{M}_{s_{n-1}} \to \mathcal{M}_{s_n}$, and to every relation symbol r of arity $s_0 s_1 \dots s_n$ a relation $r^{\mathcal{M}} \subseteq \mathcal{M}_{s_0} \times \mathcal{M}_{s_1} \times \cdots \times \mathcal{M}_{s_n}$.

□ Definition 2

Definition 3: (the type ds)

Let d be an arbitrary 1-sorted type like defined in Definition 1/(i). We expand d to a 2-sorted type ds in the following way. The set of sorts of ds is $\{\underline{d}, \underline{s}\}$; the function symbols of ds are those of type d, i.e. $domain(ds_0) = domain(d_0)$, and for every function symbol f, if $d_0(f) = n$, then $ds_0(f) = \underline{d}^{n+1}$; the relation symbols of ds are those of d together with one additional symbol $\in$ of arity $\underline{dds}$. In other words, $domain(ds_1) = domain(d_1) \cup \{\in\}$, for every relation symbol r of d, if $d_1(r) = n (\neq 0)$, then $ds_1(r) = \underline{d}^n$, and $ds_1(\in) = \underline{dds}$.

□ Definition 3

Notation 4: (Md_{ds}, $<\underline{D}, S, \in>$)

(i) Given any (1-or many-sorted) type t, we denote by Md_t the class of all models of type t.

(ii) Given a model $\mathcal{M} \in Md_{ds}$, we identify $\mathcal{M}$ with the triple $<\underline{D}, S, \in^{\mathcal{M}}>$ (or just $<\underline{D}, S, \in>$), where $\underline{D} \triangleq < \mathcal{M}_{\underline{d}}, \mathcal{M}^1 \mid (domain(d_0) \cup domain(d_1)) >$, which is the "restriction" of $\mathcal{M}$ to the type d and obviously is a model of type d, and $S \triangleq \mathcal{M}_{\underline{s}}$ (which is the universe of sort $\underline{s}$)[2].

□ Notation 4

Definition 5: (F_{ds}, F_d, L_{ds})

(i) Given a 1-sorted type d, F_{ds} denotes the set of all classical, first order, 2-sorted formulas of type ds with variables in $X = \{x_n : n \in \omega\}$ (the set of variables of sort $\underline{d}$), and in $Y = \{y_n : n \in \omega\}$ (the set of variables of sort $\underline{s}$). We assume that X and Y are disjoint.

(ii) F_d denotes the set of all (classical) first order formulas of type d with variables in X.

(iii) We have now all ingrediens together for the classical, 2-sorted language L_{ds} of type ds, namely $L_{ds} \triangleq < F_{ds}, Md_{ds}, \models>$, where $\models$ is the usual satisfaction relation. (For more on many-sorted first order logics see e.g. Monk[46].)

□ Definition 5

Notation 6: For any model $\mathcal{M} = <\underline{D}, S, \in^{\mathcal{M}}> \in Md_{ds}$, any a_1, $a_2 \in D$ and any $b \in S$, we will write $(a_1, a_2) \in^{\mathcal{M}} b$ instead of $(a_1, a_2, b) \in \in^{\mathcal{M}}$. In formulas we will replace $\in (x_1, x_2, y)$ by $(x_1, x_2) \in y$. (As indicated in Notation 4/(ii), we will simply write $\in$ instead of $\in^{\mathcal{M}}$.)

□ Notation 6

[2] $\triangleq$ denotes "equals by definition."

The language L_{ds} defined above is not yet a language of our framework logic. What is missing is that for the models $\mathcal{M}$ of L_{ds}, $\in^{\mathcal{M}}$ is not necessarily the true membership relation, moreover, we need the comprehension axioms to govern the behaviour of the objects of the "new" sort $\underline{s}$. So the final framework language FL_{ds} of type ds is going to be the restriction of L_{ds} to models of the axiom of extensionality and scheme of comprehension.

Definition 7: (Cox)

(i) The following formula Ex (of type) ds is called axiom of extensionality:

$$Ex \triangleq \forall y_1 \forall y_2 [y_1 = y_2 \longleftrightarrow \forall x_1 \forall x_2 ((x_1, x_2) \in y_1 \longleftrightarrow (x_1, x_2) \in y_2)].$$

(ii) The following set Co of formulas (of type ds) is called the comprehension scheme (and sometimes first order comprehension with parameters–see e.g. Makowsky- Sain[40]):

$$Co \triangleq \{\exists y \forall x \forall x' ((x, x') \in y \longleftrightarrow \phi(x, x', x_1, x_2, \ldots, x_n, y_1, y_2, \ldots, y_m)) :$$

$$\phi(x, x', x_1, x_2, \ldots, x_n, y_1, y_2, \ldots, y_m) \in F_{ds}\}.$$

(iii) $Cox \triangleq Co \cup \{Ex\}$

(iv) Given any type t and a set Th of formulas of type t, we denote by $Md_t(\mathrm{Th})$ the class of those models of type t, which satisfy (all formulas in) Th.

(v) Now we are ready to define our framework logic FL. In general, a logic L is a (meta)function $L = < L_t : t \in T >$, where each L_t is a language of type t and T is a class of types.

□ Definition 7

Definition 8: (FL, FL_{ds})

Our framework logic FL $\triangleq <$ FL$_{ds}$: d is a 1-sorted type$>$, and for each 1-sorted type d, FL$_{ds}$ $\triangleq$ $< F_{ds}, Md_{ds}(Cox), \models >$ is the restriction of the classical, first order, 2-sorted language L_{ds} to those models $\mathcal{M}$ which satisfy the axioms of comprehension and in which $\in^{\mathcal{M}}$ is the true membership relation (between $D \times D$ and S).

□ Definition 8

Remarks: 1) The models of our framework language FL$_{ds}$ of type ds are in fact triples $< \underline{D}, S, \in >$, where $\underline{D}$ is a model of type d, $S \subseteq P(D \times D)$ and $\in$ is the membership relation ($\subseteq D \times D \times S$).

2) Concluding this section we would like to note once again (see Introduction), that we arrived at weak second order logic as our framework logic (cf. Makowsky- Sain[39],[40] for more details and literature on weak second order logic).

3 Recursive Programs and Denotational Semantics

In this section we will first define the syntax of our programming language. This language consists of recursive programs and is strictly more powerful than the language of corresponding type with only assignments and while-loops -see section 4 for examples. Given a (recursive) program π and a model $\mathcal{M} =< \underline{D}, S, \in>$, we define the meaning $\pi^{\mathcal{M}}$ of π in $\mathcal{M}$ denotationally as the least fixed point of a function(al) on S. Finally, we prove that $\pi^{\mathcal{M}}$ exists in S and is a function (on D).

Definition 9: (the programming language P_d)

Throughout this definition, let d be an arbitrary 1-sorted type (cf. Definition 1/(i)). Further, let P be a set whose elements we call program symbols and which are all different from the symbols of d.

(i) For every program symbol $p \in$ P and variable $x \in X$ (of sort $\underline{d}$), we define the 1-sorted type dpx as follows: $(dpx)_1 = \emptyset$, and $(dpx)_0 = d_0 \cup \{< p, 1 >\} \cup \{< if\phi(x)\ then\ else,\ 2 >$: $(\phi =)\phi(x) \in F_d$ is quantifier free$\}$, i.e. dpx has no relation symbols and in addition to the function symbols in d (whose arities remain unchanged), it has one unary function symbol p, and for every quantifier-free formula $\phi(= \phi(x))$ of type d with no variable other than x occuring in it, the binary function symbol "$if\phi(x)\ then\ else$".

(ii) We denote by Tm_{dpx} the set of all terms of type dpx (in the usual sense of logic–cf. Monk[46] def. 11.1). An example of such a term would be: $if\phi(x)\ then\ else(p(x_2), p(p(x_3)))$, where ϕ is a quantifier-free formula of type d with all variables in x . We will return to the usual notation "$if\phi(x)\ then\ \tau_1\ else\ \tau_2$" instead of writing "$if\phi(x)\ then\ else(\tau_1, \tau_2)$".

(iii) We are now ready to define the programs of type d. A string of symbols is a program of type d exactly if it is of the form $p(x) := \tau(x)$, where $p \in$ P, $x \in$X and $\tau(= \tau(x)) \in Tm_{dpx}$ is a term of type dpx with no variable other than x occurring in it.

We denote by P_d the set of all programs of type d.

□ Definition 9

A typical *example* of a (type d and a) program in P_d is the following: $p(x) :=\ if\ r(x)\ then\ f(x)\ else\ h(p(g_1(x)),$

$p(g_2(x)))$, where r is a unary relation symbol, f, g_1 and g_2 are unary function symbols, and h is a binary function symbol of type d.

Theorem 4-6 of Manna[42] proves that there is no flowchart (i.e. while-) program which is equivalent to this recursive program.

Remarks: 1) As defined above, our programming language has the flaw that it does not allow composition of programs. However, we happily accept this flaw as a price we pay for simplicity.

Our aim in this version of the paper, is to concentrate on pure recursive programs which are simple enough and yet strictly more powerful than while-programs. It is very easy to get rid of this flaw by allowing $\tau(x)$ in the definition of $p(x) := \tau(x)$ to contain symbols of programs which are different from p and whose definitions have already been completed. In this case a program like $q(x) := p(r(x))$ would also be legal, assuming, say, that p and r appear on the left hand side of program definitions of the type defined in Definition 9/(iii).

2) Our definition allows only unary programs. Again, this is a price we pay for simplicity. However, in later versions we will turn to the most general kind of program definition, which is of the form $p(x_1, \ldots, x_n) = (p_1(x_1, \ldots, x_n), \ldots, p_n(x_1, \ldots, x_n))$, where $p_1(x_1, \ldots, x_n) = \tau_1(x_1 \ldots, x_n)$ $, \ldots, p_n(x_1, \ldots, x_n) = \tau_n(x_1, \ldots, x_n)$, and each τ_i may contain $p_1 \ldots, p_n$. Note that in this definition composition becomes legal. The transition from the unary to the n-ary case is in fact straightforward–see Makowsky-Sain[40], pg. 6, Remark (ii).

□ Remarks

Next, we would like to define the (denotational) semantics of our programming language(s) P_d . For this purpose, let the 1- sorted type d be arbitrary, but fixed.

Definition 10: (semantics of P_d)

Let $p(x) := \tau(x)$ be a (recursive) program of type d as defined above in Definition 9/(iii) and let us denote it by π or $\pi(x)$ to indicate that x is the only variable occurring in it. Further, let $\mathcal{M} =< \underline{D}, S, \in>$ be a model of our framework language FL_{ds} (i.e. $\underline{D}$ is a model of type d, $S \subseteq P(D \times D)$, $\in$ is the membership relation between $D \times D$ and S, and $\mathcal{M}$ satisfies the comprehension axioms). We are going to define the meaning $\pi^{\mathcal{M}}$ of π in $\mathcal{M}$.

(i) First, we define a function $F_\tau : P(D \times D) \to P(D \times D)$, uniquely determined by the right hand side $\tau(x)$ of π. For every $R \in P(D \times D)$, $F_\tau(R) \triangleq \tau^{\mathcal{M}}(p/R)$, where $\tau^{\mathcal{M}}(p/R)$ (read "the meaning of τ in $\mathcal{M}$ with p evaluated to R") is a (binary) relation on D defined recursively as follows: let $a, b \in D$ be arbitrary:

1. Suppose that $\tau(x) = x$. Then $(a, b) \in \tau^{\mathcal{M}}(p/R)$ iff $a = b$.
2. Suppose that $\tau(x) \neq x$ and that for every subterm τ' of τ, $\tau'(p/R)$ has already been defined.
   - (a) If $\tau(x) = f(\tau_1(x), \ldots, \tau_n(x))$ for some function symbol f of d of arity n, then $(a, b) \in \tau^{\mathcal{M}}(p/R)$ iff $b = f^{\mathcal{M}}(b_1, \ldots, b_n)$ for some $b_1, \ldots, b_n \in D$ such that $(a, b_i) \in \tau_i^{\mathcal{M}}(p/R)$ for $i = 1, 2, \ldots, n$.
   - (b) Suppose $\tau(x) = p(\tau'(x))$. Then $(a, b) \in \tau^{\mathcal{M}}(p/R)$ iff there is an element $c \in D$, such that $(a, c) \in \tau'^{\mathcal{M}}(p/R)$ and $(c, b) \in R$.

(c) Let $\tau(x) = if\ \phi(x)\ then\ \tau_1(x)\ else\ \tau_2(x)$. Then $(a,b) \in \tau^{\mathcal{M}}(p/R)$ iff $(a,b) \in \tau_1^{\mathcal{M}}(p/R)$ in case $\mathcal{M} \models \phi[a]$, and $(a,b) \in \tau_2^{\mathcal{M}}(p/R)$ otherwise.

(ii) Finally, we are ready to define $\pi^{\mathcal{M}}$, namely $\pi^{\mathcal{M}} \triangleq \mu_S F_\tau$, where $\mu_S F_\tau$ denotes the least fixed point of F_τ in S (i.e. a relation $R \in S$, such that $F_\tau(R) = R$ and for any $R' \in S$, if $F_\tau(R') = R'$, then $R \subseteq R'$).

□ Definition 10

Of course, the first question one raises at this point is whether the above definition makes sense, i.e.

1. Does $\mu_S F_\tau$ exist in S?
2. If it does, is it a (partial) function, as one would expect (since $\pi(x)$ is deterministic)?

The answers to these questions are given in

Theorem 11: ($\pi^{\mathcal{M}}$ exists and is a function)

Let the type d, the model $\mathcal{M} =< \underline{D}, S, \in>$, the program $\pi = \pi(x)$ and the function F_τ all be as above, in Definition 10. Then the least fixed point $\mu_S F_\tau$ in S of F_τ really exists, moreover, it is a function.

□ Theorem 11

Summing up what we have done so far, we have defined the meaning of (recursive) programs $p(x) := \tau(x)$ of type d in an arbitrarily given model $\mathcal{M} =< \underline{D}, S, \in>$ of our framework language FL_{ds} of type ds as the least fixed point $\mu_S F_\tau$ of the function F_τ (defined by $F_\tau(R) = \tau^{\mathcal{M}}(p/R)$ for every $R \subseteq D \times D$). We have then proved that this definition makes sense.

4 Examples

To gain insight into the meaning of recursive program in (nonstandard) models $\mathcal{M} =< \underline{D}, S, \in> \in Md_{ds}(Cox)$, we will look at three examples.

First example:

The type d of this example has a unary operation symbol *pred* (for predecessor), a constant symbol 10 and a binary relation symbol $<$. $\underline{D}_1$ denotes the model of type d whose universe is ω (the set of all natural numbers) augmented by a simple Z-chain $\{z_i : i \in Z\}$, with the symbols *pred*, 10 and $<$ given their obvious meanings (for all $i, j \in Z$, $pred^{\underline{D}_1}(z_i) = z_{i-1}$, $z_i <^{\underline{D}_1} z_j$ iff $i < j$ and for all $n \in \omega$, $n < z_i$).

Now let us define the following recursive program $\pi_1 : p(x) := if\ x > 10\ then\ p(pred(x))\ else\ x$. If we denote the right hand side by τ, then the meaning of π_1 in any model $\mathcal{M} =< \underline{D}_1, S, \in>$ of FL_{ds} is given by the least fixed point of F_τ in S. But first, let us examine the fixed points of F_τ in $P(D_1 \times D_1)$. It is easy to check that $F_\tau(R_{st}) = R_{st}$ for $R_{st} \triangleq \{(n,n) : n <= 10\} \cup \{(m,10) : m \in \omega, m > 10\}$, which is the usual, standard meaning of π_1 in $\underline{D}_1$. But for every element $a \in D_1$, the function $R_a \triangleq R_{st} \cup \{(z_i, a) : i \in Z\}$ is also a fixed point of F_τ, moreover, one can easily show that these are all the fixed points of F_τ which are functions. Of special interest is R_{10}, which is the only functional fixed point of F_τ definable in F_d (by the formula $((x < 10 \vee x = 10) \wedge x = x') \vee (x' = 10 \wedge x > 10))$. As a consequence, since $\mathcal{M} \models Co$, $R_{10} \in S$, *no matter what S exactly is.* (R_{10} is the least definable fixed point of F_τ in the sense of Cartwright[14].) Although we know very little about $S \subseteq P(D_1 \times D_1)$, we can nevertheless conclude, due to $\mathcal{M} \models Co$, that the meaning of π_1 in $\mathcal{M}$ is either R_{10}, if S contains none of the other fixed points R_a, or otherwise R_{st} (the standard meaning of π_1 in $\underline{D}_1$), since for any two fixed points, their intersection is R_{st} and has to be in S (because of $\mathcal{M} \models Co$).

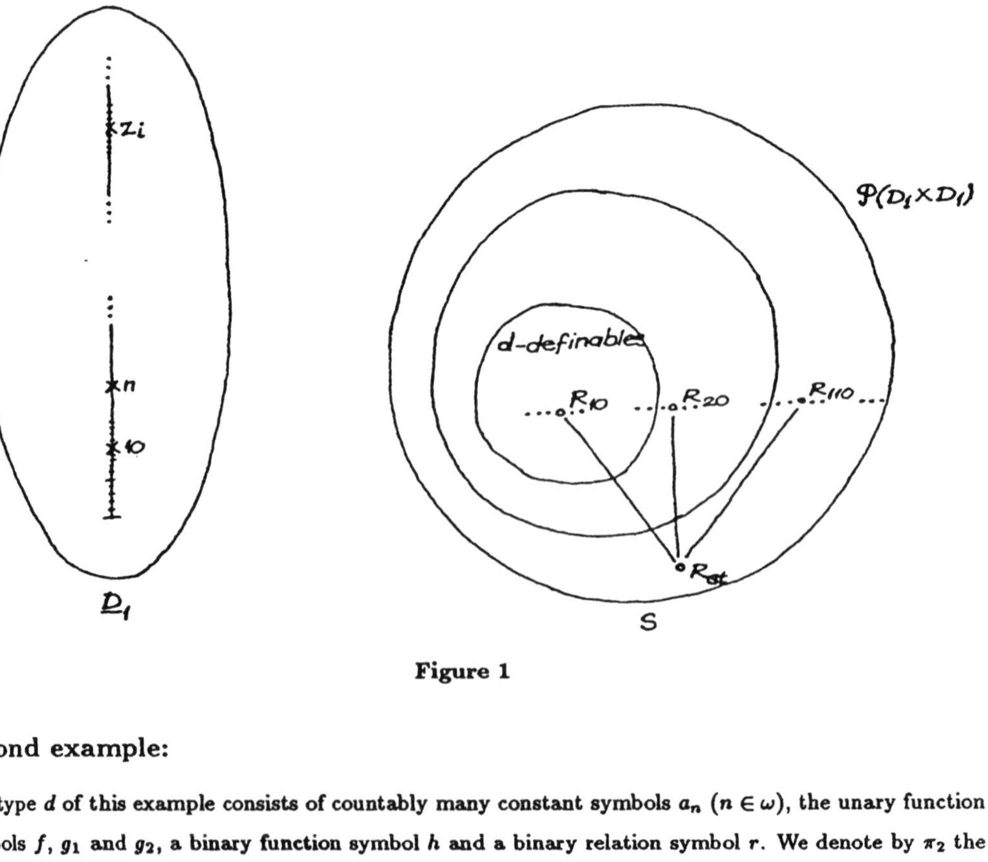

Figure 1

Second example:

The type d of this example consists of countably many constant symbols a_n $(n \in \omega)$, the unary function symbols f, g_1 and g_2, a binary function symbol h and a binary relation symbol r. We denote by π_2 the program $p(x) := \mathit{if}\, r(x)\, \mathit{then}\, f(x)\, \mathit{else}\, h(p(g_1(x)), p(g_2(x)))$. Unlike π_1, this program is not equivalent to any flowchart (i.e. while-) program (see Manna[42] Theorem 4-6). However, unlike in $\underline{D}_3$ of the third example, in $\underline{D}_2$ defined below, π_2 is going to be equivalent to a while-program. As a convention, we will sometimes drop, throughout the remainder of this section, superscripts like e.g. in $a_n^{\underline{D}_2}$ (which we may denote by a_n) or $h^{\underline{D}_2}$ (which we may denote by h), and also parentheses like e.g. in $g_1(g_2(g_1(a_n)))$ (which we may denote by $g_1g_2g_1(a_n)$). The model $\underline{D}_2$ of type d is the absolutely free (or Herbrand) algebra of the type d but without the function symbol h, generated by the constants a_n ($n \in \omega$) and an extra element b_0; in addition to this, $h^{\underline{D}_2}(a, a') = b_0$ for any $a, a' \in D_2$ and the relation $r^{\underline{D}_2}$ consists exactly of the elements of the form $g_ig_i \dots g_i(a_n)$, where g_i occurs n times and $n \in \omega$ and i is 1 or 2 (e.g. $g_1g_2g_1(a_3)$, $g_2g_1(a_2) \in r^{\underline{D}_2}$).

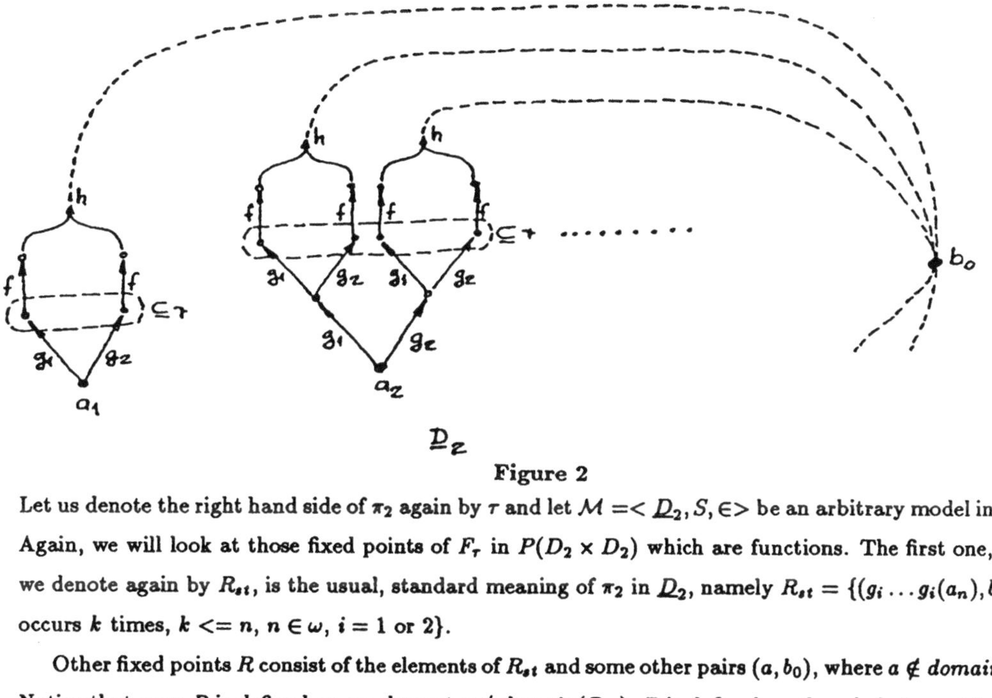

Figure 2

Let us denote the right hand side of π_2 again by τ and let $\mathcal{M} =< \underline{D}_2, S, \in>$ be an arbitrary model in FL_{ds}. Again, we will look at those fixed points of F_τ in $P(D_2 \times D_2)$ which are functions. The first one, which we denote again by R_{st}, is the usual, standard meaning of π_2 in $\underline{D}_2$, namely $R_{st} = \{(g_i \dots g_i(a_n), b_0) :\ g_i$ occurs k times, $k <= n$, $n \in \omega$, $i = 1$ or $2\}$.

Other fixed points R consist of the elements of R_{st} and some other pairs (a, b_0), where $a \notin domain(R_{st})$. Notice that once R is defined on an element $a \notin domain(R_{st})$, R is defined on the whole tree of figure 3,

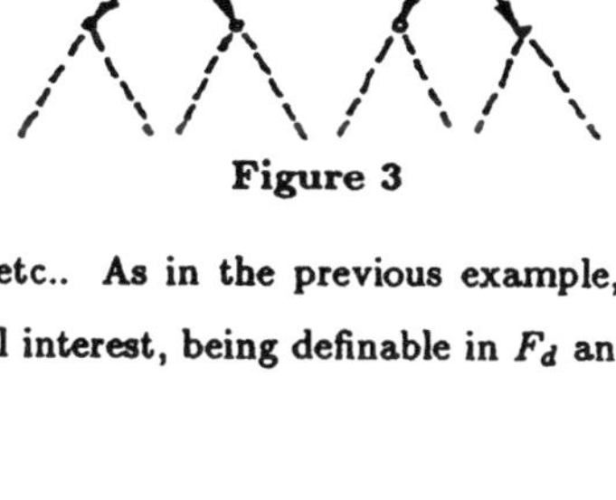

Figure 3

since $F_\tau(R)(a) = h(Rg_1a, Rg_2a)$ etc.. As in the previous example, one of these fixed points, namely $U = \{(a, b_0) : a \in D_2\}$, is of special interest, being definable in F_d and therefore being an element of any

S whatsoever (since $\mathcal{M} \models Co$). U containing all other (functional) fixed points of F_τ in $P(D_2 \times D_2)$, all we can say is that $\pi_2^{\mathcal{M}}$, i.e. the least fixed point of F_τ in S, is either U, if S contains none of the other fixed points R mentioned above, or one of the other fixed points R, in which case it is the intersection of all the fixed points of F_τ in S. This is not necessarily R_{st}.

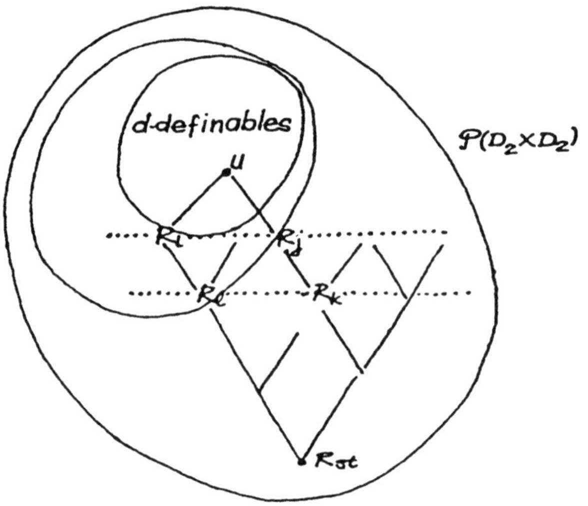

Figure 4

Third example:

The type d of this example is the same as in the second example and π denotes the program π_2 of the second example. We will give a stepwise description of the model $\underline{D}_3$ of type d. First we take a partial algebra $\underline{D}$ (i.e. the operations may not be defined everywhere) consisting of the constants a_n $(n \in \omega)$ and an infinite binary tree B whose root we denote by b_0 and with the property that every node b_m has exactly two sons b_{m0} and b_{m1} $(m \in \omega^*)$, furthermore $h(b_{m0}, b_{m1}) = b_m$. Now $\underline{D}_3$ is the free completion of type d of $\underline{D}$ augmented by the unary relation $r^{\underline{D}_3}$ which consists of exactly the same elements as $r^{\underline{D}_2}$.

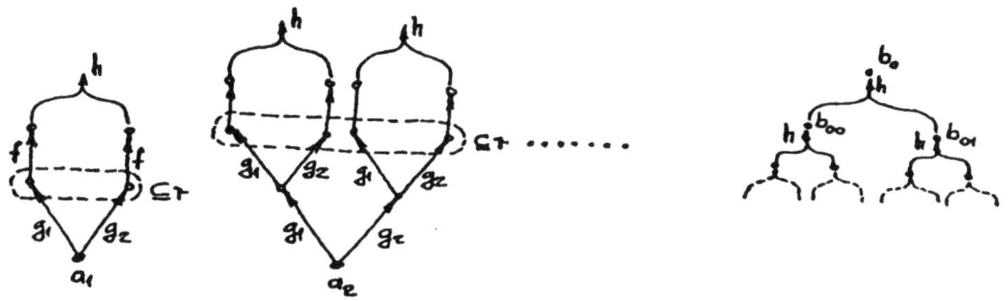

$\underline{D}_3$

Figure 5

Notice that there is no while-program whose standard meaning in $\underline{D}_3$ is the same as the standard meaning of π in $\underline{D}_3$ (cf. the proof of Theorem 4-6 in Manna[42]). Let $\mathcal{M} =< D_3, S, \in>$ be an arbitrary model in FL_{ds}. Before discussing the meaning of π in $\mathcal{M}$, we will look again at the fixed points of F_τ in $P(D_3 \times D_3)$ (τ denoting the right hand side of π).

The standard meaning of π in $\underline{D}_3$, which we denote again by R_{st}, is $R_{st} = \{(a_1, h(fg_1a_1, fg_2a_1)), (g_1a_1, fg_1a_1), (g_2a_1, fg_2a_1), (a_2, h(h(fg_1g_1a_2, fg_2g_1a_2), h(fg_1g_2a_2, fg_2g_2a_2))), (g_1a_2, h(fg_1g_1a_2, fg_2g_1a_2)), \ldots\}$. F_τ has also (uncountably many) fixed points R which contain R_{st} and assign to some elements $a \notin domain(R_{st})$ elements $b_m \in B$ in a precisely prescribed fashion: if $R(a) = b_m$, then necessarily $R(g_1a) = b_{m0}$ and $R(g_2a) = b_{m1}$ because $F_\tau(R)(a) = h(Rg_1a, Rg_2a)$ and $F_\tau(R)(a) = R(a) = b_m$. This example differs from the two earlier examples in many respects. If we denote by S_0 the set of binary relations on D_3 which are (first order) definable in $\underline{D}_3$, then F_τ has no fixed point at all in S_0. We will next assume that S (of our model $\mathcal{M}$) is the "closure of S_0 under comprehension," denoted by S_0. First we form S_1, the set of all relations, which are (first order) definable in $< \underline{D}_3, S_0, \in>$. Then e.g. $F_\tau(D_3 \times D_3)$, $F_\tau(F_\tau(D_3 \times D_3))$, and in general $F_\tau^n(D_3 \times D_3) \in S_1$ $(n \in \omega)$. Let S_2 be the set of those relations, which are definable in $< \underline{D}_3, S_1, \in>$. Then certainly $U \triangleq \cap\{R : R \in S_1,\ F_\tau(R) \subseteq R\}$ has to be in S_2 . One can see that U consists of the elements of R_{st} and of all pairs (a, b), where $a \notin domain(R_{st}$) and $b \in B$ or b is of the form $h(b_1, b_2)$, or $h(h(b_1, b_2), h(b_3, b_4))$, etc., where all $b_i \in B$. If S_3 denotes the set of those relations which are definable in $< \underline{D}_3, S_2, \in>$, then $R_{st} \in S_3$, because R_{st} is definable as the set of all $(a, b) \in U$, such that U is functional on a. Now $S = \mathrm{S}_0$ can be obtained by continuing this process (i.e. S_α is the set of relations definable in $< \underline{D}, S_\beta, \in>$ if α is a successor ordinal of the form $\beta + 1$, $S_\alpha \triangleq \cup\{S_\beta : \beta < \alpha\}$ if α is a limit ordinal, and $\mathrm{S}_0 \triangleq \cup\{S_\alpha : \alpha < \omega_1\}$, since $S_{\omega_1} = S_{\omega_1+1}$), but the point is, however, that $R_{st} \subseteq \mathrm{S}_0$, and since R_{st} is the least fixed point of F_τ in $P(D_3 \times D_3)$, R_{st} must be the least fixed point of F_τ in $\mathcal{M} =< \underline{D}_3, \mathrm{S}_0, \in>$. In other words, the meaning $\pi^{\mathcal{M}}$ of π in $\mathcal{M} =< \underline{D}, \mathrm{S}_0, \in>$ is just the standard one.

However, for arbitrary S, there is quite little one can say about $\pi^{\mathcal{M}}$, since the least fixed point of F_τ in S depends very much on S itself. For example, one can easily imagine an S for which $\pi^{\mathcal{M}}$ is $R_{st} \cup R_a^b$, where a is an arbitrary element not in the domain of R_{st} and b is an arbitrary node of B and R_a^b is defined as follows: $(a, b) \in R_a^b$, and whenever $(a',\ b') \in R_a^b$, (g_1a', c) and (g_2a', d) are as well in R_a^b, and $b' = h(c, d)$; nothing else is in R_a^b. On the other hand, if both, say $R_1 = R \cup R_a^b$ and $R_2 = R \cup R_a^{b'}$ ($b \neq b'$), are in S, then $\pi^{\mathcal{M}} = R_{st}$ again, since $R_{st} = R_1 \cap R_2 \in S$.

5 The First Order Dynamic Logic DL^{rec} of Recursive Programs. Completeness Theorem

After having defined the syntax and the semantics of the programming language P_d for every given type d, we are now ready to define the first order language DL_d^{rec} of type ds for reasoning about the programs in P_d. The logic $DL^{rec} = \{DL_d^{rec} : d \text{ a 1-sorted type}\}$ will turn out to be the absolute version of the standard logic or denotational semantics of recursive programs (cf. Introduction) we were searching for. The main result of this section (and in fact of the paper) is the (strong) completeness of DL^{rec} : for every type d there is a (decidable) proof concept $\overset{rec}{\vdash}$ for the language DL_d^{rec}, such that for every set Th of formulas of DL_d^{rec} and for every formula ϕ of DL_d^{rec}, $Th \models \phi$ iff $Th \overset{rec}{\vdash} \phi$.

Definition 12: (the language DL_d^{rec})

Let d be an arbitrary 1-sorted type.

(i) DF_d^{rec} is defined to be the smallest set satisfying the following three conditions:

1. $F_{ds} \subseteq DF_d^{rec}$. (Remember that F_{ds} is the set of all first order formulas of type ds – see Definition 5/(i).)
2. For every program $\pi \in P_d$ and every formula $\psi \in DF_d^{rec}$, $\Box(\pi, \psi) \in DF_d^{rec}$, and
3. For any formulas $\phi, \psi \in DF_d^{rec}$ and any variable $z \in X \cup Y$, $\{\neg\phi, (\phi \wedge \psi), \exists z\phi\} \subseteq DF_d^{rec}$.
   This defines the set DF_d^{rec} of dynamic formulas of type d.

(ii) Now we define the meaning of the dynamic formulas in the 2-sorted models $\mathcal{M} \in Md_{ds}(Cox)$. Let therefore $\mathcal{M} =< \underline{D}, S, \in>$ be such a model and let $e_1 : X \to D$ and $e_2 : Y \to S$ be evaluations of the variables into $\mathcal{M}$ of sort $\underline{d}$ and $\underline{s}$ respectively. We are going to define $\mathcal{M} \models \phi[e_1, e_2]$ for all $\phi \in DF_d^{rec}$.

1. If $\phi \in F_{ds}$, then $\mathcal{M} \models \phi[e_1, e_2]$ is already defined (see Definition 5/(iii)).
2. Let $\pi = \pi(x) \in P_d$ and $\psi \in DF_d^{rec}$ and assume that $\mathcal{M} \models \psi[l_1, l_2]$ has already been defined for arbitrary evaluations l_1 and l_2 of the variables into $\mathcal{M}$. Then $\mathcal{M} \models \Box(\pi, \psi)[e_1, e_2]$ iff for every $a \in D$, $(e_1(x), a) \in \pi(x)^{\mathcal{M}}$ implies $\mathcal{M} \models \psi[e_1(a/x), e_2]$, where the variant $e_1(a/x)$ of e_1 is defined as usually: for every $x' \in X$,

$$e_1(a/x)(x') = \begin{cases} e_1(x') & \text{if } x \neq x' \\ a & \text{otherwise} \end{cases}$$

   Notice that the above definition is the usual definition of the partial correctness of π in $\mathcal{M}$ with respect to the output condition ψ.
3. Let $\phi, \psi \in DF_d^{rec}$ and $z \in X \cup Y$. Then $\mathcal{M} \models \neg\phi[e_1, e_2]$, $\mathcal{M} \models (\phi \wedge \psi)[e_1, e_2]$ and $\mathcal{M} \models (\exists z\phi)[e_1, e_2]$ are defined the usual way.

4. For any $\phi \in DF_d^{rec}$, $\mathcal{M} \models \phi$ iff $\mathcal{M} \models \phi[e_1, e_2]$ for all evaluations e_1 and e_2 of sort $\underline{d}$ and $\underline{s}$ respectively into $\mathcal{M}$.

(iii) The language DL_d^{rec} of type d of the first order dynamic logic for recursive programs is defined as $DL_d^{rec} = < DF_d^{rec}, Md_{ds}(Cox), \models >$, where $\models$ is defined in (ii) above.

□ Definition 12.

Notation 13: **(i)** We will use the derived connectives $\forall, \rightarrow, \longleftrightarrow, \wedge, \diamond$ in the usual, standard way (e.g. $\diamond(\pi, \psi)$ abbreviates $\neg\Box(\pi, \neg\psi)$).

(ii) The partial correctness assertion $\{\phi\}\pi\{\psi\}$ becomes here the formula $\phi \rightarrow \Box(\pi, \psi)$.

(iii) Let $Th \subseteq DF_d^{rec}$, $\phi \in DF_d^{rec}$ and $\mathcal{M} \in Md_{ds}(Cox)$. We will use the notations $\mathcal{M} \models Th$ and $Th \models \phi$ in the usual way, e.g. $Th \models \phi$ means that for every model $N \in Md_{ds}(Cox)$, $N \models Th$ implies $N \models \phi$.

□ Notation 13.

We will now prepare our proof of the completeness of DL_d^{rec}. Throughout the remainder of this section let d be an arbitrary 1-sorted type. First recall from Definition 5/(iii) the classical, 2-sorted, first order language L_{ds}.

Definition 14: (proof concept – see Monk[46])

Let $L = < F, M, \models >$ be a language. By a proof concept for L we understand a relation $\vdash \subseteq P(F) \times F$ together with a set $Pr \subseteq P_\omega(F) \times F^* \times F$, such that for every set $Th \subseteq F$ and every $\phi \in F$, $Th \vdash \phi$ iff $< H, w, \phi > \in Pr$ for some finite set $H \subseteq Th$ and for some $w \in F^*$. The proof concept $<\vdash, Pr>$ is decidable iff the set Pr is decidable (in $P_\omega(F) \times F^* \times F$, where $P_\omega(F)$ denotes the set of all finite subsets of F). Sometimes we will say "$\vdash$ is decidable", rather than "$<\vdash, Pr>$ is a decidable proof concept".

□ Definition 14.

Notation 15: (the classical proof concept for L_{ds})

(i) We will denote by $<\vdash, Prc>$ the classical proof concept of L_{ds} (cf. Monk[46] 10.14-10.27). In other words $Prc \triangleq \{< H, w, \phi >: H \subseteq F_{ds}$ is finite, $\phi \in F_{ds}$ and w is a classical first order proof of ϕ from H in the sense of Monk[46] 10.24$\}$. We also recall that $<\vdash, Prc>$ is a decidable, sound and complete proof concept for L_{ds}, i.e. Prc is decidable and for any $Th \subseteq F_{ds}$ and $\phi \in F_{ds}$, $Th \models \phi$ iff $Th \vdash \phi$.

(ii) The proof concept $<\vdash, Prc>$ immediately provides a decidable, sound and complete proof concept for our framework langugage FL_{ds}. Namely, in FL_{ds} it is true that for every $Th \subseteq F_{ds}$

and $\phi \in F_{ds}$, $Th \models \phi$ iff $(Th \cup Cox) \vdash \phi$. We denote this proof concept for FL_{ds} by $<\vdash, Prcox>$, where $Prcox \triangleq \{<H, w, \phi>: H \subseteq F_{ds}$ is finite, $\phi \in F_{ds}$ and w is a classical first order proof of ϕ from $H \cup Cox\}$.

□ Notation 15.

To prove the completeness of DL_d^{rec}, we will pretty much follow the idea of the proof of Theorem 2 of Andreka-Nemeti-Sain[8]. The idea is to translate the language DL_d^{rec} to the complete (and compact) framework language $FL_{ds} =< F_{ds}, Md_{ds}(Cox), \models>$ by a total computable function $\Theta : DF_d^{rec} \rightarrow F_{ds}$ in such a way, that for every $\phi \in DF_d^{rec}$ and every $\mathcal{M} \in Md_{ds}(Cox)$, $(\star)\mathcal{M} \models (\phi \longleftrightarrow \Theta(\phi))$. Then, for any $Th \subseteq DF_d^{rec}$ and $\phi \in DF_d^{rec}$, we will be able to define a $\overset{rec}{\vdash}$-proof of ϕ from Th as a sequence $< H, <\Theta(H), w, \Theta(\phi)>, \phi >$ where $H \subseteq Th$ is a finite subset of DF_d^{rec} and $<\Theta(H), w, \Theta(\phi)> \in Prcox$, i.e. is a classical first order proof of $\Theta(\phi)$ from $\Theta(H) \cup Cox$. Since Θ is a total computable function, $\overset{rec}{\vdash}$ will be decidable, moreover, using the completeness of FL_{ds} and property $(\star)$ of Θ, we will have then proved the completeness theorem for DL_d^{rec}.

Remark: The proof concept $\overset{rec}{\vdash}$ for DL_d^{rec} just described above is obviously not going to be a Hilbert style inference system. But following the example of Sain[59], it would not be difficult to give a Hilbert style inference system. In this first version of the paper however, our goal is to convey the NLP-approach described in the Introduction and not so much the esthetics.

Proposition 16: (the translation function $\Theta : DF_d^{rec} \rightarrow F_{ds}$)

There is a total computable function $\Theta : DF_d^{rec} \rightarrow F_{ds}$, such that condition $(\star)$ below holds.

$(\star)$ For every $\phi \in DF_d^{rec}$ and $\mathcal{M} \in Md_{ds}(Cox)$, $\mathcal{M} \models (\phi \longleftrightarrow \Theta(\phi))$.

□ Proposition 16.

Definition 17: (The proof concept of DL_d^{rec})

By a $\overset{rec}{\vdash}$ - proof of $\phi \in DF_d^{rec}$ from $Th \subseteq DF_d^{rec}$ we understand a sequence $< H, <\Theta(H), w, \Theta(\phi)>, \phi >$, such that $H \subseteq Th$ and $<\Theta(H), w, \Theta(\phi)> \in Prcox$, i.e. is a classical proof of $\Theta(\phi)$ from $\Theta(H) \cup Cox$. In more detail: Let $Prec \triangleq \{< H, <\Theta(H), w, \Theta(\phi)>, \phi >: H \subseteq DF_d^{rec}$ finite, $\phi \in DF_d^{rec}$, $<\Theta(H), w, \Theta(\phi)> \in Prcox\}$. For any $Th \subseteq DF_d^{rec}$ and $\phi \in DF_d^{rec}$, we define $Th \overset{rec}{\vdash} \phi$ iff there is a $H \subseteq Th$ and $v \in P_\omega(F_{ds}) \times F_{ds}^* \times F_{ds}$, such that $< H, v, \phi > \in Prec$. Notice that $<\overset{rec}{\vdash}, Prec>$ is a proof concept in the sense of Definition 16 (except for the fact that $v =< \Theta(H), w, \Theta(\phi)> \in P_\omega(F_{ds}) \times F_{ds}^* \times F_{ds}$, but $P_\omega(F_{ds}) \times F_{ds}^* \times F_{ds}$ can easily be identified with F_{ds}^*).

□ Definition 17.

Proposition 18: $\overset{rec}{\vdash}$ is a decidable proof concept.

proof: using the facts that DF_d^{rec} and $<\vdash, Prcox>$ are decidable and that Θ is computable, the reader can easily complete the proof.

□ Proposition 18.

Now we are ready to state our main result.

Theorem 19: (strong completeness of DL_d^{rec})

(i) The proof concept $<\overset{rec}{\vdash}, Prec>$ is a (decidable,) sound and complete proof system for DL_d^{rec}, i.e. for any $Th \subseteq DF_d^{rec}$ and $\phi \in DF_d^{rec}$, $Th \models \phi$ iff $Th \overset{rec}{\vdash} \phi$.

(ii) The language DL_d^{rec} is compact.

□ Theorem 19

6 Discussion and the Relation to Other Logics of Programs.

Referring to the Introduction, let us recall that one of the most important objectives of the NLP-approach is to provide adequate background or frame logics for the comparative study of program verification methods for given programming languages. As a successful product of the NLP-approach let us look, for example, at Nonstandard Dynamic Logic (NDL for short). NDL is one of the many possible absolute versions (see e.g. Csirmaz[16], Makowsky-Sain[40] or Sain[63]) of Dynamic Logic (cf. e.g. Harel[33],[34]). NDL is a 3-sorted first order logic of programs with a decidable, complete proof concept (see e.g. Andreka-Nemeti-Sain[8]). As pointed out e.g. in Pasztor[49],[50] or [51], the merit of the completeness of NDL does not so much lie in itself as in the fact, that it provides enough (standard and nonstandard) models to be able to accommodate the models (i.e. semantics) for a great number of existing as well as new (interesting) program verification methods. To illuminate what we just said, let us evoke one of the first results of NDL: The partial correctness of a while-program is Floyd-Hoare-provable iff it is partially correct in exactly those models of NDL which satisfy induction on time, i.e. computational induction, restricted to formulas with no quantified variables of sort time. More generally, a typical result of NDL would read like this:

(**) The property P of a program π is provable by the (verification) method M iff π has property P in all models of NDL satisfying the (decidable!) set of axioms Ax_M of NDL.

One immediate consequence of (**), of course, is the completeness of the method M endowed with the proper semantics, i.e. all models of Ax_M in NDL. But a *far* more important consequence of (**) is that it provides a ground for *comparing* the powers of different methods M for proving property P of programs:

comparing the power of a method M with that of an other method, say, M', boils down, in view of (**), to comparing Ax_M with Ax'_M within NDL, i.e. within one and the *same* logic.

(Why one would like to compare the powers of different methods is a long story, which could and may be should be told in a separate paper, but a temporary answer would be the following: one would like to know the "price" of each method, e.g. it follows from the results mentioned above, that, although weaker than Burstall's verification method, Floyd's method is "cheaper," because it requires only a restricted type of computational induction, while Burstall's method requires full induction. So, depending on the kind of programs we want to prove correct, we can decide on the "price" we have or do not have to pay.) What is the relation of all this to the results of the present paper? Let us recall that DL^{rec} is an absolute version of standard denotational semantics of recursive program, the same way NDL is an absolute version of DL. Its proof concept $\overset{rec}{\vdash}$ provides a (syntactic) method to prove properties of recursive programs π. If π happens to be equivalent to a while-program, then obviously the question arises how $\overset{rec}{\vdash}$ relates to the other verification methods. To answer this, we need to characterize $\overset{rec}{\vdash}$ in NDL. As a result, we obtain the following instance of (**): for every recursive program π of type d, equivalent to a while-program π' of type d, the partial correctness of π with respect to a formula $\psi \in F_d$ (i.e. $\Box(\pi, \psi)$) is $\overset{rec}{\vdash}$-provable in DL_d^{rec} iff π' is partially correct with respect to ψ in all models of NDL satisfying the set DAX of axioms of NDL (defined and discussed in e.g. Sain[58],[59],[62],[65] or in Sain-Nemeti[66]). So with respect to partial correctness (and in fact it is interesting to note that with respect to total correctness, too,) the proof concept $\overset{rec}{\vdash}$ is equivalent to the proof concept $DAX \overset{NDL}{\vdash}$ (i.e. the proof concept of NDL endowed with DAX as logical axioms). It is not relevant for us here to know what exactly DAX is, but it is relevant that DAX is more powerful than any of the known partial correctness methods (like Floyd-Hoare, Pnueli, Burstall, etc.). In more detail, if $\Box(\pi', \psi)$ is provable by, say, Pnueli's method, then it is also provable by $DAX \overset{NDL}{\vdash}$, i.e. $\Box(\pi, \psi)$ is provable by $\overset{rec}{\vdash}$ in DL_d^{rec}. But there are programs π in P_d, equivalent to while-programs π', whose partial correctness with respect to some $\psi \in F_d$ is provable in DL_d (i.e. $\Box(\pi', \psi)$ is provable by $DAX \overset{NDL}{\vdash}$), but $\Box(\pi', \psi)$ is not provable by any of the above mentioned methods. (For an example for such a π—a theorem prover—see Sain[65]). The situation is very similar concerning total correctness: $\overset{rec}{\vdash}$ (i.e. $DAX \overset{NDL}{\vdash}$) is strictly stronger than, say, Manna-Cooper's Intermittent Assertions Method (see e.g. Sain[58] or [62]). As Example three of section 4 shows, there are recursive programs, which are not equivalent to any while-program. DL^{rec} is a new logic to reason about such (genuinly) recursive programs. For example, if d is the 1- sorted type of Example three, section 4, and $\underset{\sim}{D}_3$ is the data type or model of type d therein, let us denote by Th the theory of $\underset{\sim}{D}_3$, i.e. $Th = Th(\underset{\sim}{D}_3)$ is the set of all first order formulas of type d true in $\underset{\sim}{D}_3$. Let $\psi \in F_d$ be the following first order formula of type d: $\psi \triangleq \exists x_1 \exists x_2 (x = h(x_1, x_2))$. Let π be the recursive program of Example three, section 4. Recall that in $\underset{\sim}{D}_3$, π is not equivalent to any while-program. It is easy to see that in every model $\mathcal{M} =< \underline{D}, S, \in > \in Md_{d_s}(Cox)$ satisfying Th, $\mathcal{M} \models \Box(\pi, \psi)$. By Theorem 22 then, $Th \overset{rec}{\vdash} \Box(\pi_3, \psi)$. Similarly, one can prove that

$$Th \overset{rec}{\vdash} [\exists x_1(x = f(x_1) \rightarrow \Box(\pi, \exists x_1 \exists x_2 \exists x_3 \exists x_4(x = h(h(x_1, x_2), h(x_3, x_4)))].$$

Exactly like in the case of NDL, the main reason for constructing the complete logic DL^{rec} of recursive programs is to provide the ground for a *comparative study of program verification methods for recursive programs.* This defines and formulates the task of the present paper, namely to prepare and make it possible to start this comparative study. It is the task of follow-up papers to take those verification methods M of the literature which are designed for proving properties of recursive programs and, following perhaps the techniques e.g. of Sain[61] and [64], represent them in DL^{rec} (i.e. prove results analogous to instances of (**)) and then compare their reasoning powers within DL^{rec}.

7 Acknowledgements

Without the encouragement and help of Hajnal Andreka, Istvan Nemeti and Ildiko Sain, it is very likely that the present paper would not have been written. Numerous discussions during their visit to Florida International University in July-August of 1987 brought me to finally understand the essence and the real width of the NLP-approach.

References

[1] Andreka,H.: Sharpening the characterization of the power of Floyd's method. In: Logics of Programs and their Applications, ed.: A. Salwicki (Proc. Conf. Poznan 1980) Lecture Notes in Computer Science Vol 148, Springer 1983, 1-26.

[2] Andreka,H., Nemeti,I.: Completeness of Floyd logic. Bull.Section of Logic, Vol 7, Wroclaw 1978, 115-120.

[3] Andreka,H., Nemeti,I., Sain,I.: A complete first order logic. MTA MKI Preprint, 1978.

[4] Andreka,H., Nemeti,I., Sain,I.: Completeness problems in verification of programs and program schemes. In: Mathematical Foundations of Computer Science '79, ed.: J. Becvar (Proc. Conf. Olomouc Chehoslovakia 1979, Lecture Notes in Computer Science Vol 74, Springer 1979, 208-218.

[5] Andreka,H., Nemeti,I., Sain,I.: Henkin-type semantics for program schemes to turn negative results to positive. In: Fundamentals of Computation Theory '79, ed.: L.Budach (Proc.Conf.Berlin 1979) Akademie Verlag, Berlin, 1979, Band 2, 18-24.

[6] Andreka,H., Nemeti,I., Sain,I.: Program verification within and without logic. Bull.Section of Logic Vol 8, No 3, Wroclaw 1979, 124-130.

[7] Andreka,H., Nemeti,I., Sain,I.: A characterization of Floyd-provable programs. In: Mathematical Foundations of Computer Science'81. eds.: J. Gruska, M. Chytil (Proc.Conf. Strbske Pleso Csehszlovakia 1981) Lecture Notes in Computer Science Vol 118, Springer 1981, 162-171.

[8] Andreka,H., Nemeti,I., Sain,I.: A complete logic for reasoning about programs via nonstandard model theory, Parts I- II. Theoretical Computer Science 17, 1982, No 2: 193-212, No 3: 259-278.

[9] Apt,K.R., J.W.De Bakker: Exercises in denotational semantics, in Proc. 5th Symp. Mathematical Foundations of Computer Science (A. Mazurkiewicz, ed.), pp. 1-11, Lecture Notes in Computer Science 45, Springer, 1976.

[10] De Bakker,J.W.: Mathematical Theory of Program Correctness, Prentice-Hall International, London (1980).

[11] Barwise,J.: Admissible sets and structures. Springer 1975.

[12] Brookes,S.: A Fully Abstract Semantics and a Proof System for an Algol-like Language with Sharing in LNCS: Proc. Workshop on Foundations of Programming Semantics, Springer, 1985.

[13] Brookes,S.: Semantics of Programming Languages. Course Lectures, Carnegie-Mellon Univ., Comp Sci. Dept., Pittsburgh, 1983.

[14] Cartwright,R.: Recursive programs as definitions in first- order logic. SIAM J. Comput. 13, 2, 1984, 374-407.

[15] Cartwright,R., McCarthy,J.: First Order programming Logic. Proc.Sixth Annual ACM Symposium on Principles of Programming Languages, January 1979, 68-80.

[16] Csirmaz,L.: A Survey of semantics of Floyd-Hoare derivability. CL and CL (Computational linguistics and Computer Languages) 14, 1980, 21-42.

[17] Csirmaz,L.: Programs and program verifications in a general setting. Theoretical Computer Science 16, 1981, 199-210.

[18] Csirmaz,L.: On the completeness of proving partial correctness. Acta Cybernetica 5,2, 1981, 181-190.

[19] Csirmaz,L.: Theorems and problems about non-standard dynamic logic. In: Proc. of workshop on algorithms and computing theory, eds.: N. Karpinski, Z. Habasinski, Technical University Poznan, 1981, 17-19.

[20] Csirmaz,L.: On the strength of "Sometimes" and "Always" in program verification. Information and Control 57, 2-3, 1983, 165-179.

[21] Csirmaz,L.: A completeness theorem for dynamic logic. Notre Dame J. of Formal Logic 26, 1, 1985, 51-60.

[22] Csirmaz,L.: Nonstandard Logics of Programs. In: Algebra, Combinatorics, and Logic in Computer Science, eds.: J. Demetrovics, G. Katona, A. Salomaa (Proc. Conf. Gyor 1983) Colloq.Math. Soc.J.Bolyai Vol 42, North-Holland, 1986, 285-294.

[23] Csirmaz,L., Paris,J.: A property of 2-sorted Peano models and program verification. Zeitschrift fur Math. Logik u. Grundlagen der Math. 30, 1984, 325-334.

[24] Gergely,T., Szots,M.: On the incompleteness of proving partial correctness. Acta Cybernetica, Tom 4, Fasc 1, Szeged, 1978, 45-57.

[25] Gergely,T., Ury,L.: Time models for programming logics. In: Mathematical Logic in Computer Science, eds.: B. Domolki, T. Gergely (proc. Conf. Salgotarjan 1978) Colloq. Math. Soc. J. Bolyai Vol 26, North-Holland, 1981, 359-427.

[26] Gergely,T., Ury,L.: Program behaviour specification through explicit time consideration. Information Processing 80, ed.: S.H. Lavington, North-Holland, 1980, 107-111.

[27] Gonzalez,M.T.H., Artalejo,M.R.: Hoare's logic for nondeterministic regular programs: A nonstandard approach. Preprint Universidad de Madrid 84/85/cc-1, 1985. Abstracted in 12th International Colloquium on Automata, Languages and Programming, Greece, 1985.

[28] Gonzalez,M.T.H., Artalejo,M.R.: Some questions about expressiveness and relative completeness in Hoare's logic, Preprint Universidad de Madrid 84/85/cc-1, 1985.

[29] Gordon,M.: The Denotational Description of Programming Languages, An Introduction, Springer, New York (1979).

[30] Hajek,P.: Some conservativeness results for nonstandard dynamic logic. In: Algebra, Combinatorics and Logic in Computer Science, eds.: J. Demetrovics, G. Katona, A. Salomaa (Proc. Conf. Gyor 1983) Colloq. Math, Soc. J.Bolyai Vol 42, North-Holland, 1986, 443-449.

[31] Hajek,P.: A simple dynamic predicate logic. Theoretical Computer Science, 46(1986)239-259.

[32] Halpern,J., Meyer,A., Trakhtenbrot,B.: From Denotational to Operational and Axiomatic Semantics for Algol-like Languages, Lecture Notes in Computer Science, 164, (1984), pp.47-500.

[33] Harel,D.: First-order Dynamic Logic. Lecture Notes in Computer Science Vol 68, Springer, 1979.

[34] Harel,D.: Dynamic Logic. In: Handbook of Philosophical Logic, eds.: D.M.Gabbay, F. Guenthner, Reidel Publ. Co., North Holland, 1984, Vol II.10, 497-604.

[35] Henkin,L.: Completeness in the theory of types, J. Symb. Logic 15, 1950, 81-91.

[36] Kfoury,A.J., Park,D.M.R.: On the termination of program schemas. Information and Control 29 (1975) 243-251.

[37] Leivant,D.: Logical and mathematical reasoning about imperative programs. In: Proc. of the 1985 POPL (12th Annual ACM Symp. on Princliles of Progr. Languages, New Orleans, 1985) 132-140.

[38] Leivant,D.: Explicit and modal capturing of the relational semantics of programs. In: proc. of the ASL 1985 European Summer Meeting at Orsay, to appear.

[39] Makowsky,J.A., Sain,I.: On the equivalence of weak second order and nonstandard time semantics for various program verification systems. Logic in Computer Science (Proc. Conf. Cambridge USA 1986) Springer 1986, to appear.

[40] Makowsky,J.A., Sain,I.: On the equivalence of weak second order and nonstandard time semantics for Floyd-Hoare logic. MTA MKI Preprint No 70/1985. Theoretical Computer Science, to appear.

[41] Manders,K.L.: First-order Logical Systems and Set-theoretic Definability. Preprint, University of Pittsburgh, 1979.

[42] Manna, Z,: Mathematical Theory of Computation, McGraw-Hill, 1974.

[43] Meyer, A.: discussions with L. Csirmaz at Symposium on Logic in Computer Science, Cambridge, Massachusetts, June 16-18, 1986.

[44] Meyer, A: Understanding Algol: a view of a recent convert to denotational semantics, in Proc. IFIP Congress 1983, pp. 951- 962, North Holland, Amsterdam 1983.

[45] Milne,R.E., Strachey,C.: A Theory of programming Language Semantics (2 volumes), Chapman and Hall, London, and Wiley, New York (1976).

[46] Monk,J.D.: Mathematical Logic. Springer, 1976.

[47] Mosses,P.D.: Compiler generation using denotational semantics, Proc. Symposium on Mathematical Foundations of Computer Science. Gdansk, Lecture Notes in Computer Science, 45, pp. 436-41, Springer, Berlin (1976).

[48] Nemeti,I.: Nonstandard Dynamic Logic. In: Logics of Programs, ed.: D. Kozen (Proc. Conf. New York 1981) Lecture Notes in Computer Science Vol 131, Springer 1982, 311-348.

[49] Pasztor,A.: Nonstandard algorithmic and dynamic logic. J. Symbolic Computation, Academic Press, Vol 2, No 1, 1986, 59-81.

[50] Pasztor,A.: Nonstandard Dynamic Logic to Prove Standard Properties of Programs. Research Report 84-3, Carnegie- Mellon University, Pittsburgh, 1984.

[51] Pasztor,A.: Nonstandard Logics of Programs - Survey and Perspective. Research Report 85-3, Carnegie-Mellon University, Pittsburgh, 1985.

[52] Plotkin,G.D.: LCF Considered as a programming Language, Th. Comp. Science, 5, (1977), pp.223-255.

[53] Pratt,V.R.: Dynamic Logic. Logic, Methodology, and Philos. of Sci. Vol VI, North-Holland 1982, 251-261.

[54] Richter,M.M., Szabo,M.E.: Nonstandard Computation Theory. In: Algebra, Combinatorics and Logic in Computer Science, eds.: J. Demetrovics, G. Katona, A. Salomaa (Proc. Conf. Gyor 1983) Colloq. Math. Soc. J. Bolyai Vol 42, North- Holland 1986, 667-693.

[55] Sacks,G.E.: Saturated model theory, Lecture Notes Series, W.A. Benjamin Inc., Massachusetts, 1972.

[56] Sain,I.: There are general rules for specifying semantics: Observations on Abstract Model Theory. CL and CL (Computational Linguistics and Computer Languages) 13, 1979, 195-250.

[57] Sain,I.: First order dynamic logic with decidable proofs and workeable model theory. In: Fundamentals of Computation Theory'81, ed.: F. Gecseg (Proc.Conf.Szeged 1981) Lecture Notes in Computer Science Vol 117, Springer 1981, 334-340.

[58] Sain, I.: Total correctness in nonstandard dynamic logic. Bull,Section of Logic Vol 12, no 2, Warsaw-Lodz 1983, 64-70.

[59] Sain, I.: Structured nonstandard dynamic logic. Zeitschrift fur Math. Logik u. Grundlagen d. Math. Heft 3, Band 30, 1984. 481-497.

[60] Sain,I.: A simple proof for completeness of Floyd method. Theoretical Computer Science 35, 1985, 345-348.

[61] Sain,I.: The reasoning powers of Burstall's (modal logic) and Pnueli's (temporal logic) program verification methods. In: Logics of Programs, ed.: R. Parikh(Proc. Conf. Brooklyn USA 1985) Lecture Notes in Computer Science Vol 193, Springer 1985, 302-319.

[62] Sain,I.: Total correctness in nonstandard dynamic logic. Theoretical Computer Science, to appear.

[63] Sain,I.: Elementary proof for some semantic characterizations of non-deterministic Floyd-Hoare logic. Notre Dame J. Formal Logic, to appear.

[64] Sain,I.: Relative program verifying powers of the various temporal logics. MTA MKI Preprint No 40/1985. Information and Control, submitted.

[65] Sain,I.: Dynamic Logic with nonstandard model theory, Thesis, Budapest, 1986.

[66] Sain,I., Nemeti,I.: Lattice of logics of programs via nonstandard dynamic logic. Zeitschrift fur Math. Logik u. Grundlagen d. Math., to appear.

[67] Schmidt,D.A.: Denotational Semantics, A Methodology for Language Development, Allyn and Bacon, 1986.

[68] Schwarz,J.S.: Denotational semantics of parallelism, Semantics of Concurrent Computation, Proc. of the Int. Symposium, Evian, France, pp. 191-202, Lecture Notes in Computer Science, 70, Springer, Berlin (1979).

[69] Scott,D.S.: Mathematical concepts in programming language semantics, Proc. 1972 Spring Joint Computer Conference, pp. 225-34, AFIPS Press, Montvale, N.J. (1972).

[70] Scott,D.S., Strachey,C.: Towards a mathematical semantics for computer languages, Proc. of the Symposium on Computers and Automata (ed., J.Fox), pp. 19-46, Polytechnic Institute of Brooklyn Press, New York (1971); also technical monograph PRG-6, programming Research Group, University of Oxford (1971).

[71] Stoy,J.E.: Denotational Semantics: The Scott-Strachey Approach to programming Language Theory, MIT Press, Cambridge, Mass. (1977).

[72] Suchenek,M.: Floyd-Hoare implication is not uniformly axiomatizable within nonstandard dynamic logic. Preprint, Warsaw 1983. Computer Languages, Linguistics, and Logic Vol 1, to appear.

[73] Tennent,R.D.: The denotational semantics of programming languages, Comm. ACM,19(8), 437-53(1976).

[74] Trakhtenbrot, B.A., Recursive Program Schemes and Computable Functionals, Lecture Notes in Computer Science, Vol.45, (1976), pp.137-152.

[75] Trakhtenbrot,B.A., Topics in Typed programming Languages, Lecture Notes, Carnegie-Mellon University, CS Dept., Fall 1985.

Band 1

Inverse and Improperly Posed Problems in Differential Equations

Proceedings of the Conference on Mathematical and Numerical Methods, held in Halle/Saale (GDR) from May 29 to June 2, 1979

Edited by Gottfried Anger

296 Seiten, 1979 (vergriffen)

Band 2

Fundamentals of Computation Theory FCT '79

Proceedings of the Conference on Algebraic, Arithmetic and Categorial Methods in Computation Theory, held in Berlin/Wendisch-Rietz (GDR) September 17–21, 1979

Edited by Lothar Budach

576 Seiten, 1979

Band 3

Decidability and Generalized Quantifiers

By Andreas Baudisch, Detlef Seese, Hans-Peter Tuschik, and Martin Weese

XII, 235 Seiten, 1980

Band 4

Zur Elastizitäts- und Thermoelastizitätstheorie

Potentialtheoretische Methoden in Anwendung auf verschiedene Modelle

Von Lothar Jentsch und Johannes Maul

157 Seiten, 1980

Band 5

Systems Analysis and Simulation 1980

Proceedings of the International Symposium held at Berlin (GDR) September 1–5, 1980

Edited by Achim Sydow

496 Seiten, 1980 (vergriffen)

Band 6

Nichtlineare diskrete Optimierung

von Eberhard Girlich und Michail M. Kowaljow

218 Seiten, 1981

Band 7

Completeness Problems in the Structural Theory of Automata

By Jürgen Dassow

148 Seiten, 1981

Band 8

Darstellung von Lösungen linearer elliptischer Differentialgleichungen

Von Günther Wildenhain

92 Seiten, 1981

Band 9

Introduction to the Spectral Theory of Operators in Spaces with an Indefinite Metric

By I. S. Iohvidov, M. G. Krein, and H. Langer

120 Seiten, 1982 (vergriffen)

Band 10

Nonlinear Time-discrete Systems

A General Approach by Nonlinear Superposition

By Michael Gössel

115 Seiten, 1982 (vergriffen)

Band 11

Preprimal Algebras

By Klaus Denecke

162 Seiten, 1982

Band 12

Ergodic Theory and Related Topics

Proceedings of the Conference held in Vitte/Hiddensee (GDR) October 19–23, 1981

Edited by Horst Michel

232 Seiten, 1982

Band 13

Singularly Perturbed Differential Equations

By H. Goering, A. Felgenhauer, G. Lube, H.-G. Roos, and L. Tobiska

176 Seiten, 1983

Band 14

Leben wir in einer Volterra-Welt?

Ein ökologischer Zugang zur angewandten Systemanalyse

Von Manfred Peschel und Werner Mende

233 Seiten, 1983

Band 15

Markov Decision Problems with Countable State Spaces

Optimality Criteria – Algorithms – Clustering

By Hans Michael Dietz and Volker Nollau

174 Seiten, 1983

Band 16

Stochastische Verfahren zur Suche nach dem Minimum einer Funktion

Von Ryszard Zieliński und Peter Neumann

133 Seiten, 1983

Band 17

Invertibility and Asymptotics of Toeplitz Matrices

By Albrecht Böttcher and Bernd Silbermann

200 Seiten, 1983

Band 18

Structural Induction on Partial Algebras

Introduction to Theory and Application of Partial Algebras – Part II

By Horst Reichel

205 pages, 1984

Band 19

Algebraic Methods for Toeplitz-like Matrices and Operators

By Georg Heinig and Karla Rost

212 pages, 1984

Band 20

Frege Conference 1984

Proceedings of the International Conference held at Schwerin (GDR), September 10–14, 1984

Edited by Gerd Wechsung

408 pages, 1984

Band 21

Numerical Analysis of Selected Semilinear Differential Equations

Edited by Christian Großmann, Thomas Riedrich, and Hartmut Schönheinz

203 pages, 1984

Band 22

Singular Ordinary Differential Operators and Pseudodifferential Equations

By Johannes Elschner

200 pages, 1985

Band 23

Lotka-Volterra-Approach to Cooperation and Competition in Dynamic Systems

Proceedings of the 5th Meeting of UNESCO's Working Group on Systems Theory held on the Wartburg, Eisenach (GDR) at March 5–9, 1984

Edited by Werner Ebeling and Manfred Peschel

271 pages, 1985

Band 24

Convergence Structures 1984

Proceedings of the Conference on Convergence held in Bechyně (Czechoslovakia), September 24–28, 1984

Edited by Josef Novák, Werner Gähler, Horst Herrlich, and Jan Mikusiński

254 pages, 1985

Band 25

Parcella '84

Proceedings of the II. International Workshop on Parallel Processing by Cellular Automata and Arrays held in Berlin (GDR), September 25–27, 1984

Edited by Wolfgang Händler, Tamás Legendi, and Gottfried Wolf

239 pages, 1985 **(vergriffen)**

Band 26

Multiobjective and Stochastic Optimization Based on Parametric Optimization

By Jürgen Guddat, Francisco Guerra Vasquez, Klaus Tammer, and Klaus Wendler

175 pages, 1985

Band 27/28

Systems Analysis and Simulation 1985
I: Theory and Foundations
II: Applications

Proceedings of the International Symposium held in Berlin (GDR), August 26–31, 1985

Edited by Achim Sydow, Manfred Thoma, and Robert Vichnevetsky

per volume 495 pages, 1985

Band 29

Parcella '86

Proceedings of the III. International Workshop on Parallel Processing by Cellular Automata and Arrays held in Berlin (GDR), September 9–11, 1986

Edited by Tamás Legendi, Dennis Parkinson, Roland Vollmar, and Gottfried Wolf

279 pages, 1986

Band 30

Fuzzy Sets Applications, Methodological Approaches, and Results

Proceedings of the International Workshop on Fuzzy Sets Applications cosponsored by IIASA and convened in cooperation with IAEA held on the Wartburg, Eisenach (GDR), at March 3–8, 1985

Edited by Steffen Bocklisch, Sergei Orlovski, Manfred Peschel, and Yasushi Nishiwaki

233 pages, 1986

Band 31

Mathematical Methods of Specification and Synthesis of Software Systems '85

Proceedings of the International Spring School, held in Wendisch-Rietz (GDR), April 22–26, 1985

Edited by Wolfgang Bibel and Klaus P. Jantke

245 pages, 1986 (vergriffen)

Band 32

A Model Theoretic Oriented Approach to Partial Algebras

Introduction to Theory and Application of Partial Algebras – Part I

By Peter Burmeister

319 pages, 1986

Band 33

Finite Difference Method on Irregular Networks

A Generalized Approach to Elliptic Problems of Second Order

By Bernd Heinrich

206 pages, 1986

Band 34

Modern Algorithms for Large Sparse Eigenvalue Problems

By Arnd Meyer

125 pages, 1987

Band 35

Parametric Optimization and Related Topics

Edited by Jürgen Guddat, Hubertus Th. Jongen, Bernd Kummer, and František Nožička

407 pages, 1987

Band 36

Discretization in Differential Equations and Enclosures

Edited by Ernst Adams, Rainer Ansorge, Christian Großmann, and Hans-Görg Roos

250 pages, 1987

Band 37

Factorization of Measurable Matrix Functions

By Georgi Semenovich Litvinchuk and
Ilya Matveyevich Spitkovskiĭ

372 pages, 1987

Band 38

Parallel Algorithms and Architectures

Proceedings of the International Workshop
on Parallel Algorithms and Architectures
held in Suhl (GDR), May 25–30, 1987

Edited by Andreas Albrecht, Hermann Jung,
and Kurt Mehlhorn

205 pages, 1987

Band 39

Parametric Integer Optimization

By Bernd Bank and Reinhard Mandel

138 pages, 1988

Band 40

Computer Analysis of Images and Patterns 1987

Proceedings of the II. International Conference
CAIP '87 on Automatic Image Processing,
held in Wismar (GDR), September 2–4, 1987

Edited by Leonid P. Jaroslavskiĭ,
Azriel Rosenfeld, and Wolfgang Wilhelmi

282 pages, 1987

Band 41

Multivariable Control

A Graph-theoretic Approach

By Kurt J. Reinschke

274 pages, 1988

Band 42

Equilibrium Distributions of Branching Processes

By Alfred Liemant, Klaus Matthes, and Anton
Wakolbinger

240 pages, 1988

Band 43

Andreotti-Grauert Theory by Integral Formulas

By Gennadi Markovich Henkin and
Jürgen Leiterer

270 pages, 1988

Band 44

Continual Means and Boundary Value Problems in Function Spaces

By Efim Mikhailovich Polishchuk

164 pages, 1988

Band 45

Advances in Mathematical Optimization

Invited papers dedicated to Prof. Dr. Dr. h. c.
F. Nožička on occasion of his 70th birthday

Edited by Jürgen Guddat et al.

332 pages, 1988

Band 46/47

Systems Analysis and Simulation 1988

Proceedings of the International Symposium
held in Berlin (GDR),

September 12–16, 1988

I: Theory and Foundations
II: Applications

Edited by Achim Sydow, Spyros G. Tzafestas,
and Robert Vichnevetsky

426/425 pages, 1988

Band 48

Parcella '88

Proceedings of the IV. International Workshop
on Parallel Processing by Cellular Automata
and Arrays, held in Berlin (GDR),

October 17–21, 1988

Edited by Gottfried Wolf, Tamás Legendi, and
Udo Schendel

379 pages, 1988

Band 49

Algebraic and Logic Programming

Proceedings of an International Workshop
held in Gaussig (GDR), November 14–18, 1988
Edited by Jan Grabowski, Pierre Lescanne, and
Wolfgang Wechler

256 pages, 1988

The combination of algebraic and logic programming is a very promising and interesting subject in modern software development. Algebraic programming is here understood as programming by algebraic specifications and rewrite rule systems, while logic programming is meant as usually, namely Horn clause specifications and resolution principles.
The present volume contains 24 papers, including 5 invited lectures, of an international workshop with its main topics:

Equational and logic specifications,
Logic programming with types and equality,
Term rewriting,
Semantical unification,
Correctness and completeness of inference systems,
Implementation and algorithmic efficiency.

Die Kombination von algebraischer und logischer Programmierung ist ein vielversprechender und interessanter Gegenstand der modernen Softwareentwicklung. Unter algebraischer Programmierung verstehen wir hier die Programmierung mit Hilfe von algebraischen Spezifikationen und Termersetzungssystemen. Logische Programmierung wird wie üblich aufgefaßt, nämlich Hornklauseln-Spezifikationen und Resolutionsprinzipien.
Der vorliegende Band enthält 24 Arbeiten, einschließlich 5 Hauptvorträge eines internationalen Workshops, dessen Schwerpunkte die folgenden sind:

Gleichungs- und logische Spezifikationen,
Logische Programmierung mit Typen und Gleichheit,
Termersetzung,
Semantische Unifikation,
Korrektheit und Vollständigkeit von Inferenzsystemen,
Implementation und algorithmische Effektivität.

ISBN 3-05-500665-8
ISSN 0138-3019

Bei Fragen zur Produktsicherheit wenden Sie sich bitte an:
If you have any questions regarding product safety,
please contact:

Walter de Gruyter GmbH
Genthiner Straße 13
10785 Berlin
productsafety@degruyterbrill.com